T0186816

Protecting Mobile Networks and Devices

Challenges and Solutions

CRC Series in Security, Privacy and Trust

SERIES EDITORS

Jianying Zhou

Institute for Infocomm Research, Singapore

jyzhou@i2r.a-star.edu.sg

Pierangela Samarati

Università degli Studi di Milano, Italy

pierangela.samarati@unimi.it

AIMS AND SCOPE

This book series presents the advancements in research and technology development in the area of security, privacy, and trust in a systematic and comprehensive manner. The series will provide a reference for defining, reasoning and addressing the security and privacy risks and vulnerabilities in all the IT systems and applications, it will mainly include (but not limited to) aspects below:

- Applied Cryptography, Key Management/Recovery, Data and Application Security and Privacy;
- Biometrics, Authentication, Authorization, and Identity Management;
- Cloud Security, Distributed Systems Security, Smart Grid Security, CPS and IoT Security;
- Data Security, Web Security, Network Security, Mobile and Wireless Security;
- Privacy Enhancing Technology, Privacy and Anonymity, Trusted and Trustworthy Computing;
- Risk Evaluation and Security Certification, Critical Infrastructure Protection;
- Security Protocols and Intelligence, Intrusion Detection and Prevention;
- Multimedia Security, Software Security, System Security, Trust Model and Management;
- Security, Privacy, and Trust in Cloud Environments, Mobile Systems, Social Networks, Peer-to-Peer Systems, Pervasive/Ubiquitous Computing, Data Outsourcing, and Crowdsourcing, etc.

PUBLISHED TITLES

Protecting Mobile Networks and Devices: Challenges and Solutions

Weizhi Meng, Xiapu Luo, Steven Furnell, and Jianying Zhou

Real-World Electronic Voting: Design, Analysis and Deployment

Feng Hao and Peter Y. A. Ryan

Location Privacy in Wireless Sensor Networks

Ruben Rios, Javier Lopez, and Jorge Cuellar

Touchless Fingerprint Biometrics

Ruggero Donida Labati, Vincenzo Piuri, and Fabio Scotti

Protecting Mobile Networks and Devices

Challenges and Solutions

Edited by

Weizhi Meng • Xiapu Luo
Steven Furnell • Jianying Zhou

CRC Press
Taylor & Francis Group
Boca Raton   London   New York

CRC Press is an imprint of the
Taylor & Francis Group, an **informa** business

CRC Press
Taylor & Francis Group
6000 Broken Sound Parkway NW, Suite 300
Boca Raton, FL 33487-2742

© 2017 by Taylor & Francis Group, LLC
CRC Press is an imprint of Taylor & Francis Group, an Informa business

No claim to original U.S. Government works

Printed on acid-free paper
Version Date: 20160913

International Standard Book Number-13: 978-1-4987-3583-4 (Hardback)

Library of Congress Cataloging-in-Publication Data

Names: Meng, Weizhi, 1986- editor.
Title: Protecting mobile networks and devices : challenges and solutions / edited by Weizhi Meng, Xiapu Luo, Steven Furnell, Jianying Zhou.
Description: Boca Raton : CRC Press is an imprint of the Taylor & Francis Group, an Informa Business, [2017] | Series: Series in security, privacy and trust | Includes bibliographical references and index.
Identifiers: LCCN 2016026681 | ISBN 9781498735834
Subjects: LCSH: Mobile communication systems--Security measures. | Smartphones--Security measures.
Classification: LCC TK5102.85 .P76 2017 | DDC 005.4/46--dc23
LC record available at https://lccn.loc.gov/2016026681

Visit the Taylor & Francis Web site at
http://www.taylorandfrancis.com

and the CRC Press Web site at
http://www.crcpress.com

Printed and bound in the United States of America by
Edwards Brothers Malloy on sustainably sourced paper

Contents

SECTION I AUTHENTICATION TECHNIQUES FOR MOBILE DEVICES

SECTION II MOBILE DEVICE PRIVACY

SECTION III MOBILE OPERATING SYSTEM VULNERABILITIES

SECTION IV MALWARE CLASSIFICATION AND DETECTION

SECTION V MOBILE NETWORK SECURITY

Introduction

Nowadays, mobile networks and devices are developing at a rapid pace. For example, according to the International Telecommunications Union, there were more than 7 billion mobile cellular subscriptions by the end of 2015, representing a penetration rate of 97%, and 95% of the world had mobile coverage to at least the 2G level. Meanwhile, mobile broadband penetration had reached almost 50% in 2015 (a 12-fold increase since 2007). In addition to the growth of the devices and coverage, the capabilities of mobile devices have also increased considerably, with apps now leading to a wealth of uses, many of which serve to increase both the volume and sensitivity of the data that the devices store and communicate.

With the growth previously mentioned in mind, the question of how to protect mobile networks against various attacks has become a significant one. For years, security experts have predicted the rise of mobile attacks and malware, and more recent years have unfortunately proven these predications to be accurate. Mobile networks and related devices (e.g., phones or tablets) are now an increasingly significant target for various attacks. As such, the need for secure solutions in the mobile environment is increasing.

This book provides a relevant reference for researchers, students, engineers, and professionals interested in exploring recent advances in mobile security. The overall structure of the book addresses five key perspectives:

- *Authentication techniques for mobile devices*: Introducing the topic of biometric authentication on mobile devices and presenting related research on face authentication and visual authentication using visual cryptography
- *Mobile device privacy*: Investigating users' privacy issues on mobile applications and public data storage on Android-based devices
- *Mobile operating system vulnerabilities*: Illustrating how to perform security analysis on Android and iOS and how to defend Android applications against reverse engineering
- *Malware classification and detection*: Describing how to efficiently classify Android malware using robust static features and presenting a survey on Android banking malware
- *Mobile network security*: Considering a range of network-related issues, including physical layer security, how to safeguard mobile payments, and examining the cyberattack surface for next-generation mobile networks

The resulting coverage includes mobile networks, mobile phone security, and wireless security, as well as related attacks, solutions, and trends. As such, this book presents a broad view of mobile security that will be of interest in both academia and industry.

Editors

Weizhi Meng received his BE in computer science from the Nanjing University of Posts and Telecommunications, Nanjing, Jiangsu, China, and earned his PhD in the Department of Computer Science from the City University of Hong Kong (CityU), Kowloon, Hong Kong. He was known as Yuxin Meng and is currently an assistant professor in the Department of Applied Mathematics and Computer Science, Technical University of Denmark, Kongens Lyngby, Denmark. Prior to that, he worked as a research scientist in the Infocomm Security Department, Institute for Infocomm Research, Singapore, and as a senior research associate in CityU after graduation. His research interests are cyber security, including intrusion detection; mobile security and biometric authentication; malware detection; HCI security; cloud security; and intelligent security applications. He won the Outstanding Academic Performance Award during his doctoral study and is a recipient of the HKIE Outstanding Paper Award for Young Engineers/Researchers in 2014.

Xiapu Luo is a research assistant professor in the Department of Computing and an associate researcher at Shenzhen Research Institute at The Hong Kong Polytechnic University, Kowloon, Hong Kong. His research focuses on mobile networks, smartphone security, network security and privacy, and Internet measurement. Luo has a PhD in computer science from The Hong Kong Polytechnic University.

Steven Furnell is a professor of information systems security and leads the Centre for Security, Communications and Network Research at Plymouth University, Plymouth, United Kingdom. He is also an adjunct professor at Edith Cowan University in Western Australia and an honorary professor at Nelson Mandela Metropolitan University in South Africa. His research interests include mobile security, the usability of security technologies, security management and culture, and technologies for user authentication and intrusion detection. He has authored over 270 papers in refereed international journals and conference proceedings as well as books including *Cybercrime: Vandalizing the Information Society* (2001) and *Computer Insecurity: Risking the System* (2005). Prof. Furnell is the BCS representative to Technical Committee 11 (security and privacy) within the International Federation for Information Processing and is a member of related working groups on security management, security education, and human aspects of security. He is also a board member of the Institute of Information Security Professionals and chairs the academic partnership committee and southwest branch. Further details can be found at www.plymouth.ac.uk/cscan, with a variety of security podcasts also available via www.cscan.org/podcasts. Steve can also be followed on Twitter (@smfurnell).

Jianying Zhou holds a PhD in information security from Royal Holloway, University of London, United Kingdom. He is a principal scientist with the Institute for Infocomm Research and the head of the Infocomm Security Department. His research interests include applied cryptography, computer and network security, cyber-physical security, and mobile and wireless security.

Contributors

Gerd Ascheid
Institute for Communication Technologies and
Embedded Systems
RWTH Aachen University
Aachen, Germany

Man Ho Au
Department of Computing
The Hong Kong Polytechnic University
Kowloon, Hong Kong

Ari Moesriami Barmawi
School of Computing
Telkom University
Bandung, Indonesia

Lejla Batina
Institute for Computing and Information
Sciences
Radboud University
Nijmegen, the Netherlands

Bruno Bouchard
Intelligent Technologies Consultants
Montreal, Quebec, Canada

Özge Cepheli
Wireless Communications Research Laboratory
Istanbul Technical University
Istanbul, Turkey

Yang-Wai Chow
Centre for Computer and Information Security
Research
School of Computing and Information
Technology
University of Wollongong
Wollongong, New South Wales, Australia

Mauro Conti
Department of Mathematics
University of Padua
Padua, Italy

Guido Dartmann
Distributed Systems
Trier University of Applied Sciences
Trier, Germany

Robert H. Deng
School of Information Systems
Singapore Management University
Singapore, Singapore

Wenrui Diao
The Chinese University of Hong Kong
Shatin, Hong Kong, People's Republic of China

Hossein Fereidooni
Department of Mathematics
University of Padua
Padua, Italy

Igor Nai Fovino
Institute for the Protection and Security of the
    Citizen
European Commission, Joint Research Centre
Ispra, Italy

Dimitris Geneiatakis
Department of Electrical and Computer
    Engineering
Aristotle University of Thessaloniki
Thessaloniki, Greece

Ali A. Ghorbani
Faculty of Computer Science
Information Security Centre of Excellence
University of New Brunswick
Fredericton, New Brunswick, Canada

Jin Han
Twitter
San Francisco, California

Wenjun Hu
Palo Alto Networks, Inc.
Santa Clara, California

Andi Fitriah A. Kadir
Faculty of Computer Science
Information Security Centre of Excellence
University of New Brunswick
Fredericton, New Brunswick, Canada

Georgios Kambourakis
Department of Information and
    Communication Systems Engineering
University of the Aegean Karlovassi
Karlovassi, Greece

Güneş Karabulut Kurt
Department of Electronics and
    Communication Engineering
Istanbul Technical University
Istanbul, Turkey

Yan Li
School of Information Systems
Singapore Management University
Singapore, Singapore

Yingjiu Li
School of Information Systems
Singapore Management University
Singapore, Singapore

Zhou Li
RSA Laboratories
Bedford, Massachusetts

Xiangyu Liu
The Chinese University of Hong Kong
Shatin, Hong Kong, People's Republic of China

Volker Lücken
Institute for Communication Technologies and
    Embedded Systems
RWTH Aachen University
Aachen, Germany

Xiapu Luo
Department of Computing
The Hong Kong Polytechnic University
Kowloon, Hong Kong

Xiaobo Ma
MOE KLINNS Lab
Xi'an Jiaotong University
Xi'an, Shaanxi, China

Bob-Antoine J. Ménélas
Department of Computer Science
University of Quebec at Chicoutimi
Chicoutimi, Quebec, Canada

Veelasha Moonsamy
Institute for Computing and Information
    Sciences
Radboud University
Nijmegen, the Netherlands

Ricardo Neisse
Institute for the Protection and Security of the
    Citizen
European Commission, Joint Research Centre
Ispra, Italy

Riccardo Satta
Institute for the Protection and Security of the
    Citizen
European Commission, Joint Research Centre
Ispra, Italy

Filipo Sharevski
College of Computing and Digital Media
DePaul University
Chicago, Illinois

Natalia Stakhanova
Faculty of Computer Science
Information Security Centre of Excellence
University of New Brunswick
Fredericton, New Brunswick, Canada

Gary Steri
Institute for the Protection and Security of the
    Citizen
European Commission, Joint Research Centre
Ispra, Italy

Willy Susilo
Centre for Computer and Information Security
    Research
School of Computing and Information
    Technology
University of Wollongong
Wollongong, New South Wales, Australia

Florentin Thullier
Department of Computer Science
University of Quebec at Chicoutimi
Chicoutimi, Quebec, Canada

Bo Xing
Faculty of Science and Agriculture
University of Limpopo
Polokwane, South Africa

Qiang Yan
School of Information Systems
Singapore Management University
Singapore, Singapore

Kehuan Zhang
The Chinese University of Hong Kong
Shatin, Hong Kong, People's Republic of China

Zhe Zhou
The Chinese University of Hong Kong
Shatin, Hong Kong, People's Republic of China

AUTHENTICATION TECHNIQUES FOR MOBILE DEVICES

Chapter 1

Exploring Mobile Authentication Mechanisms from Personal Identification Numbers to Biometrics Including the Future Trend

Florentin Thullier, Bruno Bouchard, and
Bob-Antoine J. Ménélas

Contents

Abstract

The growing market of mobile devices forces one to question about how to protect the user's credentials and data stored on such devices. Authentication mechanisms remain the first layer of security in the use of mobile devices. However, several such mechanisms that have been already proposed were designed in a machine point of view. As a matter of fact, they are not compatible with behaviors humans have while using their mobile devices in their daily lives. Consequently, users adopted unsafe habits that may compromise the proper functioning of authentication mechanisms according to the safety aspect. In this chapter, the strength and the weakness of the current schemes, from the simpler ones such as personal identification number to the more complex biometric systems such as fingerprint, are highlighted. Besides, we offer an evaluation, based on both existing and new criteria we suggest, chiefly focused on the usability of these schemes for the user. While each of them has benefits as well as disadvantages, we finally propose a conclusion leaning toward the aftermath of authentication on mobile devices that head for new perspectives as regards *no password* and *ubiquitous* authentication mechanisms.

Keywords: Authentication, Mobile device, Smartphone, Usability

1.1 Introduction

Over the past decade, the market of mobile devices grew up exponentially. The Gartner Institute noticed that smartphone sales were over 400,000 units in the second semester of 2013 [1]. These devices take a significant place in people's everyday life. Indeed, people and more specifically young ones have their mobile devices everywhere and at any time [2] since they consider their mobile phones as an important part of their lives [3]. Nielsen [4] has pointed out that in the Q4 2013, users spent more than 30 h using applications on such devices. Besides, it should be noted that a major player of the mobile device industry used to claim that there is an application for everything. As a result, users do store private data such as pictures, videos, and secret information about their personal accounts (emails, social networks) on their mobile devices. However, they are, most of the time, not adequately wary about the safety of information they save on their devices [5].

Authentication refers to the process of an entity that has to become sure of the identity of another one [6]. Within a mobile device context, authentication remains the first entry point for security. Indeed, such a mechanism aims at protecting the sensitive information about the user that characterizes their digital identity. Over the past few years, various authentication schemes have been proposed. Hence, we divided them into three broad categories: knowledge based, token based, and biometrics [7]. Figure 1.1 illustrates all authentication mechanisms that are currently employed with mobile devices. First of all, knowledge-based authentication schemes focus on what the user

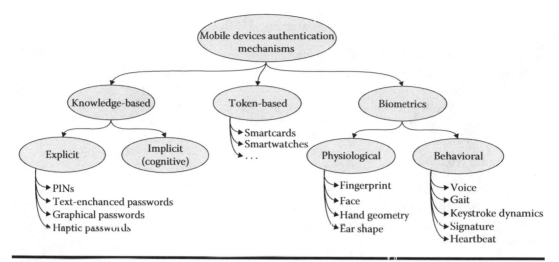

Figure 1.1   Taxonomy of authentication mechanisms employed with mobile devices.

knows. Precisely, we differentiate implicit and explicit knowledge-based mechanisms. Explicit ones imply that the user has to retain new data like a four-digit personal identification number (PIN) code or a password [8], whereas implicit knowledge-based mechanisms call upon cognitive functions of the user to exploit the data they already know [9]. Second, token-based mechanisms need the user to prove they possess a physical token that often involves a two-factor authentication process [10]. As an example, we can mention the smartcard that the user needs to authenticate himself or herself on his or her mobile phone. Finally, biometric mechanisms rely on the uniqueness of the user's physiological or behavioral trait to perform the authentication process. Consequently, we subdivide biometrics into physiological and behavioral sets. Physiological biometrics exploit singularities of the human body like fingerprints [11], while behavioral ones require users to perform some actions to prove their identity such as gait [12].

Whenever we are facing a machine that has to deal with human users, therefore, we have an interactive system [13]. As pointed out by Benyon et al. [13], a fundamental challenge with interactive systems is that human beings and machines have different characteristics. Indeed, what may be seen as strengths in a machine point of view may also be a weakness for human being. On the one hand, machines can see humans as being *vague, disorganized,* and emotional, while they are *precise, orderly, unemotional,* and *logical.* On the other hand, humans may claim to be *attentive, creative, flexible,* and *resourceful,* while machines are *dumb, rigid,* and *unimaginative.* Such differences suggest that the key challenge is first to understand the human rather than design an interactive system in the machine point of view. However, when considering the authentication systems that have been proposed over the last three decades, it seems that they have been designed without any concern for humans. As an example, it was reported that half of the population does not lock their phone at all since they estimate that entering a four-digit PIN code involves lots of trouble every time that the mobile device has to be unlocked [14]. Moreover, it is known that users have trouble remembering all the passwords they have to use nowadays [15]. Consequently, some people prefer to reuse the same password in multiple situations, while others do write down their passwords. It is clear that behaviors reported here may have a huge impact on the security of mobile devices. Accordingly, people's authentication usage may generate serious threats for the security that a system initially provides. In fact, an effective mechanism may become a weak one because it is not used as recommended.

The main contribution of this chapter is to emphasize, *via* a critical analysis, most of the weaknesses of proposed authentication schemes, when used in real-life situations. We previously pointed out that users have adopted unsafe habits in consequence of the non-user-centered design of most authentication processes. Hence, our work aims at analyzing how proposed mechanisms either suit characteristics of human users or not and why they are not entirely appropriate to their needs. The remainder of this chapter is organized as follows: First, it will review authentication schemes that have been proposed in a mobile device context, where, for each of them, both strengths and weaknesses will be highlighted. Then, we will offer an evaluation guided by criteria that were previously proposed by Jain et al. [16] in the field of authentication. In addition, we will increase such an assessment with the criteria we suggest, which stem from our critical analysis. Finally, this evaluation will lead us to conclusions on the current state of authentication mechanisms for mobile devices.

1.2 Knowledge-Based Authentication Mechanisms

Knowledge-based authentication mechanisms rely upon the user's ability to recall secret information. It is possible to make out two different kinds of knowledge-based techniques: explicit and implicit schemes. On one hand, explicit ones need the user to set and learn a piece of knowledge. On the other hand, implicit ones exploit the user memory, thanks to either, or both, personal information they already know or their everyday life preferences, for example, music they like or food they enjoy.

This section describes in detail both explicit and implicit techniques and exposes that the user's capacity to remember a secret remains a common denominator in the weakness of each knowledge-based authentication scheme.

1.2.1 Evaluation of the Strength of a Knowledge-Based Authentication Scheme

The strength of a knowledge-based authentication scheme is theoretically measurable via the evaluation of the entropy of the password space. The password space is the total number of distinct possibilities the authentication system can support. The size (S) of the password space for a system having N possible entries is given by Equation 1.1. The length of the input to retain is expressed by k. Finally, the entropy (H) can be computed by using Equation 1.2, and the result is expressed in bits:

$$S = N^k \tag{1.1}$$

$$H = \log_2(s) \tag{1.2}$$

Real use cases reveal that such an evaluation does yet not represent an accurate measure of the strength of a knowledge-based authentication mechanism. Indeed, since users have the possibility to choose their own secret input, they often refer to a familiar pattern rather than select it randomly. As an example, Yampolskiy [17] has pointed out that 47.5% of the users chose a *family-oriented* information as a secret input such as a child's name or date of birth. Therefore, a lower subset of the N possibilities is truly used, since the length of passwords is generally less than eight characters [15].

1.2.2 Explicit Schemes

1.2.2.1 Personal Identification Numbers

Example of use case scenario: Commonly, users have to choose an array of four digits that they will need to remember. Then, each time the mobile device has to be unlocked, the system prompts an input field where the user needs to fill these digits in the correct order to be authorized to access the whole content of the device.

PINs are a simple way to restrain access to an entity due to their composition from 4 to 16 digits. They appear with the growth of the ATM, and they are mostly used in the banking system. Regarding a mobile device context, using PINs currently remains the most dominant authentication method to protect the access of these devices, as concerns 2/3 of mobile device users [18]. PINs can be applied to both the device and the user's subscriber identity module—a removable token that contains the cryptographic keys required for network access. Both of the two leading mobile device operating systems, Android and iOS, provide this authentication mechanism.

However, PINs involve several issues considering memorability or human habits that may compromise the security offers by the system. In that sense, Clarke and Furnell [18] have assessed that 1/3 of mobile phone users who keep their phone locked via a four-digit PIN method consider such protection as an inconvenience in everyday life. As a result, users do need to retain a code that has a familiar signification, such as their date of birth [17]. Furthermore, Clarke and Furnell [18] also enhance the weakness of this authentication scheme; indeed, 36% of the respondents use the same PIN code for multiple services. Thus, it becomes easier for an attacker to determine the correct four-digit PIN in order to have free access to several other services where the code is used. The lack of security brought by users can also be underlined via another study that reports 26% of PIN users shared the proper code with someone else [19].

While PINs remain popular, they are also the weakest authentication mechanism on the market as they offer a theoretically low entropy—$\log_2(10^4)$. Indeed, people adopted several behaviors to cope with the large cognitive load that the system requires. All of this leads PIN authentication to be largely vulnerable to several attacks such as *code guessing* by *social engineering*, *brute force*, or *shoulder surfing* attacks [20–23].

1.2.2.2 Text-Enhanced Passwords

Example of use case scenario: Conversely, to PINs, users have to select, at least, an array of six characters that are not restricted to only digits. The whole set of characters offered by the keyboard of the mobile device is legitimate. Next, the authentication depends on the same process as PINs.

As opposed to PINs, text-enhanced passwords are more complex. Usually, they are composed of several different characters such as lower- and uppercase letters, digits, and also nonalphanumeric characters. At first, passwords were stored in plain text files without any encryption [24]. Thereby, protecting such sensitive information became crucial for numerical systems. This mechanism is also provided by both Android and iOS platforms. However, regarding authentication on mobile devices, text-enhanced passwords remain less popular than PINs among mobile device users. Indeed, authenticating users with a complex string of characters is an inherited process that comes from traditional computing, and it was not revised at all before its arrival on mobile devices.

The market of mobile applications is vast. As an example, the Apple Store counted up to 600,000 applications, and the total number of downloads surpassed 25 billion in the year 2012 [25]. Through this immense inventory, numerous applications require the user to login in order to have access to the entire set of features they offered [26]. Consequently, text-enhanced passwords also take part regarding authentication in the daily usage of mobile devices; it is now important to illustrate the weakness of this mechanism for the user's memory and how such issues affect mobile device security due to behaviors of the users. Passwords remain theoretically a strong way to secure a system. However, they are usually long and sophisticated. Hence, much more memorizing abilities are required for the user. Indeed, Yan [15] identifies that without the memorization problem, the maximally secured password would be blended with the maximum number of characters allowed by the system, randomly arranged. This is possible to do for a machine, but almost impossible to retain for a human. Moreover, text-enhanced passwords do have better entropy than PINs: $\log_2(94^k)$, where k is the password length and 94 is the number of printable characters excluding *space*. Although, because they are everywhere, and because they were designed to a machine point of view, passwords represent a mechanism not as good as claimed. Hence, a study conducted by Riley [27] shows that more than half of password users concede that they use the exact same string of characters for multiple accounts on numerical systems. Moreover, about 15% of them admit that they used to write down their list of passwords in case they forget them, while 1/3 also report using the *remember my password* function, to produce another password than the one they originally set up. The growth of the number of numerical services we use in our everyday life affects significantly the usage of passwords we have. Another study also highlights the deficiency of this mechanism as they released the *worst passwords* list, which exposed some examples as 123456, *password,* or *qwerty* that are frequently set up by users [28]. Such examples formed perfect cases of vulnerability for the security of mobile devices. Just as PINs, text-enhanced passwords are strongly exposed to *brute force, dictionary, social engineering,* and *shoulder surfing* attacks [20–23].

1.2.2.3 Graphical Passwords

Example of use case scenario: With graphical passwords, users have to recall some pieces of visual information. There are a lot of various implementations, but the most well-known is probably the one implemented in Android that appears in the earliest version of the mobile operating system. First, the user has to set up a path between dots in a matrix as shown by the gray stroke in Figure 1.2. To be granted full access to the mobile device, the user has to reproduce the path he initially set up. The order of the dots passed along the path is essential. As illustrated in Figure 1.2, if the user

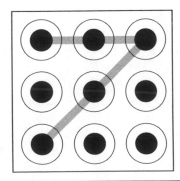

Figure 1.2 Android implementation of graphical password.

defined a path from the lower-left corner dot to the upper-left-corner dot, the inverse drawing (from upper corner to lower corner) during the authentication will not genuinely authenticate the user.

Knowing that humans have better abilities for recognizing and recalling visual information when compared to verbal or textual information [29], other mechanisms were imagined to use a graphical scheme instead of a sequence of characters as passwords. Since the patent introduced by Blonder [30] in 1996, multiple schemes have been designed. Biddle et al. [31] grouped the proposed systems into three broad categories: *recall-based systems, recognition-based systems,* and *cued recall systems.*

First of all, in a recall-based system, the first step is to choose a predefined pattern. Then, the user is presented with a selected image or a blank canvas where the secret sketch has to be reproduced each time he wants to authenticate himself. Second, in a recognition-based system the user is invited to select a sequence of predefined images among several others. The number of presented images is generally limited to ensure the usability. Finally, in a cue recall system, the user has to recall and target a specific part of a picture. In this way, such systems reduce the memory load that the user needs with recall-based systems. Biddle et al. [31] have pointed out that users are more comfortable using a graphical password than a digit or a text-based password every day. However, it is known that the Android implementation allows 389,112 possibilities [32]. Therefore, the password space is not superior to a six-digit PIN. Moreover, as reported by Uellenbeck et al. [32], users do not exploit the maximum potentiality of the security since some graphical schemes are evident to perform. Another relevant example of a graphical scheme may be the *picture gesture authentication* feature introduced by Microsoft in Windows 8 [33]. The idea behind this mechanism is to allow users to define some specific gestures (taps, circles, and lines) over either predefined pictures or user's ones. The whole set of possibilities for such a system largely relies on both the number and the nature of gestures determined. According to Sinofsky [33], when the user defines two of the most complex gesture (lines), there are 846,183 unique possibilities.

As a matter of fact, as assessed by Biddle et al. [31] it is possible to say that graphical passwords do not offer a higher level of security than PINs or text-enhanced passwords. Indeed, as regards an implementation such as the Android one, Uellenbeck et al. [32] have experienced the ability to find the most common path defined of numerous graphical schemes. They showed that it was possible to determine the right path statistically by applying a Markov model algorithm on their dataset. Concerning the gesture-based graphical authentication process, Zhao et al. [34] have demonstrated that the framework they built was able to guess a large portion of the picture passwords set of their study. Moreover, Aviv et al. [35] showed that it was possible to find the graphical scheme *via* oily residues, or smudges that users leave on the touch-screen surface. They named this vulnerability as "smudge attack." Besides, graphical passwords have also the same others vulnerabilities as all knowledge-based authentication mechanisms: *social engineering* [21,22] and *shoulder surfing* [23,36]. Recently, Gugenheimer et al. [37] have proposed a novel graphical authentication concept that claims to be robust against shoulder surfing. Through this approach, shoulder surfing attacks were reduced to 10.5% and the authors have pointed out that no participant forgot their graphical scheme. However, it is clear that their process involves a high level of memorization to recall the information due to its complexity when compared to a simple four-digit PIN code.

1.2.2.4 Haptic Passwords

Example of use case scenario: Instead of visual information, the user has to recall a sequence of kinesthetic phenomena produced by the mobile devices. The idea is to let the user define his own sequence that he has to reproduce afterward to access to the mobile device.

With the emergence of modern mobile devices, the desire to exploit haptic in numerical systems to enhance the user experience was strong. Mobile devices such as smartphones are composed of a lot of new technologies like touch screens and sensors that provide many more possibilities regarding authentication mechanisms. Consequently, several new knowledge-based authentication schemes have been designed recently. As an example, Bianchi et al. [38] suggested a novel approach through haptic passwords. The initial work of PINs is retained, but the user has to recall a sequence of vibrations instead of a sequence of single digits. As graphic passwords, haptic ones were designed to be more convenient for users than text-enhanced passwords. The implementation proposed by Bianchi et al. [39] attempts to avoid the memorability issues encountered by users and reduces behaviors that lead to security vulnerabilities. However, this study also highlights that such new authentication mechanisms still require unreasonable calls from memory. As a result, they are not the answer to fix issues provided by PINs, text passwords, or graphical passwords.

1.2.3 Cognitive Schemes

Mechanisms we described earlier are all explicit methodologies for a knowledge-based authentication. However, each person has a unique set of knowledge. Thereby, cognitive passwords aim at exploiting personal facts, opinions, and interests as a means of authentication. This process is defined as a challenge–response.

The idea behind these schemes first stems from regular computer security access where users, in addition to a conventional password, have to answer to some personal questions to be granted access. However, regarding a mobile device context, such an approach should be more carefully considered. Indeed, as we state in the first part of this chapter, users store more and more data on their devices. Thus, data such as pictures, music, and information from social media [9,40] may be exploited to build a convenient cognitive process for authentication, revised for use with mobile devices.

The experiment led by Bunnell et al. [41] regarding these authentication schemes showed that personal facts were better recalled than others. Because people socially close to the user were easily able to guess many answers, Lazar et al. [40] proposed a method to personalize cognitive passwords to individual users. Results obtained show that personalization increases the recall of cognitive passwords, but does not help in improving their secrecy.

1.3 Token-Based Authentication Mechanisms

Example of use case scenario: Token-based authentication requires the user to possess a physical piece of hardware that has first to be coupled with the mobile device. Then, the mobile device has to verify the credentials of the token to grant access to the user.

Token-based authentication mechanisms require a hardware interaction between the user and his device to complete the authentication process. Such mechanisms involve at least a two-factor authentication (*multifactor* is used when there is more than two) due to the commitment to attest that both the passwords are correct, and the user holds the token throughout the authentication process. The three major types of tokens are USB token devices, smart cards, and password-generating tokens [42]. Password-generating tokens usually imply a mobile device in the authentication process as described by Aloul et al. [43], but the purpose is not to authenticate a user directly on his mobile device.

Thereby, these old implementations—when compared to the existence of mobile devices—are no longer applicable as they were initially designed. Nowadays, we observe the growth of smart objects and connected objects as known as the Internet of Things (IoT) [44]; consequently, modern approaches regarding token-based authentication mechanisms appear to be more convenient with the use of such devices. As examples, smart watches are replacing USB devices while NFC tags will supplant smart cards overtime since it is possible to bring them everywhere (wallets, clothes). With the fifth version of Android, Google introduces a feature called "trusted devices." This feature aims at providing an automatic authentication mechanism that uses smart objects the user has to couple to his mobile device. As a result, as long as the mobile device detects a connection with the token hardware via Bluetooth, NFC, or Wi-Fi, it remains unlocked.

Regarding two-factor authentication mechanisms, Schneier [45] suggests that "they solve the security problem we had ten years ago, not the security problems we have today." The use of smart objects over the authentication process implies many cases of vulnerability issues. Indeed, whenever someone wears a smart object as a token for the authentication process, the mobile device should not be sighted off. Due to the fact that there is no need to replay the authentication process, the mobile device becomes simply accessible by anyone nearby. Another problematic situation may be observed where both devices are stolen by the same person.

1.4 Biometrics

For many years, it is known that each human exhibits a unique set of features. As an example, each human fingerprint describes a unique pattern; in the same way, blood vessels of the retina also have a unique pattern. Biometrics systems exploit these singularities in order to authenticate users. Hence, with a biometric system, there is no need for remembering or recalling any information; instead, the singularity of interest just has to be digitized and compared to the saved one. To this end, several biometric systems necessitate the use of a scanning or recording hardware that is not always adapted to suit mobile devices. When compared to other approaches, a vast majority of biometric ones have a greater cost of implementation. Moreover, due to the assessment of data that come directly from the user, biometrics raise some privacy concerns. As a result, even though these systems generally offer a high level of security and a sufficient accuracy, their usage remains limited, particularly with mobile devices.

The following review of biometric authentication that we propose does not heed sophisticated mechanisms such as blood vessels retina or iris pattern recognitions that are way too complex in hardware requirements or computational costs. Retina-based systems are currently only used in highly classified government and military facilities [46]. Although iris pattern recognition requires specific infrared hardware to authenticate users accurately [16,47,48], hence, we focus on offered mechanisms for mobile devices and those that may be materialized in the coming years.

1.4.1 Fingerprint

Example of use case scenario: First, the user has to let his device know the pattern of one or more of his fingerprints. To this end, he has to put each of the fingers he wants to use to unlock his device on the sensor, several times. Then, each time an authentication is required, the user places any previously recorded finger on the sensor.

This technology uses unique fingerprint patterns that are present in every human's fingers to authenticate users. Ridges that compose the pattern are traditionally classified into loops, arches, and whorls motifs. Figure 1.3 illustrates three examples of these patterns.

Fingerprint techniques were used for decades using ink to print the pattern onto a piece of paper [49]. However, several sensors were designed to perform the acquisition such as optical scanning,

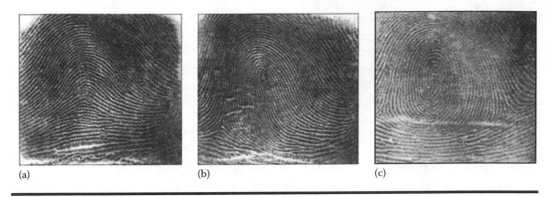

Figure 1.3 The three ridges pattern of the fingerprint: (a) arch pattern, (b) loop pattern, and (c) whorl pattern.

capacitive scanning, and ultrasound scanning [11,50,51]. Due to the maturity and the flexibility of fingerprint systems, it is now the most popular biometric system on mobile devices on the market. Indeed, fingerprint authentication delivers a high accuracy level and may be used in a wide range of environments. Moreover, with the growth of microtechnology and the emergence of mobile devices, more and more efficient fingerprint systems were integrated into these devices as means of authentication. The most well-known example of fingerprint used with mobile devices is the Apple *Touch ID* technology that comes from the patent of Bond et al. [52]. This major player of the mobile device industry builds an extra capacitive sensor in all their latest smartphones that scans subepidermal skin layers. Furthermore, they made the acquisition of the fingerprint possible up to 360° of orientation that provide a very high level of accuracy and a small error rate. Moreover, several other phone and tablet manufacturers also introduce fingerprint sensors built in their phones. The major difference between each one resides in the location of the sensors to ensure the ease for the end user (inside the power button, at the back of the phone).

Although Clarke et al. [19] assessed that 74% of mobile device users positively accept fingerprint biometrics as means of authentication, this mechanism has weaknesses that engender threats in its usage. It has been discovered that "most devices are unable to enroll some small percentage of users" [46]. The accuracy of fingerprint scanning may decrease to null when digits are either too wet or too dry and also too oily. That is the moisture effect. Fingerprints also tend to deteriorate over the time because of age, wear, or tear [46]. Furthermore, fingerprint authentication schemes suffer from confidentiality threats as well as spoofing attacks that question the security they attempt to provide. Indeed, on one hand, users may be scanned without their consent. On the other hand, Matsumoto et al. [53] proved that artificial fingers either made of silicon or gelatin were accepted by 11 fingerprint systems during the enrollment procedure.

1.4.2 Face Recognition

Example of use case scenario: In addition to the fingerprint authentication process, a user may let his device know his facial characteristics. Thereafter, when the face is recognized by the system, the user is granted access to the device.

Face recognition is the most natural mean of biometric authentication because humans also perform this evaluation in their everyday interactions. This authentication scheme may also be plainly integrated into an environment that allows image acquisition through mobile devices where a large majority of them have a frontal camera. Moreover, methods to acquire an image of the face are assessed as nonintrusive [16]. Most of the time, facial recognition systems rely on the analysis of facial features such as the position of eyes, nose, and mouth and distances between these features [54]. The evolution of both hardware and software leads facial recognition to become faster and to provide a better level of accuracy than before. Besides, it should be noted that a facial recognition may be continuously achieved. Indeed, the user may perform the authentication by his face, and then, the system may automatically verify that it is always the same face using the device when it is not in sleep mode.

Some years ago, Google offered a face-based authentication system. Although it was not genuinely successful, the improvement in the new version of the operating system still did not accurately identify the user in numerous cases such as low-lighting environment. As a matter of fact, Adini et al. [55] have pointed out this problem as a major drawback in facial recognition as well as a high complexity in the background. In addition, physical changes, such as hairstyle or beard variations and wearing hats or glasses [56], may greatly affect the matching rate of face recognition systems. Such systems also have troubles identifying very similar individuals such as twins [57] and maintaining a satisfying level of accuracy when physical changes occur owing to age [58]. Finally, the fact that a user may be scanned without his consent raises serious privacy threats.

1.4.3 Hand Geometry and Ear Shape

Example of use case scenario: In every biometric system, both hand and ear recognition need to learn from a user's unique physical features. Depending on the implementation, this process may be performed, several times, either through optical analysis or *via* the capacitive touch screen of the mobile device. Finally, the user has to repeat once the same process each time the device has to be unlocked.

Both hand geometry and ear shape biometric authentication mechanisms are based on the fact that nearly every individual's hands and ears are shaped differently. These body parts also remain practically the same after a certain age.

As regards ear shape as a means of authentication, Alphonse Bertillon's research helped to develop such biometric systems as he worked particularly on the classification of this body part [59]. Several more recent studies on the subject have shown that the acquisition of the ear was exclusively made with cameras [60–62]. In that sense, Descartes Biometric has released, recently, the most mature ear shape–based authentication system of the market—*Helix*. This software exploits the proximity sensor of the front camera on the mobile device. The user needs to place the device 6 to 12 in. away from his ear. Then, 30 images per second are recorded, processed, and finally compared to the stored template. Moreover, the company offers the possibility to configure the accuracy threshold at a higher level. However, the record of ear images with the front camera of the mobile device may be disturbing for users in their daily usage. In that sense, the Yahoo research department offers an experimental approach that handles the capacitive sensors embedded in the screen of mobile devices to record the topography of the ear. According to Hornyak [63], this system correctly identified users at 99.52% of the time. This rate is based on a test that involved 12 participants. Such a system appears to deliver a more appropriate ease of use for users since it is mimicking the act of calling.

On another side, hand geometry recognition is the ability to compare dimensions of fingers, localization of joints, shape and size of the palm, and of the phalanges. However, hand geometry is not distinctive enough to accurately identify a large set of individuals. Therefore, such a system may not be used in an authentication process, but rather in a verification process. Just as ear recognition, several studies involve a camera through the hand-record process [64–66]. By contrast, the report of Hornyak [63] noticed that the Yahoo research department also worked on a hand geometry recognition software. This experimental system allows the authentication of mobile device users in the same way as their ear recognition system. Both phalanges and palm identifications are possible, and the same matching rate of 99.52% was communicated and applied to this system.

Ear shape recognition and hand geometry appear to be an encouraging way in order to authenticate users on their mobile devices since they aim to be more usable for such devices. The example we presented earlier exposes very simple techniques, which are easy to use and which do not require any additional sensor than those already built in the mobile device. Moreover, these examples show promising results. Further, these two authentication schemes admit drawbacks. As regards ear shape, when recorded with a camera, hairs, hats, or piercings may compromise the identification process. Nevertheless, these limitations should be reduced when using the capacitive sensors of the mobile device touch screen. Regarding hand geometry, jewelry and arthritis will involve matching errors in both cases.

1.4.4 Voice: Speaker Recognition

Speaker recognition techniques are classified as a behavioral biometric. Indeed, they focus on vocal characteristics produced by the speech and not on the speech only. These features depend on the dimension of the vocal tract, mouth, and nasal cavities, but also rely on voice pitch, speaking style, and language [67] as represented in Figure 1.4.

There are two leading methods to process speaker recognition: text dependent and text independent [68,69]. Text-dependent recognition requires the user to pronounce a predefined passphrase. It is considered as a *voice password* and used both for the enrollment and the verification process. By contrast, text-independent systems are able to identify the user, for any text, properly. However, Boves and den Os [70] have identified a third type of speaker recognition technique that is a combination of the two others: the text-prompted method that randomly selects a passphrase the user needs to pronounce each time the system is used. Speaker recognition is an inexpensive solution

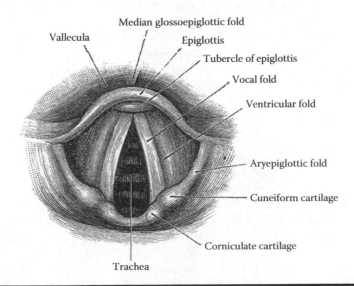

Figure 1.4　Anatomical diagram of the vocal folds or cords.

to authenticate users on their mobile devices since no additional sensor is required. In addition, speaker recognition is another mechanism that may be able to authenticate users continuously. Indeed, a first recognition may be achieved to grant the access to his owner, but such a process may also be performed each time the device either, or both, receive or emit a phone call. As an example of practical implementation, Google also introduces in *Smart Lock* the *Trusted Voice* feature. As all text-dependent-based speaker recognition, the user has to enroll his voice by pronouncing "Ok Google" three times. Then, the user must say the same passphrase. Whenever the record matches both the voice model and the passphrase, the access to the mobile device is granted.

However, speaker recognition–based authentication mechanisms demonstrate several major drawbacks. In the first place, since such systems are viewed as behavioral, the current physical, medical, or emotional condition of the user may considerably affect the accuracy. Voice is also likely to change over time due to age. Moreover, speaker recognition techniques are rarely noise resistant. Then, various loud background noises make such systems almost impossible to use in public places such as bars or public transports. Finally, speaker recognition techniques are singularly exposed to security threats. It is possible for an attacker to record or imitate the voiceprint of the user to perform a fraudulent authentication afterward [71].

1.4.5 Gait

Example of use case scenario: The process of authentication via gait analysis is independent of any action from the user. The mobile device is able to determine whether he is walking or not and then perform a recognition to unlock the device automatically in a continuous manner.

Gait recognition is a technology based on the analysis of the *rhythmic patterns associated with a walking stride* [72]. The observation that each human's walking style is different leads to the development of advanced biometric authentication systems that exploit such behavioral characteristics.

Accordingly, studies have proposed a gait analysis mechanism based on an accelerometer to collect the features that create the gait template [12,73]. It is possible for such a system to be integrated with mobile devices since they include built-in inertial sensors (accelerometer, gyroscope). As a matter of fact, gait recognition may become a convenient way to authenticate users as they always keep their mobile devices within their pocket or a bag. It is fair to say that it constitutes a human-centered system since the authentication process is wholly imperceptible to the user.

Withal, gait is not as stable over time due to changes in body weight. Such a physical change is not the only way for this human behavior to change. Indeed, brain damage, injury, and also inebriety may affect how individuals walk from a short time to a long time or permanent variation [74,75].

1.4.6 Keystroke Dynamics

Keystroke dynamics is based on the measurement and assessment of a human's typing rhythm on keyboards/keypads. This process allows the creation of a digital print upon the user's interaction with devices that are rich in cognitive qualities [76]. Characteristics of this user behavior are fairly unique to each person and hold a high potential as an authentication mechanism. With the growth of mobile devices' capacitive touch screens, keystrokes patterns are now capable of providing even more unique features for the authentication than only typing rhythm, which includes key press duration, latencies, typing rate, and typing pressure. Such characteristics may be measured up to millisecond order precision [77] and more recent studies have pointed out high accuracy levels [78,79]. It is thus nearly impossible for an attacker to replicate a defined keystroke pattern without an enormous amount of effort. The main benefit of keystroke dynamic pattern recognition is that anything except an extra software layer is required. Moreover, keystroke analysis may be employed as a continuous authentication process for free typing instead of one-time authentication such as fingerprint. However, since touch keyboards only appear when users are granted the full access, it is not possible to perform the authentication process *a posteriori*. As a result, keystroke dynamics may only be employed to verify continuously that the current user is well and truly the owner of the mobile device afterward. Thus, such a system must not be self-sufficient and require to be coupled to a noncontinuous authentication scheme as regards a mobile device usage.

1.4.7 Signature

Example of use case scenario: Generally, a blank canvas is prompted to the user where he has to make his personal signature. The analysis is based on the way the signature is produced by exploiting pressure, direction, and acceleration rather than only a comparison of the signature pattern itself. The first step resides in the definition of such a model to compare with, and then, the user has to reproduce his signature each time the mobile device has to be accessed.

Signatures were used for decades in the concrete world when people need to enact documents. The same idea is used over numerical systems for authentication purposes. Signature recognition is considered as a behavioral biometric since it is based on the dynamics of writing a signature rather than a unique comparison of the signature itself.

As the users have to reproduce their signature on a mobile device touch screen, the identification process may determine dynamics, owing the measurement of the pressure, the direction, the velocity, and the acceleration and the length of the strokes. Initially, the hardware used to record an individual's signature was not convenient enough for users [80]. However, with the advent of capacitive touch screens on mobile devices, the use of such an authentication process became more user-friendly. Moreover, a unique extra software layer is required to make it work.

As for every behavioral biometric system, despite how a person's physical and emotional condition may considerably affect an authentication mechanism based on signature recognition, it is important to note that this is the only biometric authentication scheme that can be deliberately changed by the user even though any replication requires lots of effort.

1.4.8 Heartbeat

Recently, Sufi et al. [81] have introduced the use of the electrical activity of the heart as a novel biometric authentication mechanism. Since the heartbeat is considered to be unique for each person [82], such a system requires a record of an electrocardiogram (*ECG* or *EKG*) as illustrated in Figure 1.5. Hence, as in biometric authentication schemes we described earlier, the first step of the process consists in an enrollment phase. Unique features are extracted to build a template then compared through the identification process.

Exploiting ECG as a means of authentication is suitable across a wide range of people. Indeed, heartbeat samples may be collected from any part of the body such as fingers, toes, chest, and wrist. Thus, people who suffer from large injuries may be authenticated, continuously or not, by using such a system. As we saw before, connected objects and—in a broader sense—IoT begin to take a measurable place through the numerical environment [44]. In that sense, a Canadian company patented a wristband that is fully compliant with mobile devices via a companion application. This band is able to authenticate a user by his heart signature [83].

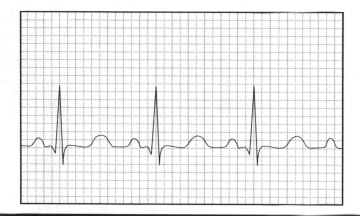

Figure 1.5   Example of ECG curve.

Such a system seems to be largely reliable, as reproducing heartbeat signature depends upon sophisticated skills and hardware. Therefore, it is nearly impossible for an attacker to spoof such authentication systems. Since it may be considered as a behavioral biometric, several factors may seriously affect its accuracy such as daily activities, nutrition, stress level, and weariness.

1.4.9 Evaluation of the Accuracy

As we have seen in earlier sections, biometric systems are not devoid of potential errors over the authentication process. These systems can make two types of errors: false acceptance rate (*FAR*) and false rejection rate (*FRR*). Figure 1.3 graphically illustrates these types of errors in detail. On one hand, *FAR*, or also false match rate, is the probability that the system incorrectly declares a successful match between the input pattern and one stored in the database. *FAR* is obtained by the following equation, where F_a is the number of false acceptances, and V_i is the number of imposter verifications:

$$FAR = \frac{F_a}{V_i} \tag{1.3}$$

On the other hand, *FRR*, or false nonmatch rate, is the probability that the system declares a failure while the input pattern matches with the template stored. *FRR* is obtained by the following equation, where F_r is the number of false rejections, and V_g is the number of genuine verifications:

$$FRR = \frac{F_r}{V_g} \tag{1.4}$$

Generally, the matching algorithm performs a decision using some parameters as a threshold. Graphically expressed, both *FAR* and *FRR* by opposition to the given threshold represent the relative operating characteristic. This plot allows finding the equal error rate (*EER*) as shown in Figure 1.6. *EER* is the rate at which both accept and reject errors are equal. This rate is commonly used to evaluate biometrics; indeed, the lower the *EER*, the more accurate the system is considered to be. The report by Mansfield and Wayman [84] describe in detail the performance evaluation for biometric systems.

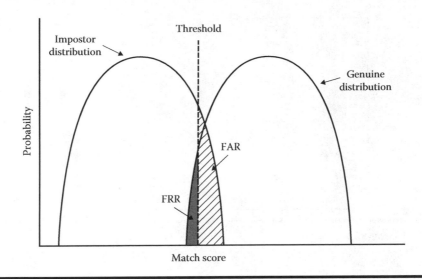

Figure 1.6 Biometric system error rates: the curves show false acceptance rate and false rejection rate for a given threshold.

Table 1.1 Evaluation of Various Biometrics Performances

	EER	FAR	FRR	Conditions
Face recognition	—	1%	10%	Varied light: indoor, outdoor
Fingerprint	2%	0.1%	2%	Rotation and exaggerated skin distortion
Hand geometry	1%	0.14%	2%	With rings and improper placement
Iris	0.01%	≈ 0%	0.99%	Indoor
Voice recognition	6%	3%	10%	Text dependent and multilingual
Keystrokes	1.8%	7%	0.1%	Data record during 6 months

Sources: Jain, A.K. et al., Biometrics: A grand challenge, in *Proceedings of the 17th International Conference on Pattern Recognition, 2004 (ICPR 2004)*, 2004, pp. 935–942; Phillips, P.J. et al., Computer, 33, 56, 2000; Jain, A. et al., *Biometric Systems Technology, Design and Performance Evaluation*, Springer Verlag, London, U.K., 2005.

Several studies [85–87] exhibit the performance of various biometric authentication mechanisms regarding evaluation criteria described earlier as illustrated in Table 1.1.

The accuracy required for biometric systems depends chiefly on the need of the application. Presently, biometric systems designers aim at reducing FARs. However, the FRRs undeniably grow. To cover *FRR* up, most of the current designs offer to bypass the biometric system that fails and suggest the user to perform another authentication scheme instead. The use of such a fallback mechanism represents a multifactor authentication scheme. Biometric was often introduced as an encouraging way to end use of passwords (knowledge-based schemes). However, since both Apple and Google recently released their respective biometric technologies *Touch ID, Nexus Imprint*, and

Smart Lock, mobile device users still have to set up a more traditional authentication mechanism such as PIN before enabling these features. As a matter of fact, we may affirm that passwords are still not dead contrary to everything said. Nevertheless, as suggested by other researchers, it seems that threats that can be thwarted by biometric authentication that operates together with rescue passwords, yet remain less secure than just only a knowledge-based authentication mechanism. Indeed, a two-factor authentication system must be treated as a conjunctive statement in opposition to a disjunctive statement. In other words, both the main system and the fallback one have to authenticate the user properly, and not just one, as offered by the main biometric solutions available on the market.

1.5 Discussion

Based on the earlier critical analysis, it is possible to say that each authentication mechanism analyzed concedes some advantages as well as drawbacks. Here we discuss the whole set of proposed mechanisms.

First, we offer an examination of knowledge-based authentication schemes via both the theoretical and real measure of the password space entropy that each of them provides—as illustrated in Table 1.2. Theoretical entropy is inevitably greater than the real one since users do choose their own piece of knowledge. The main deficiency of PINs resides in their simplicity, as they remain easy to guess by almost everyone. Text-enhanced passwords are an excessively complex solution for mobile device users that force them to choose easily-findable identification code. Finally, graphical and haptic passwords provide an adequate level of security, but yet remain easy to obtain via shoulder surfing attacks or smudge attacks. Such an observations support the idea for a more diverse ecosystem of multiple authentication schemes [88].

Second, we suggest an empirical evaluation of biometrics authentication mechanisms based on the work of Jain et al. [16] in 1999 (Table 1.3). As assessed in the previous sections, several novel studies aim at improving biometric processes to be more accurate and fast, as regards mobile devices [78,89–91]. As a matter of fact, this study requires some improvement. Hence, changes we introduced mainly focus on the performance criteria. Moreover, we include heartbeat authentication that was not discussed in their work. Our evaluation is based on the same criteria suggested by Jain et al. [16], which were defined by biometric experts. Such a guideline is described as follows:

Table 1.2 Evaluation of Knowledge-Based Authentication Mechanisms for Mobile Devices via the Password Space Entropy Metric

	Knowledge-Based		
	Explicit		
	PIN	*Text-Enhanced Passwords*	*Graphical and Haptic Passwords*
Theoretical password-space entropy	L	H	M
Real password-space entropy	L	L	M

Abbreviations: H, high; M, medium; L, low.

Table 1.3 Evaluation of Biometric Authentication Mechanisms for Mobile Devices according to the Criteria Proposed by Jain et al. [16]

	Biometrics								
	Physiological				Behavioral				
	Fingerprint	Face	Ear Shape	Hand Geometry	Voice	Gait	Keystroke	Signature	Heartbeat [91]
Universality [16]	M	H	M	M	M	M	L	L	**M**
Uniqueness [16]	H	L	M	M	L	L	L	L	**M**
Permanence [16]	H	M	H	M	L	L	L	L	**L**
Collectability [16]	M	H	M	H	M	H	M	H	**M**
Performance [16]	**H**	M [89]	H [63]	H [63]	M	M [78]	M [90]	L	**M**
Acceptability [16]	M	H	H	M	H	H	M	H	**H**
Circumvention [16]	H	L	M	M	L	M	M	L	**H**

Note: Improvements are identified by **bold.**
Abbreviations: H, high; M, medium; L, low.

1. *Universality*: Biometric solutions rely upon singularities of the human body or behavior, but the ability of such mechanisms to accurately identify the genuine user largely varies between each one. Consequently, Jain et al. [16] have suggested to quantitate the fact that each person should have that characteristic.

2. *Uniqueness*: Some physical traits of the human body (face) remain largely close in some cases (twins). Therefore, Jain et al. [16] have proposed to evaluate the probability that two individuals are potentially the same, in terms of characteristics.

3. *Permanence*: Physical or behavioral features of the human used with biometrics may gradually evolve. This criterion figures out the invariance of these characteristics with time [16].

4. *Performance*: The inability of a biometric system to identify a user with 100% accuracy leads to identifying the related performance offered by each one [16].

5. *Collectability*: Most of the time, the biometric authentication process involves additional hardware or a major computing complexity with mobile devices. This criterion refers to the evaluation of how simple a characteristic is quantitatively measurable [16].

6. *Acceptability*: All the biometric techniques still did not become the custom. Then, it is important to state the user's acceptance rate, according to such mechanisms [16].

7. *Circumvention*: The last criterion reported by Jain et al. [16] refers to the easiness of mimicking a singular trait or behavior with biometric systems. Such an evaluation delivers the strength rate of biometric systems in front of fraudulent attacks (*spoofing attack*).

Finally, we estimate that several criteria were missing to perform a proper evaluation of authentication mechanisms (Table 1.4). Consequently, we introduce four new criteria based on the

Table 1.4 Evaluation of All Authentication Mechanisms for Mobile Devices via the Criteria We Suggest

| | Knowledge-Based | | | | | Biometrics | | | | | | | | |
| | Explicit | | | Implicit | | Physiological | | | | Behavioral | | | | |
	PIN	Text-Enhanced	Graphical and Haptic	Cognitive-Based Passwords	Token-Based	Fingerprint	Face Recognition	Ear Shape	Hand Geometry	Voice Recognition	Gait Recognition	Keystroke Dynamics	Signature	Heartbeat
Confidentiality	L	L	M	H	H	M	M	M	M	M	L	H	M	L
Intrusive	N/A	N/A	N/A	M	N/A	M	H	M	M	L	L	L	L	H
Ease of use	M	L	M	M	M	H	M	M	M	M	H	H	M	H
Usage frequency	H	L	H	L	M	M	M	L	L	M	L	M	M	L

Abbreviations: H, high; M, medium; L, low.

previous critical analysis of these schemes we proposed. As a matter of fact, we define the following guidelines:

1. *Confidentiality*: Several authentication scheme designs, such as fingerprint, record or store data that relate to the secrecy of the user. Here, we target the threat that is associated with the access to such an information.
2. *Intrusive*: Several authentication mechanisms reviewed in this chapter involve providing information or features relating to the user. This criterion focuses on the ethical concern of the data that used to be collected during the authentication process. This criterion mainly focuses on cognitive schemes as well as biometric ones. Indeed, the authentication process requires handling data that directly concern users and may be performed without their broad agreement.
3. *Ease of use*: The major drawback of authentication mechanisms is predominantly due to the way that humans have to interact with these systems. In that sense, we suggest evaluating how the user is involved in the interaction process and how the system focuses on what people want to do rather than possibilities offered by the technology.
4. *Usage frequency*: Since most of the more used authentication schemes also remain the weakest ones, we offer to identify to the popularity of the mechanisms when applied to secure a mobile device.

Based on such an evaluation, it is possible to affirm that most authentication schemes raise some inconveniences for users since most of them are considered as not easy to use. Besides, proposed mechanisms that do not involve users in the process of authentication that is transparent for them, such as gait recognition or keystroke dynamics, were identified as convenient. As a matter of fact, it is clear that the future trend for authentication on mobile devices will turn to systems that focus on the users first. As an example, it is known that people do spend a considerable amount of time in a few key locations such as home or work, as assessed by Hayashi and Hong [92]. In that sense, ubiquitous mechanisms that will be able to learn about a user's habits and that will not require any passwords or tokens are close. Indeed, studies in that field of research expose promising results [93], but efforts should probably pay more attention to such a key idea in coming years.

Nowadays, authentication mechanisms remain an important field of interest for researchers and leaders of the mobile device industry. According to major players on the market, it should be noted that both Apple and Microsoft took the biometric band, respectively, with the *Touch ID* technology and the launch of Windows 10 [94]. Despite that, Google's *Advanced Technology and Projects* division is currently working on an experimental multimodal biometrics system based on behavioral analysis: *Project Abacus.* The system will identify a genuine user via a "trust score" calculated through a real-time analyzer of the user's voice recognition, keystroke dynamics and touch gestures, facial recognition, and location. The firm presented the research results at its annual conference in 2015 and claimed that the entropy of such a system is now 10 times higher than the most valuable fingerprint system on the market. While the project is still in development, Google aims to integrate such a system into the next versions of the Android operating system. Hence, this research project simply confirms the trend in the evolution of authentication for the near future.

1.6 Conclusion

The present review of mobile device authentication mechanisms leads us to maintain that each of the schemes we introduced concedes several strong and weak aspects. Since knowledge-based mechanisms were designed to a machine point of view, they first involve an enormous amount of memory efforts from users. Behaviors they consequently adopt to overcome a system they are not comfortable with yield several threats and weaknesses as regards the security of their mobile devices. Token-based authentication schemes are not devoid of weaknesses either. However, with the ongoing *IoT*, such mechanisms appear to be more convenient than knowledge-based systems and will certainly keep growing. Presently, biometric methods become more and more popular and easy to reach for everyone. Some remain just too much intrusive for users or lead them to believe that providing personal and unique features describing them is an important threat to their privacy. Nevertheless, biometric techniques are, overall, very accurate, but also accept certain dysfunctions. However, if they gather a number of biometric mechanisms together, they allow the entropy of the entire multifactor system to extend, and consequently increase its accuracy. The major drawback of biometrics resides in the fallback mechanism designed to cope with false rejections and let a genuine user proceed to his authentication properly, although most of the current biometric solutions for mobile devices available on the market such as *Touch ID* or *Smart Lock* have adopted this method.

As we state that the human factor is the fundamental drawback for authentication mechanisms since they involve lots of interaction with users, we consider that the aftermath of such systems should singularly take care of this criterion. Such mechanisms already evolve to become *no password systems.* Although they currently provide better convenience, they do not provide better security. Based on such an evaluation, we assume that the optimal solution should be able to recognize a genuine user without the need for any interaction of his part. To be really accurate and secure, such a solution should be based on the user's everyday habits (what he does throughout his day, in which order?) that involve collecting most of the possible relevant patterns via the mobile device. However, it is important to consider that such a solution implies a large set of information more than just for one user. The processing of all of this knowledge, considering each mobile device users, will undeniably increase the cost of hardware requirement. This observation now questions us about the ecological issues related to the growth of the number of brand new data centers everywhere in the world.

References

1. R. Van Der Meulen and J. Rivera. Gartner says smartphone sales grew 46.5 percent in second quarter of 2013 and exceeded feature phone sales for first time, 2013. Available at http://www.gartner.com/newsroom/id/2573415, accessed April, 2016.
2. T.-A. Wilska. Mobile phone use as part of young people's consumption styles. *Journal of Consumer Policy*, 26:441–463, 2003.
3. G. Goggin. *Cell Phone Culture: Mobile Technology in Everyday Life*. Routledge, 2012.
4. Nielsen. Smartphones: So many apps, so much time, 2014. Available at http://www.nielsen.com/us/en/insights/news/2014/smartphones-so-many-apps-so-much-time.html, accessed April, 2016.
5. H. Falaki, R. Mahajan, S. Kandula, D. Lymberopoulos, R. Govindan, and D. Estrin. Diversity in smartphone usage. In *Proceedings of the Eighth International Conference on Mobile Systems, Applications, and Services*, San Francisco, CA, pp. 179–194. 2010.
6. G. Lowe. A hierarchy of authentication specifications. In *Proceedings of the 10th Computer Security Foundations Workshop*, Rockport, MA, pp. 31–43. 1997.
7. S. Z. Li. *Encyclopedia of Biometrics: I-Z*, Vol. 1. Springer Science & Business Media, 2009.
8. L. Lamport. Password authentication with insecure communication. *Communications of the ACM*, 24:770–772, 1981.
9. M. Zviran and W. J. Haga. User authentication by cognitive passwords: An empirical assessment. In *Proceedings of the Fifth Jerusalem Conference on Information Technology, 1990, Next Decade in Information Technology*, (Cat. No. 90TH0326–9), Jerusalem, Israel, pp. 137–144. 1990.
10. J. Xu, W.-T. Zhu, and D.-G. Feng. An improved smart card based password authentication scheme with provable security. *Computer Standards & Interfaces*, 31:723–728, 2009.
11. A. Jain, L. Hong, and R. Bolle. On-line fingerprint verification. *IEEE Transactions on Pattern Analysis and Machine Intelligence*, 19:302–314, 1997.
12. D. Gafurov, K. Helkala, and T. Søndrol. Biometric gait authentication using accelerometer sensor. *Journal of Computers*, 1:51–59, 2006.
13. D. Benyon, P. Turner, and S. Turner. *Designing Interactive Systems: People, Activities, Contexts, Technologies*. Pearson Education, Addison-Wesley, 2005.
14. N. Ben-Asher, H. Sieger, D. Telekom, A. Ben-Oved, N. Kirschnick, J. Meyer et al. On the need for different security methods on mobile phones. In *Proceedings of the 13th International Conference on Human Computer Interaction with Mobile Devices and Services (MobileHCI '11)*. ACM, New York, NY, pp. 465–473. 2011.
15. J. Yan. Password memorability and security: Empirical results. *IEEE Security & Privacy*, 2(5):25–31. 2004.
16. A. K. Jain, R. Bolle, and S. Pankanti. *Biometrics: Personal Identification in Networked Society*. Springer Science & Business Media, 1999.
17. R. V. Yampolskiy. Analyzing user password selection behavior for reduction of password space. In *Proceedings 2006 40th Annual IEEE International Carnahan Conferences Security Technology*, Carnahan, pp. 109–115. 2006.

18. N. L. Clarke and S. M. Furnell. Authentication of users on mobile telephones—A survey of attitudes and practices. *Computers & Security,* 24:519–527, 2005.

19. N. L. Clarke, S. M. Furnell, P. M. Rodwell, and P. L. Reynolds. Acceptance of subscriber authentication methods for mobile telephony devices. *Computers & Security,* 21:220–228, 2002.

20. S. Hansman and R. Hunt. *A Taxonomy of Network and Computer Attack Methodologies,* Vol. 7. Department of Computer Science and Software Engineering, University of Canterbury, Christchurch, New Zealand, 2003.

21. T. N. Jagatic, N. A. Johnson, M. Jakobsson, and F. Menczer. Social phishing. *Communications of the ACM,* 50:94–100, 2007.

22. G. L. Orgill, G. W. Romney, M. G. Bailey, and P. M. Orgill. The urgency for effective user privacy-education to counter social engineering attacks on secure computer systems. In *Proceedings of the Fifth Conference on Information Technology Education,* Salt Lake City, UT, pp. 177–181. 2004.

23. F. Tari, A. Ozok, and S. H. Holden. A comparison of perceived and real shoulder-surfing risks between alphanumeric and graphical passwords. In *Proceedings of the Second Symposium on Usable Privacy and Security,* pp. 56–66. 2006.

24. R. Morris and K. Thompson. Password security: A case history. *Communications of the ACM,* 22: 594–597, 1979.

25. F. Cuadrado and J. C. Dueñas. Mobile application stores: success factors, existing approaches, and future developments. *Communications Magazine, IEEE,* 50:160–167, 2012.

26. M. A. Sasse, S. Brostoff, and D. Weirich. Transforming the 'weakest link'—A human/computer interaction approach to usable and effective security. *BT Technology Journal,* 19:122–131, 2001.

27. S. Riley. Password security: What users know and what they actually do. *Usability News,* 8:2833–2836, 2006.

28. SplashData. "123456" Maintains the top spot on our annual worst passwords *List,* Janvier 20, 2014. Available at http://splashdata.com/splashid/worst-passwords/, accessed April, 2016.

29. E. A. Kirkpatrick. An experimental study of memory. *Psychological Review,* 1:602, 1894.

30. G. E. Blonder. Graphical password. U.S. Patent 5559961, 1996.

31. R. Biddle, S. Chiasson, and P. C. Van Oorschot. Graphical passwords: Learning from the first twelve years. *ACM Computing Surveys (CSUR),*44:19, 2012.

32. S. Uellenbeck, M. Dürmuth, C. Wolf, and T. Holz. Quantifying the security of graphical passwords: The case of android unlock patterns. In *Proceedings of the 2013 ACM SIGSAC Conference on Computer & Communications Security,* Berlin, Germany, pp. 161–172. 2013.

33. S. Sinofsky. Signing in with a picture password, 2011. Available at http://blogs.msdn.com/b/b8/archive/2011/12/16/signing-in-with-a-picture-password.aspx, accessed April, 2016.

34. Z. Zhao, G. J. Ahn, and H. Hu. Picture gesture authentication: Empirical analysis, automated attacks, and scheme evaluation. *ACM Transactions on Information and System Security,* 17:1–34, 2015.

35. A. J. Aviv, K. Gibson, E. Mossop, M. Blaze, and J. M. Smith. Smudge attacks on smartphone touch screens. *WOOT,* 10:1–7, 2010.

36. A. H. Lashkari, S. Farmand, D. Zakaria, O. Bin, and D. Saleh. Shoulder surfing attack in graphical password authentication. *International Journal of Computer Science and Information Security,* 6(2): 145–154, 2009.

37. J. Gugenheimer, A. De Luca, H. Hess, S. Karg, D. Wolf, and E. Rukzio. ColorSnakes: Using colored decoys to secure authentication in sensitive contexts. In *Proceedings of the 17th International Conference on Human-Computer Interaction with Mobile Devices and Services,* Copenhagen, Denmark, pp. 274–283. 2015.

38. A. Bianchi, I. Oakley, and D. S. Kwon. The secure haptic keypad: A tactile password system. in *Proceedings of the Conference on Human Factors in Computing Systems,* Atlanta, GA, pp. 1089–1092. 2010.

39. A. Bianchi, I. Oakley, and D. S. Kwon. Counting clicks and beeps: Exploring numerosity based haptic and audio PIN entry. *Interacting with Computers*, 24:409–422, 2012.

40. L. Lazar, O. Tikolsky, C. Glezer, and M. Zviran. Personalized cognitive passwords: An exploratory assessment. *Information Management & Computer Security*, 19:25–41, 2011.

41. J. Bunnell, J. Podd, R. Henderson, R. Napier, and J. Kennedy-Moffat. Cognitive, associative and conventional passwords: Recall and guessing rates. *Computers & Security*, 16:629–641, 1997.

42. F. F. I. E. Council. Authentication in an internet banking environment. Financial Institution Letter, FIL-103–2005. Federal Deposit Insurance Corp. (FDIC), Washington, DC. Vol. 18, Retrieved March 2005.

43. F. Aloul, S. Zahidi, and W. El-Hajj. Two factor authentication using mobile phones. In *IEEE/ACS International Conference on Computer Systems and Applications, 2009 (AICCSA 2009)*, Rabat, Morocco, pp. 641–644. 2009.

44. D. Miorandi, S. Sicari, F. De Pellegrini, and I. Chlamtac. Internet of things: Vision, applications and research challenges. *Ad Hoc Networks*, 10:1497–1516, September 2012.

45. B. Schneier. Two-factor authentication: too little, too late. *Communication of the ACM*, 48:136, 2005.

46. S. Nanavati, M. Thieme, N. Raj, and R. Nanavati. 2002. *Biometrics: Identity Verification in a Networked World*. John Wiley & Sons, Inc., New York, NY.

47. J. Daugman. How iris recognition works. *IEEE Transactions on Circuits and Systems for Video Technology*, 14:21–30, 2004.

48. R. P. Wildes. Iris recognition: an emerging biometric technology. *Proceedings of the IEEE*, 85:1348–1363, 1997.

49. J. Berry and D. A. Stoney. The history and development of fingerprinting. *Advances in Fingerprint Technology*, 2:13–52, 2001.

50. K. L. Kroeker. Graphics and security: exploring visual biometrics. *Computer Graphics and Applications, IEEE*, 22:16–21, 2002.

51. A. K. Jain, A. Ross, and S. Pankanti. Biometrics: A tool for information security. *IEEE Transactions on Information Forensics and Security*, 1:125–143, 2006.

52. R. H. Bond, A. Kramer, and G. Gozzini. Molded fingerprint sensor structure with indicia regions. U.S. patent US D652332 S1, 2012.

53. T. Matsumoto, H. Matsumoto, K. Yamada, and S. Hoshino. Impact of artificial gummy fingers on fingerprint systems. In *Electronic Imaging 2002*, San Jose, CA, pp. 275–289. 2002.

54. M. Dabbah, W. Woo, and S. Dlay. Secure authentication for face recognition. In *IEEE Symposium on Computational Intelligence in Image and Signal Processing, 2007 (CIISP 2007)*, Honolulu, HI, pp. 121–126. 2007.

55. Y. Adini, Y. Moses, and S. Ullman. Face recognition: The problem of compensating for changes in illumination direction. *IEEE Transactions on Pattern Analysis and Machine Intelligence*, 19:721–732, 1997.

56. A. M. Martínez. Recognizing imprecisely localized, partially occluded, and expression variant faces from a single sample per class. *IEEE Transactions on Pattern Analysis and Machine Intelligence*, 24:748–763, 2002.

57. P. J. Grother, G. W. Quinn, and P. J. Phillips. Report on the evaluation of 2D still-image face recognition algorithms. *NIST Interagency Report*, 7709:106, 2010.

58. A. Lanitis, C. J. Taylor, and T. F. Cootes. Toward automatic simulation of aging effects on face images. *IEEE Transactions on Pattern Analysis and Machine Intelligence*, 24:442–455, 2002.

59. A. Bertillon. Anthropometric identification; descriptive instructions—Identification anthropométrique: instructions signalétiques. Imprimerie Administrative Publisher, 1893.

60. T. Yuizono, Y. Wang, K. Satoh, and S. Nakayama. Study on individual recognition for ear images by using genetic local search. In *Proceedings of the Congress on Evolutionary Computation, 2002. CEC '02*. Honolulu, HI, pp. 237–242. 2002.

61. M. Choraś. Ear biometrics based on geometrical feature extraction. *Electronic Letters on Computer Vision and Image Analysis,* 5:84–95, 2005.

62. H. Chen and B. Bhanu. Contour matching for 3D ear recognition. In *Seventh IEEE Workshops on Application of Computer Vision, 2005 (WACV/MOTIONS'05),* Vol. 1, pp. 123–128. 2005.

63. T. Hornyak. Yahoo wants you to use your ears and knuckles to unlock your phone, 2015. Available at http://www.pcworld.com/article/2913972/yahoo-explores-ears-knuckles-to-unlock-smartphones.html, accessed April, 2016.

64. A. Kumar, D. C. Wong, H. C. Shen, and A. K. Jain. Personal verification using palmprint and hand geometry biometric. In *Audio-and Video-Based Biometric Person Authentication,* Guildford, UK, pp. 668–678. 2003.

65. A. Ross and A. Jain. A prototype hand geometry-based verification system. In *Proceedings of Second Conference on Audio and Video Based Biometric Person Authentication,* Washington DC, pp. 166–171. 1999.

66. R. Sanchez-Reillo, C. Sanchez-Avila, and A. Gonzalez-Marcos. Biometric identification through hand geometry measurements. *IEEE Transactions on Pattern Analysis and Machine Intelligence,* 22:1168–1171, 2000.

67. A. Eriksson and P. Wretling. How flexible is the human voice?–A case study of mimicry. *Target,* 30:29–90, 1997.

68. G. R. Doddington, M. A. Przybocki, A. F. Martin, and D. A. Reynolds. The NIST speaker recognition evaluation–overview, methodology, systems, results, perspective. *Speech Communication,* 31:225–254, 2000.

69. B. Gold, N. Morgan, and D. Ellis. *Speech and Audio Signal Processing: Processing and Perception of Speech and Music.* John Wiley & Sons, 2011.

70. L. Boves and E. den Os. Speaker recognition in telecom applications. In *Proceedings of the IEEE Fourth Workshop on Interactive Voice Technology for Telecommunications Applications, 1998 (IVTTA'98),* Torino, Italy, pp. 203–208. 1998.

71. Y. W. Lau, M. Wagner, and D. Tran. Vulnerability of speaker verification to voice mimicking. In *Proceedings of 2004 International Symposium on Intelligent Multimedia, Video and Speech Processing, 2004,* Hong Kong, pp. 145–148. 2004.

72. M. P. Rani and G. Arumugam. An efficient gait recognition system for human identification using modified ICA. *International Journal of Computer Science and Information Technology,* 2:55–67, 2010.

73. M. O. Derawi, C. Nickel, P. Bours, and C. Busch. Unobtrusive user-authentication on mobile phones using biometric gait recognition. In *2010 Sixth International Conference on Intelligent Information Hiding and Multimedia Signal Processing (IIH-MSP),* pp. 306–311. 2010.

74. J. E. Boyd. Synchronization of oscillations for machine perception of gaits. *Computer Vision and Image Understanding,* 96:35–59, 2004.

75. S. Sarkar, P. J. Phillips, Z. Liu, I. R. Vega, P. Grother, and K. W. Bowyer. The humanid gait challenge problem: Data sets, performance, and analysis. *IEEE Transactions on Pattern Analysis and Machine Intelligence,* 27:162–177, 2005.

76. M. S. Obaidat. A verification methodology for computer systems users. In *Proceedings of the 1995 ACM Symposium on Applied Computing,* Nashville, TN, pp. 258–262. 1995.

77. C. Senk and F. Dotzler. Biometric Authentication as a service for enterprise identity management deployment: A data protection perspective. In *2011 Sixth International Conference on Availability, Reliability and Security (ARES),* pp. 43–50. 2011.

78. Y. Deng and Y. Zhong. Keystroke dynamics advances for mobile devices using deep neural network. *Gate to Computer Science & Research,* 2:59, 2015.

79. M. Trojahn, F. Arndt, and F. Ortmeier. Authentication with keystroke dynamics on touchscreen keypads-effect of different N-graph combinations. In *Third International Conference on Mobile Services, Resources, and Users (MOBILITY)*, pp. 114–119. 2013.

80. V. S. Nalwa. Automatic on-line signature verification. *Proceedings of the IEEE*, 85:215–239, 1997.

81. F. Sufi, I. Khalil, and J. Hu. ECG-based authentication. In *Handbook of Information and Communication Security*, pp. 309–331. Springer, 2010.

82. M. G. Khan, *Rapid ECG Interpretation*: Springer, 2008.

83. S. Z. Fatemian, F. Agrafioti, and D. Hatzinakos. HeartID: Cardiac biometric recognition. In *2010 Fourth IEEE International Conference on Biometrics: Theory Applications and Systems (BTAS)*, Washington DC, pp. 1–5. 2010.

84. A. J. Mansfield and J. L. Wayman. *Best Practices in Testing and Reporting Performance of Biometric Devices*. Centre for Mathematics and Scientific Computing, National Physical Laboratory Teddington, Middlesex, U.K., 2002.

85. A. K. Jain, S. Pankanti, S. Prabhakar, L. Hong, and A. Ross. Biometrics: a grand challenge. In *Proceedings of the 17th International Conference on Pattern Recognition (ICPR 2004)*, Cambridge, MA, pp. 935–942. 2004.

86. P. J. Phillips, A. Martin, C. L. Wilson, and M. Przybocki. An introduction evaluating biometric systems. *Computer*, 33:56–63, 2000.

87. A. Jain, D. Maltoni, D. Maio, and J. Wayman. *Biometric Systems Technology, Design and Performance Evaluation*. Springer Verlag, London, U.K., 2005.

88. A. Forget, S. Chiasson, and R. Biddle. User-centred authentication feature framework, *Information & Computer Security*, 23(5):497–515, 2015.

89. M. E. Fathy, V. M. Patel, and R. Chellappa. Face-based active authentication on mobile devices. In *2015 IEEE International Conference on Acoustics, Speech and Signal Processing (ICASSP)*, Brisbane, Australia, pp. 1687–1691. 2015.

90. C. Nickel, T. Wirtl, and C. Busch. Authentication of smartphone users based on the way they walk using k-NN algorithm. In *2012 Eighth International Conference on Intelligent Information Hiding and Multimedia Signal Processing (IIH-MSP)*, pp. 16–20. 2012.

91. N. Belgacem, R. Fournier, A. Nait-Ali, and F. Bereksi-Reguig. A novel biometric authentication approach using ECG and EMG signals. *Journal of Medical Engineering & Technology*, 39:1–13, 2015.

92. E. Hayashi and J. Hong. A diary study of password usage in daily life. In *Proceedings of the SIGCHI Conference on Human Factors in Computing Systems*, Vancouver, BC, Canada, pp. 2627–2630. 2011.

93. N. Micallef, M. Just, L. Baillie, M. Halvey, and H. G. Kayacik. Why aren't users using protection? Investigating the usability of smartphone locking. In *17th International Conference on Human-Computer Interaction with Mobile Devices and Services*, pp. 284–294, New York, 2015.

94. J. Belfiore. Making Windows 10 more personal and more secure with Windows Hello, 2015. Available at http://blogs.windows.com/bloggingwindows/2015/03/17/making-windows-10-more-personal-and-more-secure-with-windows-hello/.

Chapter 2

When Seeing Is Not Believing

Defeating MFF-Based Attacks Using Liveness Detection for Face Authentication on Mobile Platforms

Yan Li, Qiang Yan, Yingjiu Li, and Robert H. Deng

Contents

Abstract

Face authentication is one of the most promising biometric-based user authentication mechanisms that have been widely available in this era of mobile computing. With built-in front cameras on mobile devices such as smartphones and tablets, face authentication provides an attractive alternative to the legacy passwords for its simplicity and memory-less authentication process. Nonetheless, existing face authentication systems are inherently vulnerable to the so-called media-based facial forgery (MFF) attacks in which adversaries use victims' photos/videos to circumvent face authentication. The threat of MFF-based attacks becomes clear and present as numerous personal images and videos are made available online every day. In order to prevent such attacks, it is imperative to incorporate liveness detection mechanisms in face authentication systems. In this chapter, we provide a comprehensive overview of the latest development of liveness detection for face authentication, with a focus on mobile platforms. We categorize existing liveness detection techniques based on the types of liveness indicators being employed and evaluate their performance in terms of effectiveness, practicability, and robustness. We also comment on the future of designing a reliable liveness detection mechanism for face authentication on mobile devices.

Keywords: Face authentication, Media-based facial forgery (MFF), Liveness detection, Mobile platform

2.1 Introduction

As one of the promising biometric-based user authentication mechanisms, face authentication has been widely available on all kinds of mobile devices such as smartphones and tablets with built-in camera capability. Popular face authentication systems on mobile devices include Face Unlock [13], Facelock Pro [10], Visidon [39], and Smile-to-Pay [4]. These systems are usually used to protect screen locking/unlocking of mobile phones, mobile payment, and password managers and sensitive files on mobile devices [1,13,22,31]. The face authentication systems provide attractive alternatives of legacy passwords, as face authentication requires zero memory efforts from users and usually has higher entropy than legacy passwords as users tend to choose easy-to-guess passwords [29]. Unfortunately, face authentication has an intrinsic vulnerability against the *media-based facial forgery* (MFF) where an adversary forges or replays photos/videos containing a victim's face [24]. Previously, the major obstacle for an adversary to spoof face authentication systems was that physical proximity was required to capture a victim's facial photos/videos with sufficient quality. However, this is no longer necessary as abundant personal photos and videos are being published online. Prior research has shown that 53% of facial photos in online social networks, such as Facebook and Google+, can be simply used to spoof such face authentication systems [25]. The MFF-based attacks pose a severe threat to the existing face authentication systems on mobile platform, as mobile devices are usually used at public places in varied environments and could be accessible to adversaries.

To prevent the MFF-based attacks, liveness detection is designed to effectively distinguish between the live face and the forged face biometrics based on specific information, namely, liveness indicator [6,17,20,32]. Simple liveness detection techniques have been proposed and used in face authentication on mobile platforms for detecting the *photo-based attacks* where an adversary replays a facial photo. For example, the eyeblink-based approach relies on the detection of users' eyeblink [32]. More advanced liveness detection techniques are proposed against both *photo-based attacks* and *video-based attacks* in literature. For example, a recent research work thwarts the

MFF-based attacks by analyzing the correlation of mobile device movements and how a user's face view changes [24].

Moreover, the design of liveness detection on mobile platforms should be practical and robust in order to address specific real-world challenges of mobile devices. These challenges mainly include *limited hardware capability* and *varied usage environments*. Device manufacturers are reluctant to add extra hardware features like an infrared camera if they are not driven by strong business demand. Meanwhile, the real nonideal lighting conditions in varied usage environments may lead to low quality of input images/videos due to noisy pixels, loss of pixels, illumination, etc. The real lighting may also diminish the texture information and the geometric features of important facial landmarks such as nose and eyes and make it difficult to detect such texture information and facial landmarks in liveness detection of face authentication.

In this chapter, we present a survey of liveness detection techniques for face authentication on mobile platforms, which covers liveness detection systems deployed in practice and liveness detection methods proposed in literature. We evaluate and compare these liveness detection techniques for face authentication in terms of the effectiveness, practicability, and robustness that are crucial to address the real-world challenges of mobile devices. We further discuss the future of the reliable liveness detection techniques to protect face authentication on mobile platforms. The rest of the chapter is structured as follows. In Section 2.2, we provide background knowledge of the face authentication system. Section 2.3 presents the MFF-based attacks and threats against face authentication on mobile platforms. In Section 2.4, readers can find a comprehensive survey of the existing liveness detection techniques for face authentication. We also provide the evaluation and comparison of these liveness detection techniques and discuss the future direction of designing the reliable liveness detection in the same section. Section 2.5 concludes this chapter.

2.2 Overview of Face Authentication System

As one of the most promising biometric-based user authentication mechanisms, face authentication verifies a claimed identity of a user based on the facial features extracted from the images/videos that are directly captured from the live face of the user. A typical face authentication system usually consists of two subsystems including a face recognition subsystem and a liveness detection subsystem, as shown in Figure 2.1. In the face authentication process, on one hand, the face recognition subsystem determines whether the face in the images/videos from the camera is the same

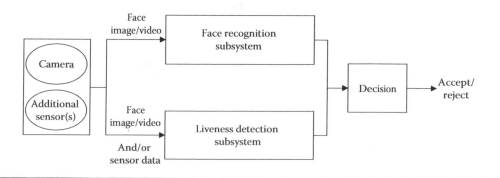

Figure 2.1   Components of a typical face authentication system.

as a user's registered face. On the other hand, the liveness detection subsystem determines if the images/videos are directly captured from a live face. Based on the decisions from the two subsystems, the face authentication system makes the final decision whether an authentication claim is accepted or rejected. In this section, we present the workflows of the face recognition subsystem and the liveness detection subsystem, respectively.

2.2.1 Face Recognition Subsystem

Taking the user's facial image/video as input through a camera, the face recognition subsystem compares the facial image/video with the enrolled face biometrics for the claimed identity and calculates a similarity score. This subsystem accepts the user if the similarity score for the input facial image/video is higher than a predetermined recognition threshold and rejects the user otherwise.

In the recognition process, two key modules are involved: the face detection module and the face matching module. The face detection module first identifies the face region in the image or selected frame in the video and removes the irrelevant parts of the image/video. Then the face detection module passes the processed image to the face matching module. The face matching module compares the image of the face region with the enrolled face template containing key features that can be used to distinguish a user from other users and imposters. The face matching module computes a similarity score for the input image. Different algorithms may be used for the modules. In particular, in face recognition, the major types of popular face recognition algorithms include holistic approaches and local landmark–based approaches [2,41]. But all face recognition systems generally have these two modules and follow this workflow. The holistic approaches, such as PCA-based algorithms and LDA-based algorithms, use the whole face region as input. Local landmark–based approaches extract local facial landmarks such as the eyes, nose, and mouth and feed locations and local statistics of these local facial landmarks into a structure classifier. In the end, a face recognition subsystem outputs the decision (i.e., accept or reject) according to whether or not the similarity score is higher than the recognition threshold (Figure 2.2). This threshold is carefully chosen so as to achieve a proper balance between false rejection rate and false acceptance rate.

As a face recognition subsystem is designed to recognize the user from an input facial image/video and cannot detect forged biometrics, it is inherently vulnerable to the MFF where an adversary forges a victim's face biometrics by using images/videos of the victim's face.

2.2.2 Liveness Detection Subsystem

Aiming at the MFF-based attacks, the liveness detection is designed to distinguish between the faces of live users and the faces forged from facial images/videos. The liveness detection subsystem

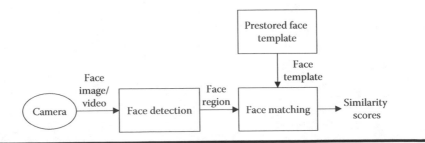

Figure 2.2 Work flow of the face recognition subsystem.

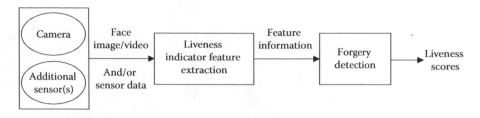

Figure 2.3 Workflow of the liveness detection subsystem.

usually uses a camera and/or other additional sensors to capture the data about a live user during the face authentication. Based on the captured data, the liveness detection subsystem computes a liveness score for the authentication process. The subsystem determines that it is a live face if the liveness score is higher than a liveness threshold. Otherwise, the face is determined to be forged.

In a liveness detection subsystem, there are two key modules, including the liveness indicator feature extraction and the forgery detection. Taking the captured raw data as input, the liveness indicator feature extraction module extracts the required feature information of the liveness indicator used by the liveness detection. For example, 3D geometric characteristics of face, texture information, and real-time response are popular liveness indicators [6,7,17,20]. Based on the extracted feature information, the forgery detection module computes a liveness score. At last, the liveness detection subsystem makes the decision according to whether or not the similarity score is higher than the liveness threshold that should be carefully chosen (Figure 2.3). In this chapter, we mainly focus on the liveness detection subsystems.

2.3 Media-Based Facial Forgery

The MFF means that users' face biometrics can be forged by photos or videos containing the users' faces. These photos or videos could be accessible to potential adversaries via malicious applications on mobile devices or in online social network sites, such as Facebook. Thus, the MFF-based attacks pose a serious threat against face authentication systems, as adversaries may replay facial photos/videos to spoof face authentication.

During the face recognition process, face authentication system has an intrinsic vulnerability against the MFF-based attacks. As shown in Figure 2.1, the face recognition subsystem recognizes a user from an input facial photo/video. However, it cannot distinguish if the input facial photo/video is directly from a live user or from a prerecorded photo/video of the same user. According to the types of the media used for facial forgery, the MFF-based attacks can be categorized into *photo-based attack* and *video-based attack*. In the photo-based attack, an adversary replays a single static photo containing a victim's face in order to spoof the face authentication system. In the video-based attack, the adversary obtains a video about the victim's face and replays the video to the face authentication system. The facial video obtained by the adversary could contain the victim's facial expressions and motions.

The MFF-based attack can be launched in two different levels including *sensor level* and *channel level*, as shown in Figure 2.4. In the sensor level, an adversary launches the attack outside the face authentication system. The adversary can print out facial photos or display the facial photos/videos on a digital screen and simply present these photos or videos to the camera connected to the face authentication system. In the channel level, the adversary should be able to intercept the channel

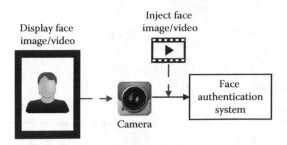

Figure 2.4 Media-based facial forgery–based attacks.

between the camera and the face authentication system. The adversary can inject the prerecorded facial photos or videos into the face authentication systems. The face authentication systems on mobile platform could be especially vulnerable to the attacks in channel level as malicious applications on mobile devices can intercept the channel between the camera and the system and inject the photos or videos.

Therefore, liveness detection is introduced into the face authentication system in order to thwart the MFF-based attacks. Liveness detection is usually based on liveness indicators that can be derived from human physiological activities or the properties of the input photos/videos. There are four major types of liveness indicators, including real-time response, texture pattern, 3D face, and multimodal [17].

Simple liveness detection approaches are designed to detect the photo-based attacks. In particular, eyeblink is a popular real-time response–based liveness indicator that has been used in popular face authentication systems such as Google's FaceUnlock [13]. These liveness detection mechanisms require no additional hardware; they work with moderate image quality and incur relatively low usability cost. They can effectively detect the photo-based attacks; however, they are still vulnerable to the video-based attacks. The video-based attacks to face authentication systems pose significant security risks as massive amounts of personal photos and videos are published online. It is likely that these photos and videos contain facial motions such as eyeblink, which is a natural physiology behavior of humans. Even worse, the facial motions can be animated and synthesized through a single static facial photo from which a dynamic 3D face model is estimated [2,41].

More sophisticated liveness detection approaches are proposed to prevent the advanced MFF-based attacks, such as video-based attacks and the channel-level attacks. For example, Alipay's Smile-To-Pay uses texture pattern analysis–based liveness detection approaches to prevent both photo-based attacks and video-based attacks [4]. These liveness detection approaches analyze the difference between the texture pattern of the media carriers and the texture pattern of a real face. However, the approaches can be vulnerable to the MFF-based attacks launched in channel level because the media for facial forgery is captured directly from a live face and injected to the face authentication without any media carrier texture pattern. The adversaries could probably launch the attacks in channel level via the malicious applications and vulnerabilities of mobile systems. Therefore, it is important for liveness detection to defend against all types of the MFF-based attacks, especially the video-based attacks and the channel-level attacks.

In addition to the effectiveness against the MFF-based attacks, the practicability and the robustness are also important to liveness detection on mobile platform so as to address the real-world challenges such as limited hardware and varied usage environments. On the one hand, the liveness detection mechanism on a mobile platform should be practical in the sense that it should not require

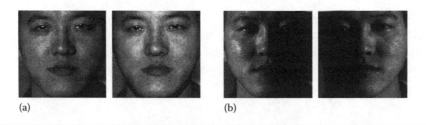

(a)　　　　　　　　　　　　　(b)

Figure 2.5   Face photos in (a) ideal lighting condition and (b) under illumination in varied environments. (From Gross, R. et al., *Image Vision Comput.*, 28(5), 807, 2010.)

any additional hardware that is uncommonly available on existing mobile devices. The hardware deployed on mobile devices is limited. Additional hardware requests may increase costs of mobile devices. On the other hand, the liveness detection mechanism should be robust to complex lighting conditions because mobile devices are usually used in varied environments. The real lighting conditions in varied usage environments may lead to low quality of input images/videos due to many reasons such as noisy pixels, loss of pixels, and illumination as shown in Figure 2.5 and further lead to difficulty of liveness indicator feature extraction.

2.4 State-of-the-Art Liveness Detection for Face Authentication

Liveness detection mechanism is designed to effectively distinguish between live faces and forged faces based on the liveness indicators including real-time response, texture pattern, 3D face, and multimodal. In order to address the real-world challenges, the effectiveness, practicability, and robustness are important to the liveness detection mechanisms for face authentication on mobile platform. In this section, we survey the latest liveness detection approaches for face authentication on mobile platform both in literature and in real-world systems. We categorize these liveness detection approaches according to the liveness indicator they use. We further provide evaluation and comparison of the liveness detection approaches based on the effectiveness, practicability, and robustness.

2.4.1 Categorization

The indicators used by liveness detection mechanisms mainly include real-time response, texture pattern, 3D face, and multimodal [17]. The real-time response can be used as a liveness indicator in the assumption that legitimate users can interact with the system in real time while it is difficult for fake faces to do so. For example, the eyeblink-based liveness detection requires users to blink their eyes several times. Second, the texture pattern can be used as a liveness indicator given the assumption that printed fake faces contain certain texture patterns that do not exist in real faces. Third, the 3D face indicator is defined based on the fact that a real face is a 3D object with depth characteristics while a fake face in a photo/video is planar (2D). For instance, different views of a live face can be observed as positions of a camera changes, while a face photo always has the same view. Finally, the multimodal usually requires a user to provide facial biometrics and additional biometric traits that are difficult to obtain by any adversary at the same time. For example, a face–fingerprint-based authentication system captures both face image and fingerprint of a user during the authentication process.

We summarize the liveness detection approaches according to the liveness indicators used. The categorizations of the liveness detection approaches mainly include real-time response–based approach, texture pattern–based approach, 3D-face–based approach, multimodal-based approach, and combination. These liveness detection approaches will be introduced and compared in the following section.

2.4.2 Literature

2.4.2.1 Real-Time Response

The real-time response–based approaches require interaction with users in real time. Among various forms of interaction, eye movement is a popular form of interaction that has been widely used by the liveness detection mechanisms in practice.

Jee et al. proposed an eye-movement-based liveness detection algorithm according to the assumption that significant shape variations of eyes can be observed on a live face due to eyeblink or unconscious movements of pupils while the shape of eyes on a single static face image cannot change [16]. By taking five sequential face images as input, the proposed algorithm mainly consists of four steps.

First, the eyes in the input images should be accurately detected since the proposed algorithm needs to use the variation of eye shape for liveness detection. In order to accurately locate the eyes in each input image, the candidate eye regions are extracted by applying a Gaussian filter to the image and using gradient descent algorithm [9]. Then the invalid eye candidates are reduced by a trained Viloa's AdaBoost classifier [38]. Second, the found eye regions in the face image are normalized since the input face could vary in size and orientation. Moreover, the negative effect of illumination in the image is reduced by applying self-quotient image that can be viewed as a high-pass filter [40]. The two eye regions in the image are extracted with a size of 10×20 pixels in this step.

Third, the extracted eye regions are binarized in order to have pixel value of 0 or 1 by applying a threshold. The threshold is adaptively obtained from the mean pixel value of each eye region. In the last step, the five binarized left eye regions and the five binarized right eye regions are compared, respectively, by using hamming distance. A liveness score is determined according to the average number of pixels that differs between the binarized eye regions. If the liveness score is higher than the threshold, the input images are considered to be from a live face since the live face always generates a significant variation of eye shape. Otherwise, it is discriminated to a fake face. In order to evaluate the performance of the proposed algorithm, the experiments are conducted based on 100 live faces and 100 fake faces. The experimental results show that the lowest *false acceptance rate* (FAR) and the *false rejection rate* (FRR) are 1% and 8%, respectively.

The eye-movement-based liveness detection mentioned earlier can detect photo-based attacks because it would be hard to produce significant eye shape variation by using a single static face image. It also has the advantage that only a common camera is required. However, the proposed algorithm can be vulnerable to the video-based attacks where an adversary displays a video containing eyeblink or pupil movements [33]. The complex lighting condition could affect the accurate localization of eyes in the images and therefore significantly harm the results of the proposed liveness detection algorithm [41].

Another liveness detection approach based on eyeblink is proposed by Pan et al. [32]. The proposed liveness detection approach required users to blink their eyes for the detection of a live face where an eyeblink is defined as a physiological activity of rapid closing and opening of the eyelid. In order to detect the eyeblink, the proposed liveness detection approach models the blink behavior as inference in an undirected conditional random field (CRF) framework [11], which enables the long-range dependencies among observations and states. More specifically, each eyeblink activity is represented by an image sequence S containing T images. In the images, eye states are defined as *open*, *close*, and *ambiguous*. Therefore, a typical eyeblink activity can be described as a state change pattern of *open* $\rightarrow$ *ambiguous* $\rightarrow$ *close* $\rightarrow$ *ambiguous* $\rightarrow$ *open*.

Give a sequence images S, the proposed approach analyzes the state change pattern in S by using the undirected CRF framework incorporated with a discriminative measure of the eye states, namely, eye closity. The eye closity measures the degree of an eye's closeness and is introduced to simplify the complex of the inference by the CRF framework and to improve the performance simultaneously. The experimental results show that the proposed approach achieves a detection rate as high as 95.7%.

This eyeblink-based liveness detection approach is practical because it only needs a generic camera with the frame rate of no less than 15 fps. It can prevent the photo-based attack. Unfortunately, this liveness detection approach is subject to the video-based attacks if the facial video used for the attacks contains the required eyeblink.

2.4.2.2 Texture Pattern

The texture pattern is another liveness indicator used by many liveness detection techniques, which assume that the printed fake faces contain detectable texture patterns due to the printing process and the material printed on.

Li et al. propose a texture pattern–based liveness detection algorithm that analyzes the 2D Fourier spectra characteristics of input face images [23]. The proposed algorithm is based on two assumptions: (1) high-frequency components of photo image must be less than those of a real face image if the size of the photo is smaller than that of the live face and the photo is 2D; (2) the standard deviation of frequency components in sequence must be small because the expressions and poses of a fake face are invariant.

In the proposed algorithm, according to the Lambertian model, a face image is described as

$$I(x,y) = \rho(x,y)n(x,y)^T s \tag{2.1}$$

where
ρ is the albedo (surface texture) of face
$n(x,y)^T$ is the surface normal (3D shape) of the face
s is the point source, which can vary arbitrarily

If the given input is a fake face printed on a 2D image, $n(x,y)^T$ should be a constant due to the 2D planar structure of the photograph. Therefore, under the same illumination, images from live faces are determined by the albedo and the surface normal, whereas those from a fake are determined by the albedo only. The difference leads to different reflectivity of light and therefore is reflected in frequency distribution of an image. In addition, the size of a fake image is assumed to be smaller than that of a live face, which causes loss of details contained in the input face image during recapturing. In order to detect these differences, the Fourier spectra analysis is applied to the input image where the high-frequency descriptor (HFD) is calculated to indicate the ratio of the energy of high-frequency components. Since the Fourier spectra of a live face have many more high-frequency components than those of a fake image, the HFD of a live face should be higher than a reasonable threshold.

However, only using HFD value may not be sufficient for the liveness detection because the differences between a live face and a fake face could be reduced if a clear and big size face image is used. Because a fake face on a single static image does not have dynamic characteristics, the frequency dynamics descriptor (FDD) of the image is calculated to represent temporal changes of the face. The FDD value of the live face should be higher than a chosen threshold while the FDD value of the fake face should be close to zero. The experimental results reported by the authors show that the proposed algorithm can correctly detect all live faces and fake faces under controlled settings.

The proposed liveness detection algorithm mentioned earlier can prevent the photo-based attacks as the FDD value in a single photo may not be high enough. Another advantage of the proposed algorithm is no requirement of additional hardware support except a generic camera. However, the proposed algorithm can be vulnerable to the video-based attacks, especially the attacks in channel level since the injected video contains the required feature information. In addition, the outdoor complex lighting condition can significantly affect the performance of the proposed algorithm that requires invariant illumination condition.

The texture pattern can also be extracted based on specular components of images. Bai et al. propose a liveness detection method that analyzes the specular components of the recaptured images printed on paper [5]. The assumption of this method is that recaptured images can be visually distinguishable from the original ones reliably by inspecting the spatial distribution of their specular component.

In particular, a surface required by the attacks to display the image or video is modeled by the bidirectional reflectance distribution function (BRDF). Based on the BRDF model, the input image is decomposed into *diffuse components* and *specular components* by

$$I(x) = \omega_d(x)D(x) + \omega_s(x)S(x) \tag{2.2}$$

where I is the color vector of the image while $D(x)$ and $S(x)$ are the diffuse components and secular components, respectively. ω_d and ω_s represent the weighting factors for the diffuse reflections, and specular reflections, respectively. The specular components are extracted because it will give us clues on deciphering the surface texture that generates the image. Microtextures are present on the paper due to uneven paper thickness and ink deposition pattern, which changes the gradient of the specular image. A natural image will have smooth transitions for the gradient in the specular components, having a gradient histogram with a peak at the lowest value and a long tail. A recaptured image with evidence of microtextures will have a gradient histogram with shorter tail and higher midvalues. The proposed method is reported to achieve the equal error rate (EER) of as low as 6.7%.

The proposed liveness detection method can detect the MFF-based attacks in sensor level by using one image/frame if the image/video is displayed on a paper/screen with a coarse surface. However, the proposed method requires that the input image be ideally sharp and clear enough, which may not be easily achievable in practice because mobile devices could be used at any place. It is hard to completely avoid images that are out of focus or contain motion blur under varied lighting conditions. Moreover, if the textures on photographic prints are finer than those compared to common paper, the effectiveness of the method could drastically decrease. The proposed method is also the MFF-based attacks in channel level as there is no display medium used in the attacks.

Another description of the texture pattern is obtained by local binary patterns (LBPs). Maatta et al. propose a liveness detection approach by adopting micro-texture analysis based on the LBP extracted from a single image [26]. The authors assume that face prints usually contain printing quality defects that can be well detected using micro-texture patterns. Human faces and prints reflect light in different ways because a human face is a complex nonrigid 3D object, whereas a photograph can be seen as a planar rigid object. This may cause different specular reflections and shades.

In order to detect the attacks, the proposed approach first detects, crops, and normalizes the input face image into a 64 × 64 pixel image. Then discriminative feature space is enhanced by applying three different LBP operators that are different in sizes of neighborhood pixels and circular radius. Each LBP operator is regarded as a micro-texton. The histograms calculated based on the LBP operators are fed to a trained SVM classifier so as to determine if it is a live face or a fake face. In the experiments, the authors show that the proposed liveness detection approach can detect photo-based attacks using paper images at the EER of as low as 2.9%.

Similar to the first two texture pattern–based liveness detection algorithms, the proposed approach has the advantage that it can prevent the MFF-based attacks in sensor level and only needs a generic camera. However, the effectiveness of the approach could decrease if the paper or screen with finer surface is used to display the image/video. The complex lighting is another negative factor that could affect the LBP operator and then harm the effectiveness of the proposed approach. There is no way for this approach to prevent any MFF-based attacks in channel level.

2.4.2.3 3D Face

The 3D face liveness indicator is based on the clue that a real face is a 3D object and has depth characteristics. The detection of the characteristics of a 3D face is usually associated with relative face movements, optical flow analysis, and depth information.

The characteristics of a live face can be detected in relative movements of the face (e.g., face view changes). Li et al. propose FaceLive, a liveness detection mechanism to protect the face authentication on mobile devices against the MFF-based attacks [24]. FaceLive detects the MFF-based attacks by measuring the consistency between device movement data from the inertial sensors and the head pose changes from the facial video captured by a built-in camera. To perform the liveness verification, a user needs to hold and move a mobile device over a short distance in front of his or her face. The video and the motion data are captured simultaneously and independently by a front-facing camera and the inertial sensors on the device, respectively. When the device is moving, the user's face is captured by the front-facing camera from different camera angles. The changes of the user's head poses correlated with the movement can be observed in the facial video if a live face is in front of the device.

FaceLive consists of three modules, including device motion estimator (DME), head pose estimator (HPE), and consistency analyzer (CA), as shown in Figure 2.6. The DME module takes the sensor data from the accelerometer and gyroscope and computes a motion vector representing the device movement over time. The HPE module estimates the head poses from the extracted frames and computes a head pose vector over time. The CA module compares the motion vector and the head pose vector and extracts the feature information. Based on the extracted features, the CA module uses a classification algorithm to distinguish between a live user and an MFF-based attack. The details of these modules are introduced in the following text.

DME uses the motion data including acceleration values and angular speeds on axis X in an inertial sensor coordinate system. The motion data is first preprocessed by applying low-pass and high-pass filters for reducing the effect of gravity and a threshold for reducing the effect of physiological hand tremor. Based on the preprocessed motion data, the device movement is estimated by a dead-reckoning algorithm [18]. In particular, given acceleration readings $a_{s,t_{i-1}}$ and a_{s,t_i} where $s \in \{x, y, z\}$, the movement distance $d_{s,t_{i-1}}$ on axis s at time t_i can be calculated as

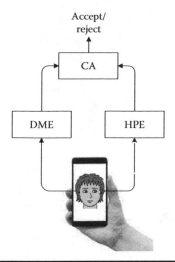

Figure 2.6 FaceLive mainly consist of device movement estimator (DME), head pose estimation (HPE), and consistency analyzer (CA).

$$d_{s,t_i} = \frac{(v_{s,t_{i-1}} \cdot \cos \varphi_{s,t_i} + v_{s,t_i}) \cdot (t_i - t_{i-1})}{2} \qquad (2.3)$$

where v_{s,t_i} is the velocity at time t_i estimated based on $a_{s,t_{i-1}}$ and a_{s,t_i}.

Given the angular speed $R_{t_i} = (r_{x,t_i}, r_{y,t_i}, r_{z,t_i})$. The rotation angle along each axis can be calculated as

$$\theta_{s,t_i} = \frac{(r_{s,t_{i-1}} + r_{s,t_i}) \cdot (t_i - t_{i-1})}{2} \qquad (2.4)$$

where $s \in \{x, y, z\}$. Note $(\theta_{x,t_i}, \theta_{y,t_i}, \theta_{z,t_i})$ is also called as Cardan angles that are typical for rotation of a 3D coordinate system [37]. Then, we have

$$\cos \varphi_{x,t_i} = \cos \theta_{y,t_i} \cdot \cos \theta_{z,t_i}$$
$$\cos \varphi_{y,t_i} = \cos \theta_{x,t_i} \cdot \cos \theta_{z,t_i} \qquad (2.5)$$
$$\cos \varphi_{z,t_i} = \cos \theta_{x,t_i} \cdot \cos \theta_{y,t_i}$$

Thus, the movement distance $d_{s,t_{i-1}}$ is estimated by combining Equations (2.3) and (2.5). Finally, DME outputs a device motion vector (D_x, D_y) between t_0 and t_m, where $D_s = (d_{s,t_1}, d_{s,t_2}, \ldots, d_{s,t_m})$ for $s = \{x, y\}$.

At the same time, HPE estimates the head poses in the facial video frames by two steps: facial landmark localization and head pose estimation. In the first step, a set of facial landmarks is located in each frame by an open-source library FaceTracker [36]. In the second step, the head pose in a frame is estimated by the Perspective-n-Point problem–based 2D head pose estimation algorithm [30,34]. Given a set of points from a face in the 3D coordinate system and the projection of these points in the 2D coordinate system, the transformation between the two systems can be estimated with the following equations:

$$\tilde{m} = A[R|T]\tilde{M} \qquad (2.6)$$

where

$\tilde{m} = (u\ v\ 1)^T$ is a 2D point (u, v) in homogenous coordinates

$A = \begin{pmatrix} f & 0 & c_x \\ 0 & f & c_y \\ 0 & 0 & 1 \end{pmatrix}$ is an internal calibration matrix

$R = \begin{pmatrix} r_{11} & r_{12} & r_{13} \\ r_{21} & r_{22} & r_{23} \\ r_{31} & r_{32} & r_{33} \end{pmatrix}$ is a rotation matrix

$T = (t_1\ t_2\ t_3)^T$ is a translation vector

$\tilde{M} = (x\ y\ z\ 1)^T$ is a 3D point (x, y, z) in homogenous coordinates

A can be obtained from the camera settings where f in A is the focal length of the camera in pixel unit while (c_x, c_y) is the center point of the 2D image in pixels.

Based on the device motion vector D_x from **DME** and the head pose change vector H_{yaw} from **HPE**, CA examines the correlation between the two vectors and uses a classifier to make the final decision based on the correlation. In order to represent the correlation, D_x and H_{yaw} are aligned by a dynamic time warping (DTW) algorithm [28]. Then the ratio values of D_x and H_{yaw} are calculated. The warping distance of DTW and the statistics about the ratio values are fed to the classifier as features for the decision of liveness. The authors collect a dataset from 73 participants and evaluate FaceLive based on the dataset. According to the experimental results, FaceLive achieves EER of as low as 4.7%. Moreover, FaceLive can be robust to partial facial landmark missing due to complex lighting conditions. The experimental results show that EER of FaceLive is as low as 7.2% when one facial landmark is missing.

FaceLive can effectively protect face authentication from the MFF-based attacks with low equal error rates. FaceLive is practical because it only requires a front-facing camera and the inertial sensors that are widely available on mobile devices. FaceLive can be robust to unsuccessful detection of partial facial landmarks that may happen in complex lighting condition.

Another liveness detection algorithm relying on the relative movement of face is proposed by Chen et al. Different from FaceLive using relative movements of a holistic face, the proposed algorithm examines the liveness of a face by using relative movement of the nose region on the face [8]. It is based on a simple idea that a real face has a 3D nose. Thus, as a mobile phone is moving in front of the 3D nose, the edge of the nose region in 2D face image captured by front camera should change accordingly. In order to determine the liveness of a user, the liveness detection mechanism compares the correlation between the direction changes of the mobile phone measured by the accelerometer and the changes of a clear nose edge observed in the video from the camera.

In particular, given a sequence of face images or video frames, the proposed algorithm first applies histogram equalization to the images or frames so as to enhance the contrast of the images or frames. The edges of the facial landmarks, including the nose, can be highlighted. Then the nose region is extracted by the Haar Cascades algorithm. Based on the extracted nose region, the edge of the nose is estimated as two straight lines l_1 and l_2 by using minimum mean square error (MMSE). The angles formed by l_1 and l_2 represent the changes of the nose edge. On the other hand, the movement of the mobile phone is estimated from the sensor data recorded by the accelerometer on the mobile phone. Based on the estimated angles of the nose edge and the estimated movements of the mobile phone, the correlation is calculated to decide if it is a live face or an attack. In the experiments, the proposed liveness detection algorithm is reported to achieve the detection rate of 97% and the false rejection rate of 3%.

The liveness detection algorithm mentioned earlier can prevent MFF-based attacks. It only requires a generic camera and an accelerometer on mobile phones. However, since the algorithm relies on accurate estimation of nose edge in order to produce the clear nose edge, a controlled lighting is required to cast the shadow of nose without any occlusion. This may be difficult to achieve in practice where the controlled lighting is not possible. The effectiveness of the liveness detection mechanism is also limited for those who have flat noses.

The second popular characteristic of a 3D face is that the motion speed of the central part of face is higher than the outer face region [17]. This characteristic can be analyzed based on the optical flow. Along this line, Bao et al. proposed a liveness detection algorithm that analyzes the differences and properties of optical flow generated from a holistic 3D face and 2D planes [6]. In this work, a motion in an image is assumed to be a combination of four basic movement types including *translation*, *rotation*, *moving*, and *swing*. In particular, the swing movement of a face (i.e., head pose changes in yaw or pitch) creates more differences between 3D live face and 2D fake face. The proposed liveness detection algorithm is basically based on the differences caused by the swing movements of face. The optical flow field is used to analyze the differences. Compared to 3D objects, the optical flow filed for 2D objects can be represented as a projection transformation. The difference between optical flow fields is calculated so as to determine whether the test region is planar or 3D. The proposed algorithm is evaluated based on three groups of sample data. The experimental results show that detection rate of the algorithm is as high as 92% with a threshold of 0.8. The proposed algorithm has advantages that no additional hardware support but a camera is required. However, the algorithm usually requires high-quality input videos with ideal lighting conditions, which may be difficult to achieve in practice. The algorithm is vulnerable to the video-based attacks if the video displayed by the adversary contains changes of face views.

The third characteristic of a live face is that it contains depth information. If a camera takes images of a live face with different focus regions on the face, the quality of the images changes because the focused regions (nose, ear) are clear and others are blurred. On the contrary, there is little difference for a fake face as the fake face is planar. Based on this idea, Kim et al. propose an approach that utilizes variation of pixel values by focusing between two images sequentially taken in different focuses of camera [19].

More specifically, two sequential images with different focuses on face regions are captured. The chosen face regions are nose and ears. This is because the nose is closest to the camera while the ears are farthest. The depth difference between them is significant. Second sum modified Laplacians (SMLs) of both images are calculated to determine the degree of focusing. The SML is used to represents degrees of focusing in images and those values are represented as a transformed second-order differential filter [15]. At last, the sums of differences of SMLs in each column are calculated and used as features for liveness detection. As reported in the work, the proposed approach achieves FAR of 2.86% and FRR of 0.00%. Therefore, this approach can prevent MFF-based attacks in sensor level as the media displayed to the camera is planar. However, it could be subject to the video-based attacks in channel level if the injected video contains face images with different focused regions of a face. It is also difficult for the proposed approach to deploy on existing mobile devices because almost all front cameras on the existing mobile phones are not equipped with zoom lenses and focal length of the front cameras is not adjustable.

2.4.2.4 Multimodal

Multimodal-based liveness detection approaches take face biometrics and other biometrics into account in user authentication. Akhtar et al. analyzed face and fingerprint-based biometric authentication systems that use different strategies to fuse face authentication and fingerprint authentication together [3]. In the face and fingerprint-based authentication systems, each user is required to present his or her face and fingerprint to camera and fingerprint scanner at the same time during the authentication process. After obtaining the face biometrics and fingerprint biometrics, the systems mainly use two strategies to fuse the two types of biometrics, which are *parallel fusion* and *serial fusion*.

In the systems with the parallel fusion mode, the face biometrics and the fingerprint biometrics are analyzed and compared with the enrolled face template and fingerprint template by a face matcher and a fingerprint matcher separately. Then, a face matching score S_{face} and a fingerprint matching score $S_{fingerprint}$ are calculated and fused through a certain fusion rule $f(S_{face}, S_{fingerprint})$ in order to produce a fused score S [35]. If S is no less than a chosen threshold T, the user is accepted as a legitimate user. Otherwise, it is rejected as an attacker.

In contrast, in the systems with the serial fusion mode [27], the user first only provides the face biometrics to the systems, which is analyzed and compared with the enrolled face template for a face matching score S_{face}. If S_{face} is higher than an upper threshold T_u^*, the user is accepted as a legitimate user. If S_{face} is no higher than a lower threshold T_l^*, the user is rejected as an attacker. If $T_l^* \leq S_{face} \leq T_u^*$, then the user needs to present his or her fingerprint biometrics to the systems. The fingerprint matcher compares the input fingerprint with the enrolled fingerprint template and determines a fingerprint matching score $S_{fingerprint}$. If $S_{fingerprint}$ is no less than a threshold $T_{fingerprint}$, the user is accepted by the system. Otherwise, the user is rejected.

The systems with the different fusion modes are evaluated in the experiments. Under the MFF-based attacks, the systems with the parallel fusion mode achieve FAR of 4.28% while the systems with the serial fusion mode achieve FAR of 3.74%. Therefore, the systems with the serial fusion mode are generally more reliable and better than those with the parallel fusion.

However, the multimodal-based authentication systems usually require additional hardware to collect the biometric information. The additional hardware may not be easily available on most existing mobile devices.

2.4.2.5 Combination

Liveness detection approaches can use more than one type of liveness indicator together in order to take advantages of the multiple liveness indicators.

AMILAB proposes a liveness detection system that combines two indicators including texture pattern and real-time interaction [7]. The team extracts different types of visual features including color, texture, and edges and uses a set of SVMs for texture analysis. Then eyeblink is analyzed and detected. For eyeblink detection, every two adjacent frames are compared with the difference and binarized in order to locate the eyes. Then the eye-open template is created with several frames. In order to detect eyeblinks, the similarity between the frames is calculated that decreases/increases as eyes close/open. Third, motion of face is analyzed to differentiate face and background. In the first step, nonchange pixels in the image stream are filtered out by using simple background and temporal differences (an input color image and a reference background image) according to interframe changes. In the second step, the pixels associated with stationary or moving objects are further classified as background of foreground based on the learned statistics of colors or color co-occurrences using the Bayes decision rule. In the third step, foreground objects are segmented by combining the classification results from both stationary and moving parts. At last, background models are updated. The experimental results show that the proposed system achieves FAR of 0% and FRR of 1.25%.

The liveness detection system mentioned earlier can prevent the MFF-based attack in sensor level. Only a generic camera is required by the system. However, it is subject to the MFF-based attack in channel level if an adversary injects into the system a prerecorded face video that could contain all the liveness indication required by the system.

2.4.3 Comparison between the Liveness Detection Approaches

Table 2.1 gives a comparison between the existing liveness detection approaches based on the liveness indicator used and the important metrics for mobile devices, including the effectiveness against

Table 2.1 Comparison between the Liveness Detection Approaches Summarized in Section 2.4.2

| Algorithm | Liveness Indicator | | | | Resist Sensor-level Attack | | Resist Channel-level Attack | | Robustness | Practicability |
	Real-Time Response	Texture Pattern	3D Face	Multimodal	Photo Based	Video Based	Photo Based	Video Based		
Jee et al. [16]	○				○		○		Low	High
Pan et al. [32]	○				○		○		Medium	High
Li et al. [23]		○			○	○			Low	High
Bai et al. [5]		○			○	○			Low	High
Maatta et al. [26]		○			○	○			Low	High
FaceLive [24]			○		○	○	○	○	Medium	High
Chen et al. [8]			○		○	○	○	○	Low	High
Bao et al. [6]			○		○		○		Low	High
Kim et al. [19]			○		○	○	○		Medium	Low
Akhtar et al. [3]				○	○	○	○	○	Medium	Medium
AMILAB [7]	○	○			○	○			Low	High

○ Possess the property.

the MFF-based attack, the robustness to complex lighting, and the practicability. We present the observation in comparison as follows:

- Among the different types of the MMF-based attacks, the liveness detection approaches generally prevent the photo-based attacks more effectively than the video-based attacks. The MMF-based attacks in sensor level can be detected more effectively than the attacks in channel level.
- The liveness detection approaches using 3D face or multimodal as a liveness indicator are generally more effective to detect the MFF-based attacks. The approaches based on the real-time interaction are vulnerable to the video-based attacks because the video used in the attacks can contain the required forms of the interaction. The approaches based on the texture pattern are subject to the MFF-based attacks in channel level since the photo/video used in the attacks needs no medium to display and Is directly injected into the system.
- Complex lighting conditions can have a significant impact on many liveness detection approaches, such as the texture pattern–based approaches, which usually require ideal lighting or controlled illumination.
- Among the liveness detection approaches, FaceLive [24], the liveness detection algorithm proposed by Chen et al. [8], and the face–fingerprint-based authentication [3] are generally better than the other liveness detection approaches in terms of the effectiveness, robustness, and practicability. The three approaches use additional hardware to enhance the effectiveness of the liveness detection. In particular, FaceLive [24] and the approach proposed by Chen et al. [8] require inertial sensors that are widely available on the existing mobile devices. The face–fingerprint-based authentication [3] require additional fingerprint scanner that is available on limited mobile devices, such as Samsung's flagship phones and iPhones.

2.4.4 Discussion

The complex lighting conditions, including illumination and dimness, are a prominent challenge in the realm of face recognition as well as the liveness detection on mobile platforms [12,21,41]. The illumination diminishes facial features and texture information in images/videos [12,41]. Dimness may also diminish facial and texture features since the luminance in images/videos is too low for liveness detection algorithm to analyze [21]. Therefore, the liveness detection algorithms should not rely on the accurate localization of only a few facial landmarks or fine texture information in the input images/videos because the mobile devices could be often used in varied environments with complex lighting conditions.

The video-based attacks and the channel-level attacks pose serious threats against the liveness detection for face authentication on mobile platforms. These attacks become realistic and common as a huge number of personal images and videos are published online and mobile devices are affected by powerful malicious applications. In order to effectively prevent these MFF-based attacks, additional hardware can be helpful in designing reliable liveness detection for face authentication. But introducing the additional hardware should not incur significant increase in cost for the practicability of the liveness detection. It would be preferable to use the hardware that are commonly available on most of the mobile devices. For example, accelerometer and gyroscope used by FaceLive [24] and the liveness detection approach proposed by Chen et al. [8] are widely available on most existing mobile devices. Some other promising hardware, such as the fingerprint scanner, are becoming cheaper and more common on mobile devices. These components could benefit the liveness detection, such as face–fingerprint-based authentication [3], in the future.

More advanced attacking techniques, such as 3D face models, could create more serious threats to face authentication systems. As more high-quality facial images and videos are published online, it could be possible for a powerful adversary to construct a targeted 3D face model with facial animations in real time based on these images and videos. The constructed face models could be displayed on a nonplanar screen or injected into the face authentication system in order to spoof the system when the techniques become practical in the future. Thus, it is crucial to design more reliable liveness detection mechanisms with combination of multiple liveness indicators or use new and more effective liveness indicators.

2.5 Conclusion

In this chapter, we provided an overview on liveness detection for face authentication on mobile platforms. We evaluate and compare the existing liveness detection techniques for face authentication from the perspective the effectiveness against the MFF-based attacks, robustness, and practicability. Although the liveness detection mechanisms still have challenges such as varied environments and limited hardware support, reliable liveness detection can be achievable with the help of more promising hardware available on mobile devices in the future.

References

1. One You. 2016. http://1uapps.com/, accessed July 30, 2016.
2. A. F. Abate, M. Nappi, D. Riccio, and G. Sabatino. 2D and 3D face recognition: A survey. *Pattern Recognition Letters*, 28(14):1885–1906, 2007.
3. Z. Akhtar, G. Fumera, G. L. Marcialis, and F. Roli. Evaluation of serial and parallel multibiometric systems under spoofing attacks. In *Proceedings of the 2012 IEEE Fifth International Conference on Biometrics: Theory, Applications and Systems (BTAS)*, Arlington, VA, pp. 283–288. IEEE, 2012.
4. Alipay. 2016. https://itunes.apple.com/us/app/zhi-fu-bao-zhifubao-kou-bei/id333206289?mt=8, accessed July 30, 2016.
5. J. Bai, T.-T. Ng, X. Gao, and Y.-Q. Shi. Is physics-based liveness detection truly possible with a single image? In *Proceedings of the 2010 IEEE international Symposium on Circuits and Systems (ISCAS)*, Paris, France, pp. 3425–3428. IEEE, 2010.
6. W. Bao, H. Li, N. Li, and W. Jiang. A liveness detection method for face recognition based on optical flow field. In *IASP 2009*, Taizhou, China, pp. 233–236. IEEE, 2009.
7. M. M. Chakka, A. Anjos, S. Marcel, R. Tronci, D. Muntoni, G. Fadda, M. Pili et al. Competition on counter measures to 2-D facial spoofing attacks. In *IJCB 2011*, pp. 1–6. IEEE, 2011.
8. S. Chen, A. Pande, and P. Mohapatra. Sensor-assisted facial recognition: An enhanced biometric authentication system for smartphones. In *MobiSys 2014*, Bretton Woods, NH, pp. 109–122, 2014.
9. R. O. Duda, P. E. Hart, and D. G. Stork. *Pattern Classification*. John Wiley & Sons, Hoboken, NJ, 2012.
10. FaceLock. 2013. http://www.facelock.mobi/facelock-for-apps, accessed July 30, 2016.
11. R. W. Frischholz and U. Dieckmann. Biold: A multimodal biometric identification system. *Computer*, 33(2):64–68, 2000.
12. A. S. Georghiades, P. N. Belhumeur, and D. J. Kriegman. From few to many: Illumination cone models for face recognition under variable lighting and pose. *IEEE Transactions on Pattern Analysis and Machine Intelligence*, 23(6):643–660, 2001.
13. Google. 2016. Ice Cream Sandwich Android. http://www.android.com/about/ice-cream-sandwich/, accessed July 30, 2016.

14. R. Gross, I. Matthews, J. Cohn, T. Kanade, and S. Baker. Multi-pie. *Image and Vision Computing*, 28(5):807–813, 2010.

15. W. Huang and Z. Jing. Evaluation of focus measures in multi-focus image fusion. *Pattern Recognition Letters*, 28(4):493–500, 2007.

16. H.-K. Jee, S.-U. Jung, and J.-H. Yoo. Liveness detection for embedded face recognition system. *International Journal of Biological and Medical Sciences*, 1(4):235–238, 2006.

17. O. Kahm and N. Damer. 2D face liveness detection: An overview. In *BIOSIG 2012*, Darmstadt, Germany, pp. 1–12, 2012.

18. I. Kamal. WFR, a dead reckoning robot: A practical application to understand the theory. IKA Logic: Electronics Solutions Online Documents, 2008. https://www.ikalogic.com/wfr-a-dead-reckoning-robot/, accessed July 30, 2016.

19. S. Kim, S. Yu, K. Kim, Y. Ban, and S. Lee. Face liveness detection using variable focusing. In *Proceedings of the 2013 International Conference on Biometrics (ICB)*, Madrid, Spain, pp. 1–6. IEEE, 2013.

20. K. Kollreider, H. Fronthaler, and J. Bigun. Non-intrusive liveness detection by face images. *Image and Vision Computing*, 27(3):233–244, 2009.

21. S. G. Kong, J. Heo, B. R. Abidi, J. Paik, and M. A. Abidi. Recent advances in visual and infrared face recognition: A review. *Computer Vision and Image Understanding*, 97(1):103–135, 2005.

22. D. M. Landau. 2015. Password management—With Intel True Key, you are your password. http://iq.intel.com/you-are-your-password/, accessed July 30, 2016.

23. J. Li, Y. Wang, T. Tan, and A. K. Jain. Live face detection based on the analysis of Fourier spectra. In *Defense and Security*, pp. 296–303. International Society for Optics and Photonics, Bellingham, WA, 2004.

24. Y. Li, Y. Li, Q. Yan, H. Kong, and R. H. Deng. Seeing your face is not enough: An inertial sensor-based liveness detection for face authentication. In *CCS 2015*, Denver, CO, pp. 1558–1569, 2015.

25. Y. Li, K. Xu, Q. Yan, Y. Li, and R. H. Deng. Understanding OSN-based facial disclosure against face authentication systems. In *AsiaCCS 2014*, pp. 413–424, 2014.

26. J. Maatta, A. Hadid, and M. Pietikainen. Face spoofing detection from single images using micro-texture analysis. In *IJCB 2011*, Washington, DC, pp. 1–7. IEEE, 2011.

27. G. L. Marcialis, F. Roli, and L. Didaci. Personal identity verification by serial fusion of fingerprint and face matchers. *Pattern Recognition*, 42(11):2807–2817, 2009.

28. M. Müller. Dynamic time warping. *Information Retrieval for Music and Motion*, Springer, Berlin, Heidelberg, pp. 69–84, 2007.

29. L. O'Gorman. Comparing passwords, tokens, and biometrics for user authentication. *Proceedings of the IEEE*, 91(12):2021–2040, 2003.

30. S. Ohayon and E. Rivlin. Robust 3D head tracking using camera pose estimation. In *Proceedings of the 18th International Conference on Pattern Recognition, 2006 (ICPR 2006)*, Hong Kong, China, Vol. 1, pp. 1063–1066. IEEE, 2006.

31. J. Osawa. 2014. Alibaba's Alipay turns to faces, fingerprints for security. http://blogs.wsj.com/digits/2014/10/16/alibabas-alipay-turns-to-faces-fingerprints-for-security/, accessed July 30, 2016.

32. G. Pan, L. Sun, Z. Wu, and S. Lao. Eyeblink-based anti-spoofing in face recognition from a generic webcamera. In *ICCV 2007*, Rio de Janeiro, Brazil, pp. 1–8, 2007.

33. J. Rice. 2016. Android Jelly Bean's face unlock "liveness check" circumvented with simple photo editing. http://www.androidpolice.com/2012/08/03/android-jelly-beans-face-unlock-liveness-check-circumvented-with-simple-photo-editing/, accessed July 30, 2016.

34. F. Rocca, M. Mancas, and B. Gosselin. Head pose estimation by perspective-n-point solution based on 2D markerless face tracking. In D. Reidsma, I. Choi, and R. Bargar (eds.), *Intelligent Technologies for Interactive Entertainment*, pp. 67–76. Springer, Switzerland, 2014.

35. R. N. Rodrigues, L. L. Ling, and V. Govindaraju. Robustness of multimodal biometric fusion methods against spoof attacks. *Journal of Visual Languages & Computing*, 20(3):169–179, 2009.

36. J. M. Saragih. Deformable face alignment via local measurements and global constraints. In M. G. Hidalgo, A. M. Torres, and J. V. Gómez (eds.), *Deformation Models*, pp. 187–207. Springer, the Netherlands, 2013.

37. S. Tupling and M. Pierrynowski. Use of cardan angles to locate rigid bodies in three-dimensional space. *Medical & Biological Engineering & Computing*, 25(5):527–532, 1987.

38. P. Viola and M. Jones. Rapid object detection using a boosted cascade of simple features. In *Proceedings of the 2001 IEEE Computer Society Conference on Computer Vision and Pattern Recognition, 2001 (CVPR 2001)*, Kauai, HI, Vol. 1, pp. I-511–I-518. IEEE, 2001.

39. Visidon. 2016. http://www.visidon.fi/en/Home, accessed July 30, 2016.

40. H. Wang, S. Z. Li, and Y. Wang. Face recognition under varying lighting conditions using self quotient image. In *Proceedings of the Sixth IEEE International Conference on Automatic Face and Gesture Recognition, 2004*, Seoul, South Korea, pp. 819–824. IEEE, 2004.

41. W. Zhao, R. Chellappa, P. J. Phillips, and A. Rosenfeld. Face recognition: A literature survey. *ACM Computing Surveys (CSUR)*, 35(4):399–458, 2003.

Chapter 3

Visual Authentication Based on Visual Cryptography Using Mobile Devices

Yang-Wai Chow, Willy Susilo, Man Ho Au,
and Ari Moesriami Barmawi

Contents

Abstract

Authentication refers to the process of confirming the identity of the authenticating entity. The use of passwords for user authentication has become ubiquitous in our everyday lives. Despite its widespread usage, password-based authentication has numerous deficiencies. For instance, password theft is becoming a common occurrence due to a variety of security problems associated with passwords. As such, many organizations are moving toward adopting alternative solutions like one-time passwords (OTPs), which are only valid for a single session. Nevertheless, various OTP schemes also suffer from a number of drawbacks in terms of their method of generation or delivery. In this chapter, we present a challenge–response visual OTP authentication scheme that is to be used in conjunction with the camera on a mobile device. The main feature of the proposed scheme is to allow the server to send a challenge over a public channel for a user to obtain a session key while safeguarding the user's long-term secret key. We present the authentication protocol, its security analysis, the various design considerations, and the advantages provided by our system.

3.1 Introduction

In the digital age, passwords are widely used in our daily lives. We authenticate to numerous electronic services over the Internet via username and password. Despite the development of many alternatives, text passwords are still the most dominant form of web authentication due to their convenience and simplicity [30]. However, the use of passwords has been shown to be plagued by various security problems [4,5]. In addition, over the years, many security attacks, such as spyware and phishing attacks, have been developed to extract sensitive information from computers, emails, fraudulent websites, etc. Consequently, password theft has become a common occurrence.

For this reason, many business companies and organizations are moving toward adopting alternative solutions to the traditional static password approach. Static password approaches are particularly vulnerable as these passwords can easily be stolen by an adversary via a variety of means (e.g., keyloggers, phishing attacks, Trojans) and used without the password owner's knowledge.

This has led to the increasing popularity of two-factor authentication (2FA), a technology that confirms the identity of an entity by the combination of two components. An example of 2FA is Google's authentication framework, which uses a two-step verification [11]. In Google's two-step verification, the first step involves the user using the traditional username and static password authentication. In the step second, the user will be asked to input a six-digit verification code for which the user can obtain through a variety of different means. Examples of these means include a Short Message Service (SMS) text message, a voice call to a preregistered phone number, a list of pregenerated one-time codes, or an offline application preinstalled on the user's smartphone. The second step is often known as a one-time password (OTP) scheme, where the verification code (the OTP) is only valid for a single session.

There are a number of ways in which OTPs can be generated and distributed. In the following text, we briefly discuss these methods and their limitations:

- *OTP via SMS*: One common method of distributing the OTPs is to transmit it via an SMS text message. Mulliner [22] argues that this method cannot be considered secure. First, the security of SMS OTP relies on the confidentiality of the SMS messages, which in turn relies on the security of the cellular networks. However, this may not be the case as there are already several potential attacks that can be conducted on these services. Second, specialized mobile phone trojans have been created to compromise the security of SMS OTP approaches.

Finally, this approach also has its usability issue since it can be problematic if the user is in a location with poor mobile phone reception.

■ *OTP via Security Token*: This method is commonly used by banks and financial institutions. A security token that generates OTPs is supplied to the user. The token is capable of generating OTPs using a time-dependent algorithm to generate pseudorandom numbers to be used as the OTPs. This approach relies on accurate time synchronization between the token and the authentication server, as the OTPs generated using this approach are only valid for a short period of time. As such, this approach suffers from synchronization issues and the potential for clock skew [27].

■ *OTP from hash chains*: As a variant of OTP via security token, the OTPs can be generated based on a one-way function in the form of a hash chain. However, hash chains are known to have storage and computational complexity issues [28].

The growing popularity of smartphones has opened up a new possibility in the support of two-factor authentication since the smartphone itself can be treated as the second component in the process. In this chapter, we propose a challenge–response visual OTP authentication scheme that uses the camera on a mobile device to obtain the OTP. The design goal of the scheme is to support the transmission of the challenge over a public channel for a user to obtain a session key while safeguarding the user's long-term secret key. Our approach is based on the concept of visual cryptography and as such does not rely on mobile phone network reception, or having to establish a network link between a computer and a mobile phone. We present the construction of our scheme and its security analysis, along with various practical issues that had to be considered in the design of the visual OTP scheme.

3.1.1 Our Contribution

This chapter presents the design of a visual OTP authentication scheme. The proposed scheme is a challenge–response approach that relies on a camera on a mobile device to receive the challenge and to present the response on the mobile device's display. The advantage of this approach is that it does not suffer from common OTP issues such as poor reception, hash chain complexities, or requirement on time synchronization. In addition, unlike SMS-based approaches, our approach is not restricted to mobile phones and can be used on any mobile device with a camera and a display, including tablet computers. In the proposed scheme, the challenge can even be sent on printed media instead of via electronic means. This chapter is an extension of our scheme published in [6] and includes security analysis and concrete instantiations of the algorithm.

3.2 Related Work

A number of authentication mechanisms exists. In this section, we review the research in the area of authentication that is relevant to our system.

3.2.1 Visual Authentication

The notion of using human–computer cryptographic approaches for identification and authentication has been around for many years. Typically, these approaches implement a challenge–response mechanism that requires a human user to interact with a computer in some manner in order to perform authentication. For example, Matsumoto [20,21] investigated human–computer cryptographic schemes that presented challenges to users in the form of visual images. The approach that was examined in his study was based on the capability of human beings in memorizing and processing images to solve some simple challenges. Since then, other researchers and practitioners have also proposed and developed various graphical password schemes. Graphical passwords attempt to leverage human memory for visual information with the shared secret being related to images [2]. This capitalizes on the natural human ability to remember images, which is believed to exceed memory for text [5]. However, graphical passwords are not immune to security attacks. For example, graphical password schemes may suffer from shoulder surfing attacks where credentials are captured through direct observation of the login process or by recording the process using a recording device [2].

Other schemes that have been proposed in this area are based on using the human visual system to solve the challenge. Naor and Pinkas [23] proposed an authentication and identification approach that is based on visual cryptography. Visual cryptography was introduced by Naor and Shamir [24] as a means of using images to conceal information. The main idea behind visual cryptography is to divide a secret image into a set of shares, each to be printed on a separate transparency. Individually, the shares look like random black and white pixels that reveal no information about the secret image. When the appropriate number of shares are stacked together, the human visual system averages the black and white pixel contributions of the superimposed shares to recover the hidden information. Thus, the concealed information can be decrypted by the human visual system without any need of a computer to perform decryption computations [7]. Figure 3.1 depicts an example of Naor and Shamir's visual cryptography scheme. The secret image, shown in Figure 3.1a, is divided into two shares, which are shown in Figures 3.1b and c, respectively. The secret can be recovered by superimposing the two shares, as shown in Figure 3.1d.

In the scheme proposed by Naor and Pinkas [23], the user is required to carry a small transparency, small enough to be carried in a wallet, and the authentication and identification process simply involves the user overlaying the transparency on the message sent by an informant in order to view the concealed information. However, in their scheme, unless the user carries a stack of transparencies, which would be impractical, a single transparency will have to be used for multiple authentication sessions. It has been highlighted that since basic visual cryptography schemes are equivalent to one-time pads, an observer can eventually learn the user's secret by repeated observation [5]. In addition, Naor and Shamir's visual cryptography scheme suffers from the pixel

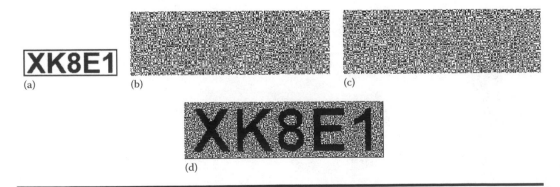

(a) (b) (c) (d)

Figure 3.1 Example of Naor and Shamir's visual cryptography scheme. (a) secret image; (b) share 1; (c) share 2; and (d) result of superimposing shares 1 and 2.

expansion problem, as illustrated in Figure 3.1, where each pixel in the secret image is split into four subpixels in the shares and the recovered image. As such, the shares are four times the size of the secret image.

To overcome a number of drawbacks with this scheme, Tuyls et al. [31] proposed a scheme where every user was to be given a small decryption display. Their approach was similarly based on visual cryptography where the small decryption display was used to replace the need for transparencies. The small decryption display required very limited computing power to perform authentication and security and since the user was required to carry his/her own trusted decryption display, it would be impossible to be contaminated by Trojans or viruses. However, this approach requires the user to use a special authentication device.

A commercially available scheme called PassWindow [26] uses a similar approach where a small transparent display is embedded in an ID card or some form of payment card. The pattern on the transparent display changes periodically based on a pregenerated sequence of patterns. To perform authentication, the user has to overlay the transparent display of the card over a patterned image sent from the server and to visually identify the digits that form as a result of superimposing the card's display onto the image. However, it should be noted that an image on screen can potentially appear at different sizes depending on the user's display settings. This approach requires that the size of the image that is displayed on screen be exactly the same as the size of the card's transparent display.

3.2.2 *Authentication Using a Personal Device*

A number of other authentication approaches that have been proposed make use of personal devices that a user usually carries around (e.g., a cellphone). In a study on how to provide a user with authenticated communication when using an untrusted computer, Clarke et al. [8] proposed a method of using a trusted personal device equipped with a camera to monitor the screen of the untrusted computer. All communication is then authenticated by a trusted proxy. This approach is quite costly in terms of computational resources required to monitor the communication.

Mannan and Oorschot [19] proposed a protocol that they called MP-Auth (*Mobile Password Authentication*), which uses a mobile device to protect user passwords from easily being recorded. In their approach, the mobile device is assumed to be free from malware as the user will enter the password into the mobile device rather than into an untrusted computer. In another approach proposed by Jeun et al. [15], the user uses an application to store his encrypted password in his smart phone and that application program is used to send the password from the smart phone itself, instead of requiring the user to enter his password via a computer's keyboard.

Phoolproof is another scheme that uses mobile phones for authentication. Phoolproof is a mutual authentication protocol used to prevent phishing using a trusted mobile phone [25]. To use the system, the user must establish a shared secret with the server using an out-of-band channel. This long-term secret is stored on the mobile phone. In order to use this protocol, the mobile phone must establish a secure Bluetooth connection with the web browser where mutual authentication occurs between the mobile phone and the website.

3.2.3 *One-Time Passwords*

To overcome some of the problems associated with static passwords, OTP approaches are increasingly being used for authentication. There are various techniques for generating and distributing OTPs. In addition, several approaches were devised to use OTPs in conjunction with mobile devices.

Paterson and Stebila [27] examined an approach of using OTPs in conjunction with OTP authentication key exchange protocols in order to ensure more secure use of OTPs. In a scheme called oPass proposed by Sun et al. [30], a trusted cellphone is used to communicate with the web server (via SMS) and the web browser (via Wi-Fi or Bluetooth). The user does not input his password into the web browser, but rather is required to enter his long-term password into the oPass program that will generate an OTP that will be sent by way of an encrypted SMS to the server.

Mulliner et al. [22] investigated attacks against SMS-based OTPs and state that attacks against cellular networks and mobile phones have shown that SMS messages cannot be deemed to be secure. They proposed a virtual dedicated OTP channel inside the mobile phone operating system to secure OTP SMS messages from being intercepted by Trojans by removing these messages from the general delivery process and redirecting them to a special OTP application.

Instead of using a mobile device, Huang et al. [13] proposed a scheme where the OTP is delivered via an instant messaging service. This approach assumes that the website that adopts the OTP authentication method must join an instant messaging network and use the network to communicate with the users.

3.3 Syntax and Security Definition of the Visual One-Time Password Authentication System

In this section, we first define a visual OTP authentication system and its application scenario. We extend this definition from the visual authentication scheme proposed by Naor and Pinkas [23].

3.3.1 Syntax

A visual OTP authentication system consists of two protocols, namely, Init and Auth, executed between entity H (Henry, aka human) and S (Sandra, aka server).

- Init: We assume both parties have a common input k, the security parameter. Upon completion of the protocol, both parties obtain a long-term secret key lk.
- Auth: We assume both H and S have lk as the common input. For practicality, we require that Auth to be a two-move protocol of the following form:
  - S generates a short-term secret key sk. We use S_{sk} to denote a visual representation of sk. S generates a random value r and produces a visual representation S_1, which is a function of r and lk. S computes $S_2 = S_{sk} - S_1$.* S sends (r, S_2) to H.
  - Upon receiving (r, S_2), H reconstruct S_1 from r and lk. Next, H computes $S_{sk} = S_1 + S_2$ to acquire S_{sk}. Using the visual capability, H can acquire sk from S_{sk}.

---

* We abuse the notation and assume visual representation can be "added" and "subtracted." Any concrete instantiation must specify the details of these operations.

3.3.2 Application Scenario and Threat Model

Without the loss of generality, we assume that there are three entities involved, namely, H (Henry), S (Sandra), and an adversary E (Evan). H is a human, and therefore, H has human visual capabilities. The purpose of the visual OTP authentication system is to enable Sandra to attest whether Henry is present in the protocol in the presence of Evan. Note that Evan can observe the channel used between Henry and Sandra. The security parameter k is involved, such that storage capacities and computing power of Sandra and Evan are polynomial in k.

There are two main stages in the visual OTP authentication scenario. The first stage is the initialization stage, where Sandra engages Henry in protocol Init on a secure channel, which is inaccessible to Evan.

In the second stage, Sandra communicates with Henry in protocol Auth via a public channel, in which Evan can also access. Upon completion of the protocol, both Sandra and Henry have a local copy of the value sk.

The threat model assumes that Evan has access to the public channel, and therefore, he has access to both r and S_2. Informally speaking, the goal is to ensure that Evan cannot obtain any information about sk.

The following game between a challenger and the adversary formally defines the security of requirement.

- *Initialization*: The challenger executes Init on behalf of both H and S to set up the long-term key lk.
- *Query Phase 1*: The adversary E is allowed to eavesdrop as many communication between H and S as it wishes. Upon invocation of this query, the challenger executes Auth on behalf of H and S with common input lk and gives the whole communication transcript, including the outcome, to E. That is, E will be given r, S_2, S_{sk}, and sk.
- *Challenge*: The challenger flips a fair coin $b \in \{0, 1\}$ and executes Auth on behalf of H and S to obtain (r, S_2) and the resulting key sk. The challenger sets $r^* = r$, $S_2^* = S_2$. If $b = 0$ it sets $sk^* = sk$. Otherwise, it sets sk^* to be a random value. Finally, the challenger gives (r^*, S_2^*) and sk^* to E.
- *Query Phase 2*: After receiving the challenge, E can still issue the same query as in Query phase 1.
- *Guess*: Finally, E outputs a bit b'.

We say that E wins the game if $b = b'$. Define the advantage of E in the game mentioned earlier as the probability that E wins minus 0.5.

Definition 3.1 Security. A visual OTP authentication protocol is *secure* if for all PPT algorithm E, the advantage of E is negligible.

3.4 Proposed Visual OTP Scheme

3.4.1 Overview

The overall architecture of the proposed visual OTP authentication system is presented in Figure 3.2. We would like to remark that the private channel for the registration can be established

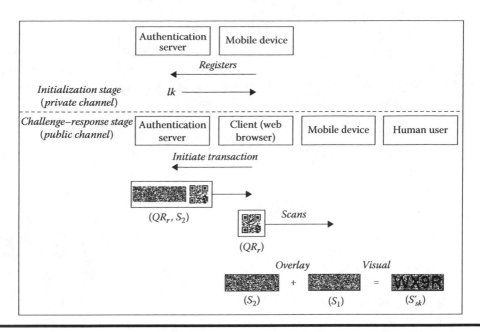

Figure 3.2 Overview of the communication between the various components in the visual one-time password scheme.

via traditional public key cryptography (assuming H is equipped with a public key) or an identity-based encryption [3] (assuming the identity of H is known—in a smart phone scenario, this could be via the phone's International Mobile Station Equipment Identity [IMEI] or phone number).

The figure depicts an example of a practical scenario where the visual OTP scheme can be implemented for conducting an online transaction. In the scenario, the user must first register his mobile device with the authentication server via a secure private channel. The server will in turn generate lk and send this to the user's mobile device. Registration only happens once for the server and mobile device to establish a long-term secret key. Subsequently, whenever the user initiates an online transaction from a web browser, the server will generate and send (QR_r, S_2) (i.e., the challenge) that will be displayed on the web browser. Upon receiving (QR_r, S_2), the user will use the camera on his mobile device to scan QR_r. With the value of lk and r, the user's mobile device will be able to generate S_1. On the mobile device's display, S_1 will be overlaid on S_2 to produce S'_{sk} (i.e., the visual reconstruction of S_{sk} on the mobile device's display), and the user will be able to visually obtain sk (i.e., the response/OTP). Only the server can generate a valid S_2, and only the user can obtain sk using S_1 that is generated on the mobile device.

3.4.2 Algorithm Details

Initialization Stage (Init): On input a security parameter k, we let S choose a long-term secret key $lk \in_R \{0, 1\}^k$. S transmits lk to H via a secure and authenticated channel. Upon successful completion of the protocol, both S and H store lk.

Challenge–Response Stage (Auth): In this stage, S selects a random number $r \in_R \{0, 1\}$ and conduct the following computation.

- Produce a quick response (QR) code that contains r, QR_r.
- Generate a short-term secret key sk, and its visual representation S_{sk}.
- Execute Algorithm 3.1 with input lk, r and S_{sk} and obtain S_1, S_2.
- Present QR_r and S_2 to H via a public channel.

Upon receiving the challenge (QR_r, S_2), H conducts the following:

- Scan the QR code to retrieve r.
- Reproduce S_1 from lk and r based on Algorithm 3.2.
- Use H's visual capability to retrieve sk from the superposition of S_1 and S_2.
- Output sk.

Note that the value of sk is obtained visually and is never stored anywhere.

Remarks: For efficiency considerations and to allow a concrete security analysis, we instantiate the function randomPixel(lk, r, i, j) employed in Algorithms 3.1 and 3.2 as $H(lk, r, i, j) \bmod 2$ where H is a hash function. Looking ahead, we would model it as a random oracle in the security analysis.

3.4.3 Security Analysis

In the following, we prove that our protocol is secure under Definition 3.1 in the random oracle model [1]. Assume there exists E that can win the game with probability ε. We use a technique known as game-hopping [10], a commonly employed approach in cryptography, to show that ε is negligible.

- **game$_1$**: We consider **game$_1$**, which is the same as the original security definition except that in the query phase 1 and 2, the transcript component r is replaced by a random value $r' \in_R \{0,1\}^k$. In other words, the value r transmitted to E is no longer related to S_2. In the random oracle model, the advantage of E remains unchanged except E makes a hash query on input $H(r', lk, i, j)$ for some i and j within the image width. This happens with probability $1/2^k$ since lk is hidden from the view of E. If E makes q hash queries, the probability that this happens is bounded earlier by $q/2^k$.
- **game$_2$**: We consider **game$_2$**, which is the same as **game$_1$** except that $r^a st$ is set to be a random value $r' \in_R \{0,1\}^k$. Now that the transcript contains no information about sk. Again, by the same token, the advantage of E in **game$_2$** is the same as **game$_1$** unless a query is made on $H(r^*, lk, i, j)$ and this happen with negligible probability.
- Now, consider the probability that E wins in **game$_2$**. Since the value r^* is no longer related to sk, E can only guess the value of b with probability $1/2$.
- To sum up, the difference in advantage of any E between the original game and **game$_1$** is bounded by $O(q/2^k)$ and that the advantage difference between **game$_1$** and **game$_2$** is also bounded by $O(q/2^k)$. The advantage of any E in **game$_2$** is 0. Thus, the advantage of any E in the original game is bounded by $O(q/2^k) + O(q/2^k) = O(q/2^k)$ that is negligible for any PPT E whose running time is polynomial in the security parameter k since q must be polynomial in k.

3.4.4 Practical Issues

In the proposed scheme, r has to be sent to H over a public channel. While it is not necessary to encode and transmit r within a QR code, we find that this is the most appropriate and convenient method of delivery. The QR code is a two-dimensional code that was invented by the company Denso Wave [9]. These days, QR codes are ubiquitous on the Internet, and the information contained within a QR code can easily be scanned by a mobile device with a camera. In addition, QR codes have a built-in error detection and correction mechanism that can be used to correctly decode corrupted QR codes, which may contain certain errors. Furthermore, QR codes contain a number of patterns to determine rotational orientation and alignment. Since (QR_r, S_2) is sent to H as a single image, QR_r can be used to facilitate the alignment of S_1 and S_2.

It is well known that traditional visual cryptography suffers from the alignment problem, in that when stacking shares, it is difficult to align the shares [18,34]. Practical approaches typically suggest the use of some reference frame to align the transparencies [23]. However, unlike traditional approaches that use physical transparencies or tokens, our approach relies on the use of a mobile device like a smart phone or a tablet. As such, using the camera's video stream to capture (QR_r, S_2), this can be used in conjunction with image processing techniques to overlay S_1 over S_2. This is akin to techniques using in augmented reality to overlay virtual content onto elements of the real world [32]. Adopting this method will allow the mobile device to appropriately scale and rotate S_1 in order for it to align with S_2.

Another problem with traditional visual cryptography when displaying a share on a computer screen and trying to place the corresponding share, which is printed on a transparency, on top of the screen, is that monitors can differ greatly and the computer can be set to different display settings. As such, the image of the share on screen may not be displayed at the same size as the share printed on the transparency. This will prevent the shares from being correctly superimposed, and thus the secret cannot be recovered. In the approach proposed in this chapter, we rely on the mobile device to virtually overlay S_1 over S_2. This means that it does not matter what size S_2 is displayed at, as long as the mobile device can accurately capture the image of S_2, because the mobile device can scale S_1 to the appropriate size. To facilitate this, the size of the squares in S_2's image should not be too small.

As previously shown in Figure 3.1, traditional visual cryptography suffers from the pixel expansion problem that significantly increases the size of the resulting shares. While there are a number of size invariant visual cryptography schemes like the probabilistic approaches proposed by Ito et al. [14] and Yang [33], these schemes do not produce the ideal visual quality required for the visual OTP. Therefore, for the purpose of generating S_1 and S_2, a random grid visual cryptography approach was deemed to be the most suitable approach. Random grid visual secret sharing was first proposed by Kefri and Keren [16], and over the years a number of random grid approaches have been investigated [12,29]. Using a random grid visual cryptography scheme, it is possible to produce shares with no pixel expansion.

In the proposed visual OTP scheme, the shared image S_1 will be generated from $lk||r$ and a pseudorandom number generator. Thus, S_1 is a random grid. S_1 can be used in conjunction with the secret image S_{sk} to generate the corresponding challenge image S_2. Algorithm 3.1 gives an example of a random grid visual secret sharing method that was adapted from Shyu [29], which can be used in the proposed visual OTP scheme. In this approach, black pixels in S_{sk} are reproduced at 100% in S'_{sk} and white pixels (i.e., transparent pixels) are reproduced at 50%. Figure 3.3 shows the results of using Algorithm 3.1 on a secret image. The secret image, shares 1 and 2, along with the reconstructed image is shown in Figures 3.3a through d, respectively.

Algorithm 3.1 An Algorithm for Generating S_2 from S_{sk}, lk and r

---

function GENERATESHARES(S_{sk}, lk, r)
    $imgWidth \leftarrow S_{sk}$ width
    $imgHeight \leftarrow S_{sk}$ height
    for $i = 1$ to $imgWidth$ **do**
        for $j = 1$ to $imgHeight$ **do**
            /* Generate S_1 as a random grid */
            $S_1[i,j] \leftarrow$ randomPixel(lk, r, i, j) /* randomPixel() outputs 0 or 1 */
            /* Generate S_2 */
            if $S_{sk}[i,j] = 0$ **then**
                $S_2[i,j] \leftarrow S_1[i,j]$
            else
                $S_2[i,j] \leftarrow \neg S_1[i,j]$
            end if
        end for
    end for
end function

Algorithm 3.2 An Algorithm for Generating S_1 from lk, r, $imgWidth$, $imgHeight$

---

function RECOVERY($imgWidth$, $imgHeight$, lk, r)
    for $i = 1$ to $imgWidth$ **do**
        for $j = 1$ to $imgHeight$ **do**
            /* Re-compute S_1 from lk and r */
            $S_1[i,j] \leftarrow$ randomPixel(lk, r, i, j) /* randomPixel() outputs 0 or 1 */
        end for
    end for
end function

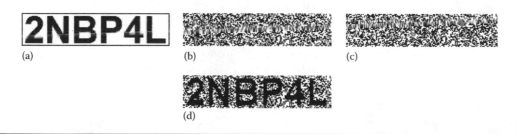

(a)   (b)   (c)

(d)

Figure 3.3 Random grid visual cryptography approach. (a) Secret image; (b) share 1; (c) share 2, and (d) result of superimposing shares 1 and 2.

Another practical issue to consider when implementing the visual OTP scheme is how clearly the user will be able to perceive the OTP in the visual reconstruction of the secret image. For this, we should consider the color of the text and the background. There are two possible variations as depicted in Figure 3.4, where Figure 3.4a shows the reconstructed secret using black text on a white background, and Figure 3.4b shows the reconstructed secret using white text on a black

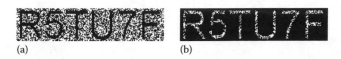

(a)   (b)

Figure 3.4   Text and background color. (a) Black text on a white background and (b) white text on a black background.

background. It has been argued that using white contents on a black background gives rise to better perceived visual quality in the reconstructed image for images with thin lines [17].

3.5 Discussion

3.5.1 Advantages of the Visual OTP Scheme

The fundamental goal of the proposed visual OTP scheme is allow the challenge to be sent over a public channel for the user to obtain a session key that can be used as an OTP, while safeguarding the user's long-term secret key. In this scheme, the user also does not have to remember any passwords.

As mobile devices are ubiquitous in this day and age, the proposed approach does not require the user to carry around a specialized authentication card or device, or a printed list of OTPs. In addition, unlike authentication schemes like SMS OTP–based approaches, authentication in the proposed method does not require any form of communication with a mobile phone network. As such, mobile phone network reception is not an issue. This also means that the visual OTP scheme can be applied to any mobile device that has a camera and is not restricted to only be usable on smart phones. The user simply has to install the visual OTP software and register it with the authentication server.

While the OTP can be used to authenticate the user, another feature provided by the scheme is that the user can also verify that the message containing the challenge was sent by a legitimate party. This is because in the proposed scheme, without knowledge of the long-term secret key, an adversary cannot generate a valid challenge. This also prevents an adversary from tampering with the challenge image, as changing QR_r will mean that the mobile device will not be able to generate the correct visual pattern (i.e., S_1) to solve the challenge, and changing S_2 will not produce a valid result when overlaying S_1 over S_2. Furthermore, such an event would raise the suspicion of the user, as it would indicate that the challenge may have been tampered with.

Another advantage of the proposed scheme is that the challenge does not have to be transmitted via electronic means. For example, if a bank wants to send a letter to an individual who has registered with the bank, the bank can send the person a letter with the challenge printed on paper. The person can verify that the letter was indeed sent from the bank (as only the bank can generate a valid challenge) and also receive the OTP that can be used for authentication purposes with the bank.

In other authentication approaches that involve the user having to overlay a transparency or an authentication token on top of another pattern, the size of the patterns has to perfectly match. Otherwise, the user will not be able to recover the secret by superimpose the different-sized patterns. This is not an issue in the proposed approach as the mobile device will be responsible for scaling and aligning the patterns. Therefore, the challenge can be displayed in any size as long as it can be captured by the mobile device's camera. The mobile device will then use augmented reality techniques to overlay the virtual pattern onto the image of the challenge pattern.

It should be noted that the OTP is obtained by the human user via the visual channel and the OTP is never stored on any device. This prevents malicious software like keyloggers or even software designed to monitor the user's activities from obtaining the OTP. Furthermore, the OTP is only valid for a single use. Additionally, the video stream on the mobile device will be used to overlay the visual patterns and present this to the user in real time. If there is any software designed to monitor the user's activities, this will require a huge amount of information to be streamed to the adversary, which will significantly degrade the system's performance and alert the user of suspicious activity.

In addition, unlike traditional graphical passwords, which may suffer from shoulder surfing attacks, this is not an issue in the proposed visual scheme. Shoulder surfing attacks are where an adversary standing behind the user, and possibly even recording the user's interactions, maybe able to observe and detect some pattern in the image or from the user's interactions, which will compromise the security of the visual password. In the proposed visual OTP scheme, the visual pattern generated on the mobile device to solve a challenge can only be used for that particular challenge. The mobile device will generate different visual patterns for different challenges.

3.5.2 Limitations

In this section, we discuss some of the limitations of the proposed visual OTP scheme.

As with all visual challenges or passwords, the proposed scheme relies on the human visual system. This means that it does not cater for the blind or visually impaired and cannot be used by an individual with a visual disability. Another potential disadvantage is that the challenge image will have to be displayed at a certain size in order for the mobile device's camera to be able to accurately capture the information contained within the challenge image. While this is not seen as a major problem, it may adversely affect the layout or aesthetics of a message, document, or webpage.

It should be noted that the proposed scheme does not deal with man-in-the-middle or similar attacks. To handle such attacks, the scheme can be combined with other security protocols that are designed to handle man-in-the-middle attacks. In addition, this approach also does not address the situation where the authentication server is hacked. The server is responsible for its own security, and it is assumed that all the necessary security, mechanisms are in place.

In the proposed scheme, the mobile device captures the challenge image using its video stream and is responsible for overlaying the virtual image on top of the challenge image. As such, it is assumed that the mobile device has the computational capabilities required to process augmented reality techniques in real time. Additionally, since the mobile device has to be used to visually present the solution to the challenge, a separate means of displaying the challenge has to be employed. In other words, if the user wants to conduct an online transaction via a web browser, this cannot be done using the mobile device's web browser as the mobile device itself will have to be used in conjunction with the challenge's display to obtain the OTP. However, this requirement is no different from several other authentication schemes that were previously presented in Section 3.2, which also require using a web browser and a separate mobile phone to perform authentication.

3.6 Conclusion

In this chapter, we presented the design of a challenge–response visual OTP authentication scheme. Using this scheme, a challenge is sent to a registered individual, which can be via a web browser or even printed media, and the user can use the camera and display of his mobile device to obtain the solution to the challenge. This approach can be implemented on a variety of mobile devices, such as mobile phones and tablets, with the main requirement being that the device must have a camera. The challenge itself can be transmitted over a public channel without the threat of it being compromised by an adversary, as the adversary can neither correctly generate nor solve the challenge. As such, the scheme does not suffer from the common issues affecting the generation and delivery of OTPs such as mobile phone reception, hash chain complexities, or time synchronization mechanisms. In addition, this scheme does not suffer from security issues like shoulder surfing attacks or keyloggers, as the mobile device will generate the specific visual pattern required to solve a particular challenge and will generate a different visual pattern when presented with a different challenge.

References

1. M. Bellare and P. Rogaway. Random oracles are practical: A paradigm for designing efficient protocols. In D. E. Denning, R. Pyle, R. Ganesan, R. S. Sandhu, and V. Ashby, eds., *Proceedings of the First ACM Conference on Computer and Communications Security (CCS 93)*, Fairfax, VA, November 3–5, 1993, pp. 62–73. ACM, New York, NY, 1993.
2. R. Biddle, S. Chiasson, and P. C. van Oorschot. Graphical passwords: Learning from the first twelve years. *ACM Computing Surveys*, 44(4):19, 2012.
3. D. Boneh and M. K. Franklin. Identity-based encryption from the Weil pairing. In J. Kilian, ed., *Proceedings of the 21st Annual International Cryptology Conference on Advances in Cryptology (CRYPTO 2001)*, Santa Barbara, CA, August 19–23, 2001, Lecture Notes in Computer Science, Vol. 2139, pp. 213–229. Springer, Berlin Heidelberg, 2001.
4. J. Bonneau, C. Herley, P. C. van Oorschot, and F. Stajano. The quest to replace passwords: A framework for comparative evaluation of web authentication schemes. In *IEEE Symposium on Security and Privacy*, San Francisco, CA, pp. 553–567. IEEE Computer Society, Washington, DC, 2012.
5. J. Bonneau, C. Herley, P. C. van Oorschot, and F. Stajano. The quest to replace passwords: A framework for comparative evaluation of web authentication schemes. Technical Report 817, University of Cambridge Computer Laboratory, Cambridge, U.K., 2012.
6. Y.-W. Chow, W. Susilo, M. H. Au, and A. M. Barmawi. A visual one-time password authentication scheme using mobile devices. In L. C. K. Hui, S. H. Qing, E. Shi, and S. M. Yiu, eds., *Proceedings of the 16th International Conference on Information and Communications Security (ICICS 2014)*, Hong Kong, China, December 16–17, 2014, Revised Selected Papers, Lecture Notes in Computer Science, Vol. 8958, pp. 243–257. Springer, Switzerland, 2014.
7. Y.-W. Chow, W. Susilo, and D. S. Wong. Enhancing the perceived visual quality of a size invariant visual cryptography scheme. In T. W. Chim and T. H. Yuen, eds., *Proceedings of the 14th International Conference on Information and Communications Security (ICICS 2012)*, Hong Kong, China, Lecture Notes in Computer Science, Vol. 7618, pp. 10–21. Springer, Berlin Heidelberg, 2012.
8. D. E. Clarke, B. Gassend, T. Kotwal, M. Burnside, M. van Dijk, S. Devadas, and R. L. Rivest. The untrusted computer problem and camera-based authentication. In F. Mattern and M. Naghshineh, eds., *Proceedings of 1st International Conference on Pervasive Computing*, Zurich, Switzerland, Lecture Notes in Computer Science, Vol. 2414, pp. 114–124. Springer, Berlin Heidelberg, 2002.
9. Denso Wave Incorporated. QRcode.com, http://www.qrcode.com/en/, accessed August 3, 2016.
10. A. W. Dent. A note on game-hopping proofs. *IACR Cryptology ePrint Archive*, 2006:260, 2006.
11. E. Grosse and M. Upadhyay. Authentication at scale. *IEEE Security & Privacy*, 11(1):15–22, 2013.
12. Y.-C. Hou, S.-C. Wei, and C.-Y. Lin. Random-grid-based visual cryptography schemes. *IEEE Transaction Circuits and System Video Technolgy*, 24(5):733–744, 2014.
13. C.-Y. Huang, S.-P. Ma, and K.-T. Chen. Using one-time passwords to prevent password phishing attacks. *Journal Network and Computer Applications*, 34(4):1292–1301, 2011.

14. R. Ito, H. Kuwakado, and H. Tanaka. Image size invariant visual cryptography. *IEICE Transactions on Fundamentals of Electronics, Communications and Computer Sciences*, 82(10):2172–2177, 1999.
15. I. Jeun, M. Kim, and D. Won. Enhanced password-based user authentication using smart phone. In R. Li, J. Cao, and J. Bourgeois, eds., *Advances in Grid and Pervasive Computing*, 7th International Conference, GPC, Hong Kong, China, Lecture Notes in Computer Science, Vol. 7296, pp. 350–360. Springer, Berlin Heidelberg, 2012.
16. O. Kafri and E. Keren. Encryption of pictures and shapes by random grids. *Optics Letters*, 12(6): 377–379, 1987.
17. F. Liu, T. Guo, C. K. Wu, and L. Qian. Improving the visual quality of size invariant visual cryptography scheme. *Journal Visual Communication and Image Representation*, 23(2):331–342, 2012.
18. F. Liu, C. K. Wu, and X. J. Lin. The alignment problem of visual cryptography schemes. *Designs Codes Cryptography*, 50(2):215–227, 2009.
19. M. Mannan and P. C. van Oorschot. Using a personal device to strengthen password authentication from an untrusted computer. In S. Dietrich and R. Dhamija, eds., *Financial Cryptography*, Lecture Notes in Computer Science, Scarborough, Trinidad and Tobago, Vol. 4886, pp. 88–103. Springer, Berlin Heidelberg, 2007.
20. T. Matsumoto. Human identification through insecure channel. In D. W. Davies, ed., *EUROCRYPT*, Brighton, UK, Lecture Notes in Computer Science, Vol. 547, pp. 409–421. Springer, Berlin Heidelberg, 1991.
21. T. Matsumoto. Human–computer cryptography: An attempt. *Journal of Computer Security*, 6(3): 129–150, 1998.
22. C. Mulliner, R. Borgaonkar, P. Stewin, and J.-P. Seifert. SMS-based one-time passwords: Attacks and defense (short paper). In K. Rieck, P. Stewin, and J.-P. Seifert, eds., *DIMVA*, Lecture Notes in Computer Science, Berlin, Germany, Vol. 7967, pp. 150–159. Springer, Berlin Heidelberg, 2013.
23. M. Naor and B. Pinkas. Visual authentication and identification. In B. S. Kaliski Jr., ed., *Advances in Cryptology—CRYPTO '97*, 17th Annual International Cryptology Conference, Santa Barbara, CA, Lecture Notes in Computer Science, Vol. 1294, pp. 322–336. Springer, Berlin Heidelberg, 1997.
24. M. Naor and A. Shamir. Visual cryptography. In *EUROCRYPT*, Perugia, Italy, pp. 1–12, 1994.
25. B. Parno, C. Kuo, and A. Perrig. Phoolproof phishing prevention. In G. Di Crescenzo and A. D. Rubin, eds., *Financial Cryptography and Data Security*, 10th International Conference, FC 2006, Anguilla, British West Indies, Lecture Notes in Computer Science, Vol. 4107, pp. 1–19. Springer, Berlin Heidelberg, 2006.
26. PassWindow. http://www.passwindow.com/, accessed August 3, 2016.
27. K. G. Paterson and D. Stebila. One-time-password-authenticated key exchange. In R. Steinfeld and P. Hawkes, eds., *Information Security and Privacy: 15th Australasian Conference*, ACISP 2010, Sydney, Australia, Lecture Notes in Computer Science, Vol. 6168, pp. 264–281. Springer, Berlin Heidelberg, 2010.
28. Y. Sella. On the computation-storage trade-offs of hash chain traversal. In R. N. Wright, ed., *Financial Cryptography*, FC 2003, Guadeloupe, French West Indies, Lecture Notes in Computer Science, Vol. 2742, pp. 270–285. Springer, Berlin Heidelberg, 2003.
29. S. J. Shyu. Image encryption by random grids. *Pattern Recognition*, 40(3):1014–1031, 2007.
30. H.-M. Sun, Y.-H. Chen, and Y.-H. Lin. OPASS: A user authentication protocol resistant to password stealing and password reuse attacks. *IEEE Transactions on Information Forensics and Security*, 7(2):651–663, 2012.
31. P. Tuyls, T. A. M. Kevenaar, G. J. Schrijen, T. Staring, and M. van Dijk. Visual crypto displays enabling secure communications. In D. Hutter, G. Müller, W. Stephan, and M. Ullmann, eds., *1st Annual Conference on Security in Pervasive Computing*, Boppard, Germany, Lecture Notes in Computer Science, Vol. 2802, pp. 271–284. Springer, Berlin Heidelberg, 2003.

32. D. Wagner and D. Schmalstieg. Making augmented reality practical on mobile phones, part 1. *IEEE Computer Graphics and Applications*, 29(3):12–15, 2009.
33. C.-N. Yang. New visual secret sharing schemes using probabilistic method. *Pattern Recognition Letters*, 25(4):481–494, 2004.
34. C.-N. Yang, A.-G. Peng, and T.-S. Chen. MTVSS: (M)isalignment (t)olerant (v)isual (s)ecret (s)haring on resolving alignment difficulty. *Signal Processing*, 89(8):1602–1624, 2009.

MOBILE DEVICE PRIVACY

Chapter 4

Dealing with User Privacy in Mobile Apps

Issues and Mitigation

Ricardo Neisse, Dimitris Geneiatakis, Gary Steri,
Georgios Kambourakis, Igor Nai Fovino, and Riccardo Satta

Contents

Abstract

Mobile platforms, such as Android, iOS, and Windows, are gaining more and more relevance within end users' applications, thanks to their usability, flexibility, and low cost. As a result, mobile Internet traffic is about to overwhelm landline traffic. Mobile platforms do not only provide end users with services similar to legacy computers but also extend their experiences exploiting the additional hardware features (e.g., sensors) incorporated in the mobile device, without the need of supplementary devices. Moreover, mobile devices are becoming the sources and repositories of sensitive information, from running performances to positioning data, from travel information to friendship preferences, from personal photos to financial data, and so on.

Data loss, modification, or exposure that a mobile device might face, on one hand, could directly impact end users' safety and privacy, but, on the other hand, they could seriously damage the trust in the rising mobile economy. A typical example of these threats could be the case in which a mobile application that monitors some end user's body parameters is exploited by an adversary to gain access to sensitive data and infer the end user's health status. The possible damages caused by this data breach can impact both the psychological and physical spheres of an end user. Although these problems and flaws already existed in the traditional information and communication technologies (ICT) systems, their magnitude increased exponentially in the case of mobile devices due to their stronger link with the owner. Indeed, as already mentioned, mobile devices embed several sensors and functionalities capable of collecting a huge amount of sensitive information. As a consequence, any vulnerability in the host platform can have a high impact on end user's privacy and security.

In this chapter, we analyze the different characteristics of the Android platform that can be manipulated and exploited by a malicious app to gain access to an end users' private data. Android, the dominant operating system (OS) in the mobile world, is indeed taken as a use case to illustrate threats, which, in the reality, affect the majority of mobile OSs. Moreover, by elaborating on these security flaws and misconfigurations, we describe different threat examples that influence end users' privacy and anonymity in a risk assessment fashion. Furthermore, this chapter reviews the different existing solutions that can be employed to mitigate the described threats and to empower end users in regaining control on their sensitive information and on the behavior of the mobile applications installed in their mobile devices.

4.1 Introduction

The Android operating system is currently the dominant choice among mobile devices and is also receiving growing attention in other device categories including smart watches, tablets, and smart televisions. One of the reasons behind its success is the large number of available applications (apps) that can be easily installed mostly for free by the end users. These apps help users accomplishing their daily tasks including banking, point-to-point navigation, and web browsing. Since users are mostly inexperienced and have little knowledge about their risks when installing these apps, malicious developers and hackers are more and more producing apps with hidden functionalities mainly aimed at collecting users' personal data or harming the user in some way.

When apps are published in the official Android app market (Google Play), a security service named Google Bouncer is in place to scrutinize these apps and identify malicious behavior that may possibly harm end users. Even though Google Bouncer is mostly successful, it is not perfect, and in many cases, apps with malicious or hidden functionality still manage to appear legitimate and are made available to a large number of users. Since the Android OS also allows users to install apps from any other source, even apps received attached to e-mail messages, a large number of users are also vulnerable to malicious apps distributed through these alternative markets. Common examples are app markets that distribute modified versions of paid apps that can be freely installed by users, but they also include additional malicious functionalities users are totally unaware of.

Malicious apps can only successfully harm the user because the Android OS has security vulnerabilities, lack of end user control over the app behavior, and access to sensitive resources. Unknown vulnerabilities or bugs are an inherent problem of any software system, and a simple way to resolve them is by updating the software with a new version that corrects the problem. In many cases, updating the Android OS is not an option since many different models of phones are available and users must rely on the manufacturers to provide these updates. As a consequence, in many situations, users are left vulnerable to attackers due to the lack of updates.

From a user control perspective, in the Android OS, users are supported only by install-time permission requests. When users install an app, they may decide to allow access to a set of sensitive resources, and in case they decide to not allow it, the app will not be installed. The information to the users about the sensitive resources being requested is mostly unclear and difficult to understand, and users have no deep insight about the risks of installing any given app. Whenever access to sensitive resources such as user data is granted, apps may do whatever they want with it, without further evaluation of the user privacy preferences and possibility of further restrictions by end users.

Our focus in this chapter is on the analysis of the Android security model, the threats to the end users, and the existing solutions proposed in the literature to address these threats. The remainder of this chapter is organized as follows. Section 4.2 overviews the high-level architecture of the Android OS including details about its security mechanisms and focusing on its permission model. Section 4.3 describes the different ways a malicious actor could gain access to otherwise private information stored in an Android device. Section 4.4 proposes a classification of existing defensive approaches that can be employed to eliminate the impact on the end users. Section 4.5 elaborates on the effectiveness of existing countermeasures. Finally, Section 4.6 concludes the chapter, giving also some ideas for future work. It is worth emphasizing that while this chapter focuses mainly on Android OS, as today it is the predominant one for mobile devices, similar privacy flaws apply to other mobile OSs as well.

4.2 Android High-Level Architecture

This section provides an overview of the Android OS architecture focusing on those features governing the way the OS grants a permission to an app. As Figure 4.1 illustrates, Android OS employs a multilayer software architecture on the basis of a Linux-based kernel. On the top of the architecture, apps are running on a virtualized environment in the context of Dalvik virtual machine (DVM), named as Android Application Sandbox. This sandbox is set up in the kernel, thus propagating the isolation on all the layers above it. As a result, all apps running in the Android OS are assigned a low-privilege user ID, are only allowed access to their own files, cannot directly interact with each other, and have a limited access to the OS resources. This way Android isolates an apps' execution, that is, each app has its own security environment and dedicated filesystem with the aim to eliminate possible interference that might affect the app's reliability.

Further, in Android OS, the functionalities an app can use are categorized and grouped in application programming interfaces (APIs) that among others grant access to sensitive resources normally accessible only by the OS. For example, among the protected APIs, there are functions for SMS and MMS management, access to location information, camera control, network access, etc.

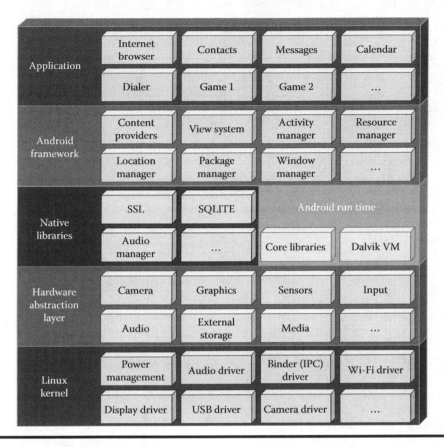

Figure 4.1   Android software stack.

The access to the protected APIs is regulated by a permission mechanism, in which a specific permission defined in the corresponding app's manifest should be granted to the app during installation in order to allow access to a given API at run time. Unprotected APIs do not require any special permission to be executed by the app. More specifically, permissions in Android OS are bundled in four different levels considering the risk level introduced to the user:

1. *Normal permissions* are considered as low risk to other apps, the system, or the end user*.
2. *Dangerous permissions* have a high risk of negative consequences for the users' personal data and experience.
3. *Signature permissions* are used to protect exported interfaces accessible only by apps signed with the same developer key.
4. *Signature-or-system permissions* are used to protect core resources available only to trusted system apps signed with the firmware key.

When installing an app, a user is notified only about the dangerous permissions required by it. The user must grant the app the requested permissions, otherwise the installation fails. Further details on app permission management in Android are given in the next sections.

---

* http://developer.android.com/guide/topics/manifest/permission-element.html.

4.3 Privacy Flaws in Mobile World

In this section, we delve into different scenarios in which end user privacy violations could take place by manipulations or misconfigurations on the Android OS security model. In our analysis, we focus on cases where root access to the mobile device is not required, while data exfiltration could be accomplished by means of different communication channels incorporated to the mobile device (e.g., WiFi, near field communication [NFC]).

Specifically, Android OS protects access to sensitive resources such as camera, and Internet connection through a permission-restricted model. Therefore, if an app wishes to have access to one or more of these resources, it should have declared in a manifest file the corresponding permission, for example, Android.Permission.CAMERA. By using this model, one is able to restrict app access to sensitive resources, but there is no guarantee that a malicious app will not manage to manipulate the corresponding manifest file. Note that the user is left with no other choice apart from accepting all the requested permission access to sensitive resources; otherwise, the app will not be installed. This is true up to version 6, which currently is a huge user base. Furthermore, any permission such as INTERNET or BLUETOOTH that has been classified by the Android OS itself as normal is automatically granted to apps without informing the user. However, this a priori categorization of permissions into low or high-risk ones may be perceived differently by individual users.

Furthermore, the permission request access notification procedure does not provide any inside information in regard to access to sensitive resources. For instance, a mobile app gaining access to the CAMERA permission is allowed to also use the different system resources for taking pictures (using the takePicture() API) as well as to record media (e.g., voice using the MediaRecorder API.) Consequently, everyday users often tend to ignore these requests. This problem has been highlighted by other researches as well [86].

In any case, the majority of users are not aware of the possible risks the Android permission system introduces to their personal sphere. As a result, to gain access to user's private information, an app is able to request access to one or several permissions that it does not actually need for providing its services. This stands true not only for malicious apps but sometimes for benevolent ones too. Putting it another way, even well-known and trusted apps stealthily manipulate access to users' data without explicitly notifying the user as demonstrated by TaintDroid [22] and other similar techniques [44,51]. For instance, apps may covertly gain access to unique identifiers, including Android ID, Android Advertising ID, Google Advertising ID, UDID, IDFV, and IDFA, which in turn constitute users vulnerable to monitoring and data correlation attacks. To exemplify this situation, Listing 4.1 shows a case in which a well-known app sends out the unique identifier of the device along with other information when a crash occurs. Although typically users get informed (in a very

```
$r4 = virtualinvoke $r8.<TelephonyManager: getDeviceId()>();

if $r4 == null goto label2;

$r2 = $r0.<class: java.util.Map mDeviceSpecificFields>;

$r3 = <class: class.ReportField DEVICE_ID>;

interfaceinvoke $r2.< Object put(Object,Object)>($r3, $r4);
```

Listing 4.1   An example of a simple code residing in a real Android application that gets access to device unique ID.

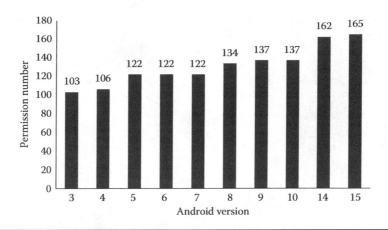

Figure 4.2   SDK Android permission evolution in numbers from version 3 to version 15.

similar way to an access to specific permission) about possible access to such identifiers (in this case, READ_PHONE_STATE), they have no insight knowledge about how these IDs and the associated information will be used. Such instances of abuse could be identified by reverse engineering the app.

The observations mentioned earlier are verified by various research works [4,24,77] so far, demonstrating that a great mass of apps manipulate their access to sensitive resources without notifying the user. In our opinion, this situation is escalated by the evolution of the Android permission model between different versions; that is, this version progression increases the Android OS privacy attack surface providing additional attack vectors to malicious users against end users privacy (Figure 4.2). This is mainly due to the additional sensors and sensitive operations incorporated to each new Android version.

As already pointed out, some apps may also request permissions they not actually need for accomplishing their tasks. These apps are characterized as overprivileged [26]. Also, such an app may be able to trigger a silent privilege escalation in terms of accessing sensitive resources after an OS update [77]. This is because Android does not inform users about the existence of a new permission that does not apply to the OS version in which the app is currently installed. Privilege escalation may also lead to a so-called confused deputy attack [24]. Namely, when an app has been granted a specific permission, it is then exploited by other apps with the aim to gain access to specific sensitive permissions and perform sensitive tasks. A typical example of this situation is concerned with an app that is granted with the permission to send SMS messages and allows another app to use its interface to read SMS messages or send SMS to premium numbers. At this point, we should mention that these types of attacks capitalize on Android's interprocess communication (IPC) component in which unlike other IPC does not support remote communication through network but only communication through other "apps" using higher level abstraction such as Intents, Messengers, and content providers. Figure 4.3 provides a graphical representation of the aforementioned attack.

Furthermore, to achieve their goals, evildoers may be able to exploit vulnerabilities in Android native libraries. For instance, Stagefright [87] is Android's media playback native library used to process, record, and play multimedia files. This means that any given app using a multimedia resource is built against this library. Consequently, if a vulnerability resides in this library, all apps using it become straightforwardly vulnerable to the very same attack vectors. This is the case for

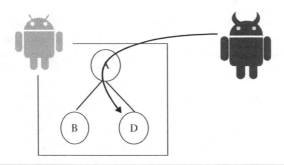

Figure 4.3 Example of vulnerability exploitation by a malware on method D to execute method U that is "out of the scope" of the app. By the term "out of the scope," we mean that the malware is taking the advantage of the app's access to resources that are not actually required to fulfill its tasks.

a vulnerability found in the Stagefright library where evildoers were able to access resources that otherwise only Stagefright library components have, for example, to the Media Server. As a direct result, an attacker exploiting this vulnerability may gain access to all services and components, such as recording audio and video, in which normally only the Media Server is entitled to. It is worth noting that an aggressor does not even need the end user to install any particular app to trigger such an attack. There are various ways for taking advantage of this vulnerability, say, through a specific crafted MMS or by exploiting other remote attack vectors as reported in [88].

Even worse, some methods in the Android API are still not protected by specific permission, and this could introduce possible backdoors in a system for a malicious entity to gain access to private information. For instance, an app might use the *exec(string) API* method to execute the process *prog* passed as a parameter. This means that apps could silently execute unprotected system commands in order to read system- and users-related information, say, reading the SD card and retrieving the list of installed apps.

Moreover, Android apps have to deal with legacy network vulnerabilities and attacks, including man in the middle, malformed messages, and other more specialized attack vectors. For instance, the work in [89] has demonstrated different cases where apps fail to protect the exchanged data in a given session even if the running app mandates Transport Layer Security (TLS). Similarly, attackers could craft specific Javascript input that in turn can grant access to otherwise private user's information in the mobile device [90].

The aforementioned discussion shows that apps themselves, either legitimate or malware, are characterized by a certain degree of privacy invasiveness, which is directly related to the permissions they request and to which use make out of the protected methods. A typical example of a category of apps showing privacy invasive behavior is the mobile games. That is, several games request access to unique identifiers or user location they do not actually need for their proper function. To cope with this threat, Android provides security services to verify apps before installation and to periodically scan the OS for harmful apps. Unfortunately, these security services themselves are also privacy invasive. Surprisingly, this is mentioned in Android documentation,* saying that the device may transmit information to Google such as log information, URLs related to the app, device ID, OS version, and IP address. In this direction, preinstalled apps are automatically granted every required permission and are considered trusted since they are part of OS firmware per se. It is therefore clear

---

* https://support.google.com/accounts/answer/2812853?hl=en.

that end users are not informed about which exact permissions these apps require. This means that users do not have any indication which specific resources these apps access in the course of time, and thus they are left totally unprotected or have poor privacy protection in place.

4.4 Countermeasures

On the basis of what was described in the previous section, it is evident how the Android ecosystem is today far from being considered a complete privacy-friendly and secure environment. The scientific and engineering communities are indeed hugely investing efforts towards this perspective. An extensive survey, classification, and analysis of Android security approaches proposed since 2011 were presented by Enck [21]. In this section, we build upon the aforementioned analysis and also succinctly describe additional approaches proposed after 2011 with the aim of building a more fine-grained classification. In our analysis, we consider the following broad classes of related work:

- Risk analysis and reputation mechanisms (C1) include approaches that analyze the risk of specific mobile app features for end users and reputation models to rank mobile apps according to their risk level for end users.
- App analysis (C2) embraces approaches that analyze an app's static features and dynamic behavior with the aim to detect improper behavior. By inappropriate behavior, we mean legitimate apps that are privacy invasive or malware apps with hidden malicious functionality.
- Information flow tracking (C3) are approaches that analyze the flow of information accessed by an app (sources, e.g., microphone or camera) and the target of the information published by the app (sinks, e.g., network connection or other apps). These approaches may analyze the app as a black box communicating with other local apps or cloud services or to consider the internal fine-grained information flow between the internal app components.
- Specification and enforcement of security policies (C4) consist of approaches that verify statically or dynamically the conformance of an app with a given security policy. Policies may be simple detective properties that must be satisfied by an app (e.g., personal information is not sent to remote servers) or preventive policies that modify the behavior of an app (e.g., close all network connections if personal data are sent over them).
- Privacy-by-design extensions (C5) accommodate approaches that introduce new features to the Android security model to protect users' privacy. These approaches propose drastic changes to the Android security model, for example, by anonymizing the user identity over all the apps running in the mobile device or by providing a private-browsing option.

Table 4.1 overviews all approaches that we have identified in the literature addressing each respective class of related work, while the following subsections describe in detail the existing work according to the classification mentioned earlier.

4.4.1 Risk Analysis and App Reputation

Users of mobile devices are expected to choose their privacy and security preferences in order to be informed or to prevent apps from accessing their personal information stored in the device. This is of course in accordance with the introduced risk as the case may be. In addition to personal information privacy, according to a user study performed by Jorgensen et al. [34], when selecting their security preferences, users need to also take into serious consideration several risks, including

Table 4.1   Approaches Proposed in Each Broad Class of Related Work

C1	WHYPER [47], Papamartzivanos et al. [48], AutoCog [54], AppProfiler [60], RiskMon [80]
C2	*Static analysis*: DREBIN [2], Barrera et al. [9], Geneiatakis et al. [26,27], CHABADA [28], RiskRanker [29,31], Liang et al. [38], Sbirlea et al. [66], [61–63,65], Shahzad et al. [71], Xuetao et al. [75], DroidMat [76], Xing et al. [77], Permlyzer [79], Droid Analytics [83], Yajin et al. [84], TatWing et al. [85]
	Dynamic runtime analysis: DREBIN [2], Neuner et al. [43], CopperDroid [58,74]
C3	FlowDroid [4], ScrubDroid [5], Droidel [14], Taintdroid [22], DidFail [35], IccTA [36,37], Mann and Starostin [39], Rasthofer et al. [56], Epicc [58]
C4	*Static policy verification*: Pegasus [18], Authzoid [40], Song and Touili [73]
	Monitoring access to sensitive data: Berthome et al. [13], AppTrace [53]
	Runtime policy enforcement: Boxify [6], AppGuard [7], DroidGuard [8], Bartel et al. [10], Androlyzer [11], MockDroid [12], SEDalvik [15], CRePE [19], Kirin [23], Feth and Pretschner [25], AppFence [30], DeepDroid [33], DroidForce [57], Kynoid [67], Constroid [68], XRay and PatchDroid [41], Apex [42], Saint [45], Porscha [46], SELinux in Android [70], Aurasium [78], TISSA [81], MOSES [82]
	Privilege escalation-focused solutions: XManDroid [16], QUIRE [20], IPC Inspection [24], pWhitelist and PDP [49], Rangwala et al. [55], RGBDroid [50]
C5	TrustDroid [17], Jinseong et al. [32], ACGs [59], Shekhar et al. [72]

monetary, device availability/stability, and data integrity ones. In any case, few would disagree that users are not well informed about the reasons behind the requested permissions by the apps they install. This is mainly due to the lack of information and documentation in the Android permissions and the associated description provided by the app itself.

RiskMon [80] introduces an automated service to assess the security and privacy risk of a given app taking into account the legitimate user's normal behavior. RiskMon leverages on (a) machine learning and (b) trusted apps' different run-time features to build the user's legitimate model.

Pandita et al. proposed WHYPER [47], which is a framework using natural language processing techniques, to process app keywords and descriptions in order to identify the need for the requested permissions. This research results show that it is feasible to automatically analyze apps to better inform users about the risks associated with the permissions granted to the apps they install. Similarly to WHYPER, AutoCog proposed by Qu et al. [54] is a system to automatically assess the description-to-permission fidelity of apps, focusing on the semantic gap and lack of documentation in the description of the allowed functionality for apps requesting the specific permissions.

AppProfiler proposed by Rosen et al. [60] defines a low-level knowledge base along with a mapping between API calls and privacy invasive behaviors that may be exhibited by apps using these API calls. More specifically, when a user installs an app, AppProfiler employs static analysis to generate high-level behavior profiles for any given app. Users are then shown a list of privacy impacts and are asked for feedback in order to check whether the impacts are what the user expects and whether they are fine. Moreover, this feedback can be used to create a reputation score of the app and also to have a choice of uninstalling the app. In this way, the users' opinions about how apps affect their privacy are analyzed and are informed about the impact and their understanding of those apps.

Papamartzivanos et al. [48] go a step further and propose a cloud-based crowdsourcing architecture where users share any locally logged information about the app of interest. The authors' goal is to use the exchanged logs to calculate the app's privacy exposure level considering the exchanged information between various participants in the system. The authors use the Cydia Substrate, which can only be installed in rooted devices to hook code in method invocations and object creations. A user may decide to always allow, deny, or be asked about what to do every time a hooked method is invoked by the running app.

4.4.2 App Analysis

Android apps can be analyzed statically or dynamically in order to detect malicious or privacy invasive behavior. On one hand, many approaches focus on the static analysis of an app bytecode evaluating the requested permissions with respect to the corresponding API method calls used by the app. The main goals of these approaches are (1) to evaluate whether an app requests more permissions than those needed for its execution, which is named in the literature as an overprivileged app or an app with a permission gap (see Chapter 3), (2) to assess whether an app exhibits a privacy invasive behavior, or (3) to assess whether an app is a malware and implements covert malicious functionality.

On the other hand, the analysis of apps may be also performed dynamically at run time using *instrumented* apps, app execution sandboxes, or a modified OS that monitors app traces at execution time.

In this section, we elaborate on approaches that perform static and run-time analysis.

4.4.2.1 Static Analysis

Grace et al. [29] propose RiskRanker, an automated system that detects zero-day-related Android malware, by analyzing statically whether a particular app exhibits dangerous behavior (e.g., sending background SMS messages). In their empirical evaluation, they analyzed 120,000 apps, from which around 3,000 were classified as risky. In this sample of risky apps, they successfully identified 719 malware samples where 322 were zero-day malware not encountered before. These results show that static analysis techniques are effective and efficient to support the analysis of Android apps in order to discover the ones incorporating malicious code.

Xing et al. identified, using static analysis techniques in [77], a new type of vulnerability in the Android update mechanism called pileup. Due to pileup vulnerability, all the permissions requested by an app installed in a low-version OS are automatically granted when the OS is upgraded. This happens only for permissions that are requested and were not defined in the lower-version OS that the app was installed at first. A recommendation to fight against this vulnerability is to scan all apps in the device after the OS upgrade in order to detect if any of the installed apps show signs of this vulnerability.

The CHABADA prototype described by Gorla et al. [28] identifies outliers in clusters of apps according to their functionality and API usage. For example, apps that send SMS messages in a cluster of weather service apps are clearly requesting permissions not needed to perform their core functionality. They empirically evaluated using their prototype around 22,000 apps and identified a 56% of new malicious apps. This prototype can work for apps that have a clear defined functionality; however, for apps that have a more general coverage of permissions like social networking, a bigger set of permissions may be requested and used rightfully.

Droid Analytics [83] develops a solution to scrutinize Android apps at the byte code level and generates the corresponding signatures that can be used by antivirus software. In the same direction, Shahzad et al. [71] rely on bigram sequences of op-codes retrofitted in machine learning classifiers to detect malware, while Permlyzer [79] analyzes an app's permission usage based on both static and dynamic analyses.

Barrera et al. [9] accomplish a permission analysis based on a self-organizing map, while Xuetao et al. [75] study the evolution of Android's permissions. Similarly, Zhu et al. [85] build a permission-based abnormal model leveraging on app description and its permission set.

Zhou et al. [84] introduce a tool for the systematic study of apps that might passively leak private information due to vulnerabilities stemming from built-in Android components, such as read/write operations to content provider. Apps are statically analyzed to identify such data flows. Analogously, Liang et al. [38] propose a malware detection engine that relies on the semantic analysis of an examined app. Sbirlea et al. [66] develop techniques for statically detecting Android app vulnerabilities aiming at obtaining unauthorized access to permission-protected information.

The work proposed by Genciatakis et al. [26] focuses on the static analysis techniques in order to identify overprivileged apps. In their study, the authors use as input the app bytecode and manifest files to check whether all the declared permissions are indeed used by the app. Apps that request permission and do not use it are classified as overprivileged. That is, overprivileged apps do not respect the least-privilege principle and essentially leave an open backdoor for malware in case the app is exploited or updated to a malicious version. Another work [27] by the same authors examines the performance of three machine learning algorithms in the detection of malicious apps. At a high level, this approach is complimentary and performs very similar to DroidMat [76] and DREBIN [2]. However, the authors experimented with additional machine learning classifiers combining also the extracted features to achieve better accuracy by eliminating false alarms. Analogous techniques to those given in this section are followed by the works in [31,61–63,65].

4.4.2.2 Dynamic Run-Time Analysis

A complete analysis of around 20 existing sandboxes for dynamic malware analysis was done by Neuner et al. [43]. The main goal of these sandboxes is to execute apps in a restricted environment and analyze all app interactions with the Android framework in order to evaluate any possible intrusive behavior. In practice, their analysis shows that many of the sandboxes are not maintained, are not capable of detecting even well-known malware, and are usually developed by integrating one or more approaches/sandboxes from third parties. In the following, we refer to a couple of recent works not included in [43].

CopperDroid [58,74] is a sandbox that focuses on the dynamic behavioral analysis of Android malware by observing OS calls including native code. The sandbox was implemented using the QEMU virtualization environment and produces as output an abstract behavioral profile, which excludes low-level events and considers only the high-level semantics of the malware activities.

DREBIN [2] contributes a solution incorporating dual-mode functionality. That is, DREBIN can be used not only for detecting malicious apps on Android platform based on their built-in static features and machine learning algorithms but also for revealing in detail those features that constitute a given app as a malware. Although previously published works such as [1,31,61–63,65, 76] propose solutions in the same direction, all of them use mainly permissions and APIs as a static features, while their evaluation is limited in comparison with DREBIN. Nevertheless, most of the existing related work concentrates on malware detection, while very few focus on enhancing users' privacy when they interact with mobile apps.

4.4.3 Information Flow Tracking

Information flow tracking approaches focus on the tracking of sensitive information after access is allowed to an app by granting a particular permission, as at least from a user's viewpoint is of fundamental importance to support app monitoring and control. The tracking of information flow is done considering a set of sources and sinks of information and taint marks that are propagated when the app bytecode is executed. This process builds on the advantages of static and dynamic analysis as given in Section 4.2. A taint mark identifies a specific data item and can be used to know where specific data items may be stored during the execution of the app. An important aspect is the classification of sources and sinks because they are the main focus for the specification of the corresponding access policies.

The following list briefly describes the different classes of information flow tracking approaches defined in the literature. Also, in this section, we use these classes as a reference to explain the existing approaches.

- *Implicit and/or explicit flow coverage*: Some solutions track only explicit variable assignments when propagating flow tracking taint information or also include conditional branches that use tainted variables in the conditional expressions.
- *Interapp, Intercomponent, Intra-app focus*: Since apps consist of components, namely, activities, broadcast receivers, content providers, and services whose communication interfaces are clearly defined, some solutions proposed by researchers are based on the analysis of inter-app and intra-app information flow, aimed at isolating information and malware at the inter-component level. Figure 4.4 shows the structure of an Android app including the flow of information from outside the app (source), inside of the app (intra-app), and to the outside (sinks).
- *Reflection and native code support*: Many static analysis information flow solutions are limited in their support for tracking flow when reflection is used to execute the dynamic code. Furthermore, support to track information flow when native code is invoked by apps is also limited due to the complexity and low-level nature of the instructions executed.

From a practical point of view, the main difference of existing approaches is the achieved precision and recall, the capability of detecting implicit or explicit information flow, the support for apps using reflection and native code, and the capability of analyzing obfuscated apps. When analyzing an app from an information flow perspective, the first step is the identification of sources

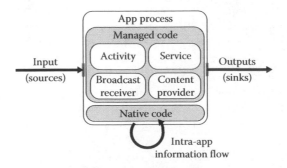

Figure 4.4 Android app structure and information flow.

and sinks. Sources are incoming channels from which the app receives external information, while sinks are outgoing channels to which the app sends information to external components. Most approaches manually identify sources and sinks, while some specific approaches aim at identifying them automatically.

Rasthofer et al. [56] propose a machine learning approach named SUSI to classify and categorize sources and sinks. Categories that are considered both sources and sinks are account, browser, calendar, contact, file, network, NFC, and sync. Exclusive source categories identified are Bluetooth, database, settings, and unique identifier. Exclusive sink categories identified are audio, log, phone connection, phone state, SMA/MMS, sync, system, and VOIP. The authors also define a no-category class for methods that do not fit in any of the aforementioned categories. The SUSI approach misses important sources, for example, there is no mention of the camera in the source categories or analysis. Furthermore, some categories are subsets of other categories and do not define a clear abstraction, for example, database is also a file, Bluetooth is a network type, and phone state may contain unique identifiers. Finally, user input may also include sensitive information that should be tracked.

After sources and sinks are identified and classified, the next step is tracking the flow of information from sources to sinks. Approaches for tracking information flow focus on the communication of apps and the OS, and the communication between the apps, mostly those that could enable collusion attacks or covert communication between them.

Taintdroid [22] is one of the first system-centric approaches to track information flow using dynamic taint analysis technique to detect unauthorized leakage of sensitive data and run-time attacks. It proposes a modified DVM that enforces its taint propagation semantics by instrumenting an app DEX bytecode to tag every variable, pointer, and IPC message that flows through the system with a taint value. The number of taint marks is limited to a set of sensitive resources, for example, device ID and location. Taintdroid does not consider implicit information flow and has been shown to be insecure to many types of attacks in [64] (e.g., timing, covert channels).

Mann and Starostin [39] propose a bytecode analysis framework that uses a secure-type system to analyze explicit information flows. Their framework takes a privacy policy as input and verifies whether the policy is violated or satisfied by an app. They also propose a set of sources consisting of location data, unique identifiers, call state, authentication data, contact, and calendar data. The set of sinks they consider is SMS communication, file output, network communication, intents, and content providers/resolvers.

FlowDroid proposed by Arzt et al. [4] is a static analysis tool that performs taint analysis of apps. FlowDroid claims to be precise and considers the app life cycle and callbacks from the framework, in contrast to other approaches that use simply a coarse-grain approximation. By considering fine-grained context, flow, field, and object sensitivity information, the authors state that they are able to reduce the number of false alarms.

DidFail (Droid Intent Data Flow Analysis for Information Leakage) [35] uses the static analysis to detect potential leaks of sensitive information within a set of Android apps. DidFail combines and augments FlowDroid (which identifies intracomponent information flows) and Epicc [58], which identifies properties of intents such as its action string to track both intercomponent and intracomponent data flow in a set of Android apps. IccTA [36,37], which was developed at roughly the same time as DidFail, uses a one-phase analysis and is more precise than the two-phase analysis of DidFail.

Droidel proposed by Blackshear et al. [14] is a general approach for modeling the Android framework that generates Java source code representing the entry point and call graphs. The authors show with a practical evaluation of seven apps that while Droibel misses only 6% of the reachable methods, FlowDroid [4] misses around 30%. Both tools have slightly different purposes; however, FlowDroid misses far too many methods in order to be considered a secure solution for taint analysis.

Some approaches show patterns that can be used to circumvent existing information flow tracking tools and attack their limitations with respect to implicit information flow. For instance, the ScrubDroid approach proposed by Babil et al. [5] concentrates on Taintdroid limitations. The authors demonstrate timing, side channel, counting, and indirect information flow type of attacks that could also be used to evade other tracking systems similar to Taintdroid. In this point we should remark that information flaw tracking is a valuable tool for enhancing end-users' privacy, however, it has its own inefficiencies [64].

4.4.4 Specifying and Enforcing Security Policies

Many solutions to provide more fine-grained policy enforcement capabilities for Android have been proposed including context information and complex conditions. Essentially, the focus of these solutions is to detect or prevent malicious behavior according to a specified policy. Policy-based approaches may be used statically to verify if an app potentially violates a security policy or at run time to detect or prevent/constrain a malicious behavior. Some approaches also focus on information flow properties or suspicious behavior that may be characterized as malware. The general problem of monitoring security policies using run-time verification in Android is described in [3].

Another class of solutions provides extensions to enforce security policies in apps, perhaps integrated with information flow tracking techniques. These solutions differ according to the expressiveness of the policy language used, the abstraction level where the policy is enforced, and the type of authorization actions allowed (detective or preventive). Expressive policy languages can be also used from a detective enforcement approach in order to discover malicious behaviors, even if, in most cases, real malwares that abuse vulnerabilities cannot be detected a priori because their signatures are not known beforehand. However, the strong point of security policy solutions is that they mostly focus on privacy-invasive behavior that, as already said, can be performed also by legitimate apps and only by recognized malwares.

4.4.4.1 Static Policy Verification

In the context of static policy verification, May et al. [40] proposed Authzoid and Chen et al. [18] propose Pegasus. Both these approaches use a formal model to verify app behavior to check for violations to authorization policies with respect to invocation of sensitive API methods. More specifically, Pegasus uses Event Graphs and model checking to detect or prove the absence of malicious behavior in apps. Authzoid, on the other hand, uses a ProVerif model to verify and point out attacks and goes one step further than Pegasus because it also includes enforcement of authorization policies in addition to static verification only, although their main strength is in the verification of apps.

Song and Touili [73] propose a model-checking approach and tool to identify malware apps. In their approach, apps are modeled using a PushDown System and malware behavior is succinctly specified using computation tree logic or linear temporal logic (LTL). In a practical evaluation of 1260 apps, they identified 10 apps that were leaking private data. The authors correctly point out that it is difficult to classify apps as malicious or malware just by analyzing a set of LTL properties since some apps have a privacy invasive behavior by design and also the precise specification of a privacy invasive behavior changes from user to user. Furthermore, the tool is as good as the set of LTL properties specified.

All approaches described in this section are clearly important and are able to verify apps automatically. However, to be really useful, they should be made available to end users, and usable extension mechanisms for user-centric specification of properties to be verified should be enabled. This is needed because users may have different requirements and perceptions about the privacy level that they would like to enjoy for each app.

4.4.4.2 Monitoring Access to Sensitive Data

Berthome et al. [13] propose an approach that deliberately injects code in an app to monitor the access to the users' data. The injected code simply logs the access and shows it in a dedicated screen. Qiu et al. [53] propose AppTrace, a dynamic instrumentation system, to help analyze the behavior of an app at run time. AppTrace has been evaluated with a set of 58 apps, and the authors' results show the feasibility of their system, which is able to notify users when sensitive APIs are invoked. Both approaches to monitor sensitive data access are useful but do not help users to control the access to their sensitive data. Putting it another way, the end users are only informed about possible privacy violations and after that they may decide to stop using the app.

4.4.4.3 Run-Time Policy Enforcement

In this section, we overview the existing related works devoted to the security policy enforcement in Android for enhancing users' security and privacy levels against apps with a privacy invasive behavior. We present these solutions in chronological order, briefly analyzing the functionality proposed, the implementation details, and the existing limitations.

Kirin, proposed by Enck et al. [23], is a security service running on the mobile device. This service is in charge of analyzing the requested permissions of an app and detecting potential security flaws. When an app is about to be installed, Kirin evaluates, using a rule-based engine, whether there is a match between the set of requested permissions and the signatures defined in the rule engine. Note that the signatures defined in the rule engine represent possible attack vectors, for example, RECORD AUDIO, and INTERNET permissions define a rule in the signature set. On the downside, Kirin produces a high number of false positives since legitimate apps that follow the defined signature pattern are often characterized as malicious.

Ongtang et al. propose Saint [45] as an extension of Kirin. In addition to the analysis performed by Kirin at install time, Saint monitors the intercomponent communication (ICC) flows also at run time, which include flows related to activities initialization, components binding to services, and access to content providers. The policies in Saint are static and define conditions to control the run-time behavior. For example, a specific activity can bind with a specific content provider considering the allowed permissions, signature, or package name. Saint is implemented as a modified Android middleware.

Similarly, Apex introduced by Nauman et al. [42] also focuses on policies enforcement for regulating ICC flows. In contrast to Saint, Apex supports more complex policy conditions using dynamic attributes including cardinality and time dimension constraints, that is, restricting the maximum number of SMS messages sent by an app. Policy rules must be defined to manage the initialization, updating, and resetting of dynamic attributes. Both Saint and Apex support authorization actions to allow or deny an ICC flow without the possibility of modifying or obfuscating a flow. In the same direction, Porscha, proposed by Ongtang et al. [46], introduces a Digital Rights

Management (DRM) framework for Android smartphones that mediates the access to protected content between different Android components. For example, it can regulate the access of an app to the content of an SMS message. The Porscha mediator supports constraints on devices, apps, and on the use (e.g., cardinality) of the protected data. Also, Porscha mediates ICC flows, with extensions including a policy filed and has been implemented as a modified Android firmware that is considered to be trusted.

CRePE introduced by Conti et al. [19] is a customized Android OS system able to enforce fine-grained security policies considering time and location features. Policies in this system intercept authorization requests before the standard Android permission checks; however, if the request is allowed by CRePE, the standard permission check may still deny it. In addition to the standard permission checks, CRePE also intercepts and enforces policies when activities are started. Policies in CRePE consisted of propositional conditions of allow or deny actions.

Shabtai et al. [70] first proposed the use of SELinux in Android order to implement low-level mandatory access control (MAC) policies. From Android 4.3 and later, SELinux is used by default to further define apps in permissive mode, only logging permission denials, while since Android 5.0 release SELinux is used in full enforcement mode, in which permission denials are logged and enforced according to the specified policies.

Batyuk et al. [11] introduced Androlyzer, a server-based solution that focuses mainly on informing users about apps potential security and privacy risks. To do so, Androlyzer first reverse-engineers the app of interest and second accomplishes a static analysis to determine the possible flaws. In addition, Androlyzer provides an approach for mitigating the identified flaws by modifying the examined app based on user's preferences. However, Androlyzer does not use an expressive policy language to employ the user's security preferences into the app.

Beresford et al. [12] propose MockDroid, which is a complementary solution to AppFence proposed by Hornyack et al. [30]. Both these approaches are implemented as a modified OS using TaintDroid as a starting point. In addition to the information flow tracking support implemented by Taintdroid, these solutions also provide shadowing of user data and filters to protect private user data according to user's preferences.

Feth and Pretschner [25] employ a usage control policy framework that leverages also in Taint-Droid for information flow tracking. Their framework uses an expressive policy language for describing users' preferences and monitors at run-time permission checks, queries to content providers, intents, and certain data sinks like the network, file system, and IPC to eliminate access to private data.

In an alternative approach, Xu et al. [78] introduce an additional sandbox security service, namely Aurasium for protecting users' against apps malicious behavior. In contrast to other similar approaches, Aurasium enforces its security policies on Android libc level through interposition as a middleware between Anrdoid kernel and user space layer. This means that the original apps are repackaged to a new app that includes the appropriate code enabling Aurasium to control access to sensitive resources.

Constroid introduced by Schreckling et al. [68] also defines a management framework for employing data-centric security policies of fine granularity. To do so, Constroid adopts the UCONABC model [69]; however, only the abstract model is detailed and no concrete example of policy is provided.

In an alternative approach, SEDalvik introduced by Bousquet et al. [15] proposes a MAC mechanism to regulate information flows between apps' objects building on the advantages of Dalvik internal debugger. The specified policies define which interactions are allowed to take place in a given context or not.

Schreckling et al. [67] introduce Kynoid, a solution that extends Taintdroid with security policies at the variable level. Kynoid retains the taint propagation performed by Taintdroid and maintains a dependency graph where a direct edge represents a security requirement (aka policy) between two objects. This way, Kynoid provides a fine-grained control to sensitive flows.

AppGuard, introduced by Backes et al. [7], is an app instrumentation framework that runs directly in the users' device and allows user-centric security policy customizations. AppGuard computes a risk score for each app considering the number of dangerous permissions and provides the option of instrumenting the app to control the access to potentially dangerous calls. For example, if an app granted access to NETWORK permission, a user may choose to enable/disable the corresponding functionality. The same direction is followed by the solution presented by Bartel et al. [10].

Zhauniarovich et al. [82] proposed MOSES, which enforces context-based policy specification at kernel-level, meaning that MOSES requires a modification to the underlying OS. In this approach, users can define a security profile that could be applied in a specific context, that is, at a specific time and location for a given app. Note that if a security profile is not linked to an app, then MOSES does not allow access to any sensitive resource since by default employs a negative authorization policy. MOSES security profile consists of allow or deny rules according to the user's requirements.

TISSA proposed by Zhou et al. [81] introduces a privacy mode functionality in Android with coarse-grain control over the behavior of an app. By using TISSA, users can have more fine-grained control over private information, for example, location, phone identity, contacts, and call log. TISSA is implemented as a modified OS with proxy content providers for each controlled information that are responsible for retrieving and enforcing the corresponding policies. TISSA's policies are hard-coded and restricted to a static set of authorization options without support for complex conditions.

DroidForce proposed by Rasthofer et al. [57] relies on the Soot framework for analyzing and instrumenting an app to enforce a security policy. This approach considers PEPs injected in multiple applications with a single policy decision point (PDP) running as an app in the mobile device, addresses information flow intra-app statically and inter-app at run time, and uses an expressive policy language with cardinality and temporal constraints. Their policies allow or deny an activity, while do not support modification/obfuscation of values. Complementary, Jing et al. [33] propose DeepDroid, which in contrast to DroidForce performs instrumentation at the native level with the possibility of intercepting system calls in addition to app API calls for regulating access to sensitive resources. However, DeepDroid neither considers information flow tracking nor uses any expressive policy language for enforcement.

Bagheri et al. [8] introduce DroidGuard, a framework for modeling inter-app vulnerabilities and employing the appropriate protection mechanism to enhance the user's privacy and security. Briefly, DroidGuard analyzes statically a set of given apps to foresee security flaws realized through apps intercommunication. The generated model is used as a policy to be employed as proactive countermeasure.

Backes et al. [6] propose Boxify, a full-fledge app sandboxing for stock Android. Finally, XRay and PatchDroid introduced by Mulliner and Oberheide [41] are two approaches to detect vulnerabilities and eliminate them from Android apps.

4.4.4.4 Privilege Escalation-Focused Solutions

Felt et al. [24] proposed the IPC Inspection OS mechanism for protection against app permission redelegation. Rangwala et al. [55] present taxonomy of privilege escalation attacks in Android apps.

Dietz et al. [20] proposed QUIRE, which is a lightweight provenance system that prevents privilege escalation attacks via confused deputy attacks (see Section 4.3). The QUIRE framework tracks the IPC flow and modifies the privileges of an app in case it executes operations on behalf of other apps, while allowing an app to exercise its full privileges when acting autonomously. In addition to tracking IPC, it also allows apps to create signed statements that will be only accessible to other apps running in the same mobile device. Technically, QUIRE requires changes to the OS's IPC subsystem, which is also used by native libraries and demonstrates less overhead than approaches that rely on bytecode instrumentation to propagate taint values. One drawback of QUIRE is that it does not address privilege escalation of maliciously colluding apps, but its focus is on the exploitation of legitimate apps by malware apps.

The XManDroid (eXtended Monitoring on Android) [16] security framework proposed by Bugiel et al. detects and prevents privilege escalation and confused deputy attacks at run time using a system-centric security policy enforcement solution. The enforcement is done at kernel level using as input a graph representing the IPC calls between apps and policy rules that sandbox a set of permissions to direct and indirect use by apps. The policy language used is a simple list of predicates connected by propositional logic operators considering the permissions of two apps that are interacting.

Park et al. [49] propose pWhitelist and PDP also as a scheme to detect and respond to privilege escalation attacks. Their scheme focuses on malware apps that illegally acquire root-level privileges by exploiting vulnerabilities of trusted programs. In fact, pWhitelist is a kernel module that only authorizes a specific set of trusted programs to run with root privileges. On the other hand, PDP works in parallel applying the least privilege principle and restricting the access to the app's data only to the app themselves. RGBDroid (Rooting Good Bye on Droid) also proposed by Park et al. [50] adds the Criticallist, which is a list of critical resources that can affect operations and behaviors in Android that not even privileged process are able to modify.

4.4.5 Privacy-by-Design Extensions

Some researchers propose innovative privacy-by-design approaches (e.g., private-browsing mode and anonymization mode) to enhance user protection in addition to the specification of verifiable/enforceable security policies. Other solutions focus on new mechanisms for privacy or security by design for user protection, including separation of advertising libraries, which can be simulated by malware and leak users' data or tracking information, from core app functionalities to allow an app and its advertising services to run separately [72]. As a result, apps do not have to request permissions on behalf of their advertising libraries. This approach could be extended also to other services and libraries incorporating access to sensitive resources.

Overall, these solutions show innovative ideas that could be adopted or embedded in Android OS. Jinseong et al. [32] propose a new mechanism to break down Android permissions in subclasses of more fine-grained permissions. For each permission of this kind, wrapper components are implemented to control access to the specific resource. Roesner et al. propose Access Control Gadgets (ACGs) [59], which are permission-granting graphical user interface (GUI) components introduced in apps. Particularly, apps that require specific permissions should embed an ACG (e.g., camera access) in the GUI, and the user can explicitly grant the permission by interacting with the ACG. By using ACG abstractions, user-driven access control can be implemented into apps.

Bugiel et al. [17] propose the TrustDroid framework to provide strong isolation between apps. Their main focus is on the isolation of business apps and data from other apps when they are installed in mobile devices that may be also used for private purposes whether they are running in personal or company-owned devices. These types of solutions can also benefit nonbusiness apps that handle personal data of users in order to isolate critical apps, for example, banking or mobile payment apps from games downloaded from unknown sources that may probably contain malware functionality.

4.5 Discussion

Today, Android is a well-established platform for the mobile world. However, mainly due to its innate properties, Android is still vulnerable to different attack vectors that target users' privacy among others. For instance, an app might be overprivileged, meaning that does not follow the least privilege principle. In other words, this could be translated to a silent escalation access to sensitive resources. Furthermore, Android also has to take into account security flaws inherited by the existing x86 platforms. It is worth noting that some of these flaws are recognized by Android OS designers as of high importance. For instance, the latest Android v6 introduced a module in which users are (every time) notified upon a permission, newly classified as dangerous, is used. This is a clear improvement over the older versions in terms of user notification. However, the users still remain vulnerable to overprivileged apps, and they are not informed about the usage of normal-labeled permissions.

Furthermore, it should be mentioned that existing mobile devices are locked down for Android update because of hardware restrictions, and thus their user basis remains vulnerable to existing flaws. This situation must not be regarded as unimportant as currently there are millions of users (the majority) relying on previous Android versions. Thus, researchers continue to focus on security enhancements not only for new Android versions but also for previous ones. In this direction, Table 4.2 overviews the existing security protection classes with the aim to mitigate apps privacy intrusive behavior that are the source of different types of threats. Recall that all these protection classes have been succinctly discussed in Section 4.4. It can be argued that these approaches are complementary to each other and can be combined to provide a complete protection against the different threats as well as to improve users' awareness for app access to sensitive resources.

In any case, it should be noted that while researchers are developing various protection schemes to protect users' privacy, attackers are continuously seeking new methods that allow them to evade detection. In this direction, static analysis can be defeated by code obfuscation and unpacking malicious code only during execution. Dynamic analysis has its own weaknesses in that the majority

Table 4.2 Overview of Protection Classes That Can Be Used to Enhance Android Security and Mitigate Privacy Invasion Behaviors on the User Side

	Protection Class				
	C1	*C2*	*C3*	*C4*	*C5*
Overprivileged	No	Yes	No	No	No
Security flaws	No	Yes	Yes	No	No
Unprotected resources	No	Yes	Yes	Yes	Yes
User's awareness	Yes	No	No	No	No

of the popular sandbox environments can be fingerprinted with minimal effort. While lately taint analysis has received a lot of attention and its methods continue to get improved, there are still leaks that produce undetected and false positives that would make it an incomplete solution for securing the user against potential data leaks.

In this direction, the framework proposed by Qiu et al. [52] focuses on the understanding of the internal logic of malware that implements antianalysis defense techniques. This framework uses information flow to analyze self-check, summing-based, antitampering defenses, and timing-based emulator detection. Self-check summing refers to the procedure when an app computes a hash of its own code to verify its integrity and possibly further use the same hash value to trigger malware execution. In case a malicious app detects modifications to itself, it behaves normally and hides its malicious behavior. Timing-based emulator detection is a technique employed to detect if a malware is running in a dynamic analysis framework. By detecting it is being analyzed, the malicious app does not execute any malicious activity and may evade detection in this way. Overall, it can be said that the study of antidefense techniques is an important step toward the improvement of malware detection techniques that may not work for certain malware behaviors.

Generally, despite the limitations shown by the existing protection techniques, we believe that all these approaches seem to be promising if they are diffused and used in a wide scale for enhancing user protection and user awareness in terms of privacy.

4.6 Conclusions

In this chapter, we presented an extensive analysis of the Android OS security architecture, the threats that the users of potentially every smart device may face, and the existing approaches proposed in the literature to counter these threats. We proposed a classification of the current defensive approaches considering their main functionalities and goals from an end user's perspective. In addition to the description of threats and classification of approaches, we also included a short but insightful discussion of the limitations and effectiveness offered by each of them.

From our analysis, it is clear that in the current security model of Android apps, which relies on permissions controlling the access to sensitive resources through the corresponding API methods, users have to trust the apps for not misusing their data after access to a sensitive resource is granted. Many approaches focus on advanced control solutions but they lack the integration with effective information flow tracking in order to be able to control misuse of user data even after the app possibly modifies it. For example, if an app reads the user location, encrypts it, and sends it to a remote server, users have no way using the existing approaches to both identify and block this privacy-invasive behavior. Furthermore, all approaches that propose to solve partially this problem are academic prototypes and are not usable in their current state by end users. There is a need for adopting and embedding these solutions by the Android OS developers.

Based on our classification of existing approaches, we can identify the main functionalities that should be present in the Android OS, or in any other mobile device OS, to protect and empower users. There is a need to identify risks and present them in an understandable way to end users, to automatically analyze apps to detect malicious or privacy invasive behaviors and make users aware of it, to empower users with information flow tracking combined with expressive run-time policy enforcement solutions, and finally to implement where possible privacy-by-design approaches in a way that users can be protected by default from apps. Overall, only a holistic approach considering all these elements can be effective to improve the current mobile device security situation for end users.

References

1. Y. Aafer, W. Du, and H. Yin. Droidapiminer: Mining api-level features for robust malware detection in Android. In *9th International ICST Conference on Security and Privacy in Communication Networks, SecureComm 2013*, Lecture Notes of the Institute for Computer Sciences, Social Informatics and Telecommunications Engineering, Sydney, Australia, Vol. 127, pp. 86–103. Springer International Publishing. 2013.

2. D. Arp, M. Spreitzenbarth, M. Hubner, H. Gascon, and K. Rieck. DREBIN: Effective and explainable detection of Android malware in your pocket. In *21st Annual Network and Distributed System Security Symposium (NDSS 2014)*. San Diego, CA, February 23–26, 2013, 2014.

3. S. Arzt, K. Falzon, A. Follner, S. Rasthofer, E. Bodden, and V. Stolz. How useful are existing monitoring languages for securing Android apps? In *Software Engineering 2013—Workshopband (inkl. Doktorandensymposium)*, Fachtagung des GI-Fachbereichs Softwaretechnik, pp. 107–122. Aachen, Germany, Februar 26–Marz 1, 2013.

4. S. Arzt, S. Rasthofer, C. Fritz, E. Bodden, A. Bartel, J. Klein, Y. L. Traon, D. Octeau, and P. McDaniel. Flowdroid: Precise context, flow, field, object-sensitive and lifecycle-aware taint analysis for Android apps. In *Proceedings of the 35th ACM SIGPLAN Conference on Programming Language Design and Implementation (PLDI'14)*, Edinburgh, UK, pp. 259–269. ACM, New York, 2014.

5. G. S. Babil, O. Mehani, R. Boreli, and M.-A. Kaafar. On the effectiveness of dynamic taint analysis for protecting against private information leaks on Android-based devices. In *International Conference on Security and Cryptography (SECRYPT)*, Lisbon, Portugal, pp. 1–8. July 2013.

6. M. Backes, S. Bugiel, C. Hammer, O. Schranz, and P. von Styp-Rekowsky. Boxify: Full-fledged app sandboxing for stock Android. In *24th USENIX Security Symposium (USENIX Security 15)*, pp. 691–706. USENIX Association, Washington, DC, August 2015.

7. M. Backes, S. Gerling, C. Hammer, M. Maffei, and P. von Styp-Rekowsky. Appguard enforcing user requirements on Android apps. In *Tools and Algorithms for the Construction and Analysis of Systems*, Rome, Italy, Lecture Notes in Computer Science, Vol. 7795, pp. 543–548. Springer Berlin, Heidelberg, 2013.

8. H. Bagheri, A. Sadeghi, R. Jabbarvand, and S. Malek. Automated dynamic enforcement of synthesized security policies in Android. Technical Report Technical Report GMU-CS-TR-2015–5, George Mason University, Fairfax, VA, 2015.

9. D. Barrera, H. Güne Kayacik, P. C. van Oorschot, and A. Somayaji. A methodology for empirical analysis of permission-based security models and its application to Android. In *Proceedings of the 17th ACM Conference on Computer and Communications Security (CCS'10)*, Chicago, IL, pp. 73–84. ACM, New York, 2010.

10. A. Bartel, J. Klein, M. Monperrus, K. Allix, and Y. L. Traon. Improving privacy on Android smartphones through in-vivo bytecode instrumentation. 2012. Available online: https://hal.archives-ouvertes.fr/hal-00700319v1/document.

11. L. Batyuk, M. Herpich, S. A. Camtepe, K. Raddatz, A.-D. Schmidt, and S. Albayrak. Using static analysis for automatic assessment and mitigation of unwanted and malicious activities within Android applications. In *Proceedings of the 2011 Sixth International Conference on Malicious and Unwanted Software (MALWARE'11)*, Puerto Rico, pp. 66–72. IEEE Computer Society, Washington, DC, 2011.

12. A. R. Beresford, A. Rice, N. Skehin, and R. Sohan. Mockdroid: Trading privacy for application functionality on smartphones. In *Proceedings of the 12th Workshop on Mobile Computing Systems and Applications (HotMobile'11)*, Phoenix, AZ, pp. 49–54. ACM, New York, 2011.

13. P. Berthome, T. Fecherolle, N. Guilloteau, and J.-F. Lalande. Repackaging Android applications for auditing access to private data. In *2012 Seventh International Conference on Availability, Reliability and Security (ARES)*, Prague, Czech, pp. 388–396. August 2012.

14. S. Blackshear, A. Gendreau, and B.-Y. Evan Chang. Droidel: A general approach to Android framework modeling. In *Proceedings of the Fourth ACM SIGPLAN International Workshop on State Of the Art in Program Analysis (SOAP 2015)*, Santa Barbara, CA, pp. 19–25. ACM, New York, 2015.

15. A. Bousquet, J. Briffaut, L. Clevy, C. Toinard, and B. Venelle. Mandatory access control for the Android dalvik virtual machine. In Presented as part of the *2013 Workshop on Embedded Self-Organizing Systems*. USENIX, Berkeley, CA, 2013.

16. S. Bugiel, L. Davi, A. Dmitrienko, T. Fischer, and A.-R. Sadeghi. Xmandroid: A new Android evolution to mitigate privilege escalation attacks. Technical Report TR-2011–04, Technische Universität Darmstadt, Darmstadt, Germany, April 2011.

17. S. Bugiel, L. Davi, A. Dmitrienko, S. Heuser, A.-R. Sadeghi, and B. Shastry. Practical and lightweight domain isolation on Android. In *Proceedings of the First ACM Workshop on Security and Privacy in Smartphones and Mobile Devices (SPSM'11)*, pp. 51–62. ACM, New York, 2011.

18. K. Chen, N. Johnson, V. D'Silva, S. Dai, T. Magrino, K. Macnamarra, E. Wu, M. Rinard, and D. Song. Contextual policy enforcement in Android programs with permission event graphs. In *Proceedings of the Conference on Networked and Distributed System Security*. San Diego, CA, 2013.

19. M. Conti, V. T. Nga Nguyen, and B. Crispo. Crepe: Context-related policy enforcement for Android. In *Proceedings of the 13th International Conference on Information Security (ISC'10)*, Boca Raton, FL, pp. 331–345. Springer-Verlag, Berlin, Heidelberg, 2011.

20. M. Dietz, S. Shekhar, Y. Pisetsky, A. Shu, and D. S. Wallach. Quire: Lightweight provenance for smart phone operating systems. In *20th USENIX Security Symposium*. San Francisco, CA, August 2011.

21. W. Enck. Defending users against smartphone apps: Techniques and future directions. In *Proceedings of the Seventh International Conference on Information Systems Security (ICISS'11)*, Calcutta, India, pp. 49–70. Springer-Verlag, Berlin, Heidelberg, 2011.

22. W. Enck, P. Gilbert, B.-G. Chun, L. P. Cox, J. Jung, P. Mc-Daniel, and A. N. Sheth. Taintdroid: An information-flow tracking system for realtime privacy monitoring on smartphones. In *Proceedings of the Ninth USENIX Conference on Operating Systems Design and Implementation (OSDI'10)*, Vancouver, Canada, pp. 1–6. USENIX Association, Berkeley, CA, 2010.

23. W. Enck, M. Ongtang, and P. McDaniel. On lightweight mobile phone application certification. In *Proceedings of the 16th ACM Conference on Computer and Communications Security (CCS'09)*, Chicago, IL. ACM, New York, 2009.

24. A. P. Felt, H. J. Wang, A. Moshchuk, S. Hanna, and E. Chin. Permission re-delegation: Attacks and defenses. In *Proceedings of the 20th USENIX Conference on Security (SEC'11)*, San Francisco, CA, pp. 22–22. USENIX Association, Berkeley, CA, 2011.

25. D. Feth and A. Pretschner. Flexible data-driven security for Android. In *Proceedings of the 2012 IEEE Sixth International Conference on Software Security and Reliability (SERE'12)*, pp. 41–50. IEEE Computer Society, Washington, DC, 2012.

26. D. Geneiatakis, I. N. Fovino, I. Kounelis, and P. Stirparo. A permission verification approach for Android mobile applications. *Computer Security*, 49(C):192–205, March 2015.

27. D. Geneiatakis, R. Satta, I. N. Fovino, and R. Neisse. On the efficacy of static features to detect malicious applications in Android. In *Proceedings of the 12th International Conference on Trust, Privacy and Security in Digital Business, TrustBus 2015*, Valencia, Spain, Lecture Notes in Computer Science, Vol. 9264, pp. 87–98. Springer, September 1–2, 2015.

28. A. Gorla, I. Tavecchia, F. Gross, and A. Zeller. Checking app behavior against app descriptions. In *Proceedings of the 36th International Conference on Software Engineering (ICSE 2014)*, Hyderabad, India, pp. 1025–1035. ACM, New York, 2014.

29. M. Grace, Y. Zhou, Q. Zhang, S. Zou, and X. Jiang. Riskranker: Scalable and accurate zero-day Android malware detection. In *Proceedings of the 10th International Conference on Mobile Systems, Applications, and Services (MobiSys'12)*, Low Wood Bay, Lake District, UK, pp. 281–294. ACM, New York, 2012.

30. P. Hornyack, S. Han, J. Jung, S. Schechter, and D. Wetherall. These aren't the droids you're looking for: Retrofitting Android to protect data from imperious applications. In *Proceedings of the 18th ACM Conference on Computer and Communications Security, CCS'11*, Chicago, IL, pp. 639–652. ACM, New York, 2011.

31. C.-Y. Huang, Y.-T. Tsai, and C.-H. Hsu. Performance evaluation on permission-based detection for Android malware. In *Proceedings of International Computer Symposium (ICS)*, Hualien, Taiwan. 2012.

32. J. Jeon, K. K. Micinski, J. A. Vaughan, A. Fogel, N. Reddy, J. S. Foster, and T. Millstein. Dr. Android and Mr. Hide: Fine-grained permissions in Android applications. In *Proceedings of the Second ACM Workshop on Security and Privacy in Smartphones and Mobile Devices (SPSM'12)*, Raleigh, NC, pp. 3–14. ACM, New York, 2012.

33. J. Jing, K. Sun, Y. Wang, and X. Wang. Deepdroid: Dynamically enforcing enterprise policy on Android devices. In *Proceedings of the Network and Distributed System Security Symposium*, San Diego, CA, 2015.

34. Z. Jorgensen, J. Chen, C. S. Gates, N. Li, R. W. Proctor, and T. Yu. Dimensions of risk in mobile applications: A user study. In *Proceedings of the Fifth ACM Conference on Data and Application Security and Privacy (CODASPY'15)*, San Antonio, TX, pp. 49–60. ACM, New York, 2015.

35. W. Klieber, L. Flynn, A. Bhosale, L. Jia, and L. Bauer. Android taint flow analysis for app sets. In *Proceedings of the Third ACM SIGPLAN International Workshop on the State of the Art in Java Program Analysis (SOAP'14)*, Edinburgh, UK, pp. 1–6. ACM, New York, 2014.

36. L. Li, A. Bartel, T. F. Bissyand'e, J. Klein, Y. L. Traon, S. Arzt, S. Rasthofer, E. Bodden, D. Octeau, and P. McDaniel. IccTA: Detecting inter-component privacy leaks in Android apps. In *Proceedings of the 37th International Conference on Software Engineering (ICSE 2015)*. Florence/Firenze, Italy, 2015.

37. L. Li, A. Bartel, J. Klein, Y. L. Traon, S. Arzt, S. Rasthofer, E. Bodden, D. Octeau, and P. Mcdaniel. I know what leaked in your pocket: Uncovering privacy leaks on Android apps with static taint analysis. Technical Report 9782–87971–129–4 TR-SNT-2014–9, Interdisciplinary Centre for Security, Reliability and Trust [Luxembourg] -SnT, Computer Science and Communications [Luxembourg] CSC, EC SPRIDE, Department of Computer Science and Engineering, April 2014.

38. S. Liang, M. Might, and D. Van Horn. Anadroid: Malware analysis of Android with user-supplied predicates. In *Proceedings the Workshop on Tools for Automatic Program Analysis*, Seattle, WA, 2013.

39. C. Mann and A. Starostin. A framework for static detection of privacy leaks in Android applications. In *Proceedings of the 27th Annual ACM Symposium on Applied Computing (SAC'12)*, Riva (Trento), Italy, pp. 1457–1462. ACM, New York, 2012.

40. M. J. May and K. Bhargavan. Towards unified authorization for Android. In *Proceedings of the Fifth International Conference on Engineering Secure Software and Systems (ESSoS'13)*, Paris, France, pp. 42–57. Springer-Verlag, Berlin, Heidelberg, 2013.

41. C. Mulliner and J. Oberheide. The real deal of Android device security: The third party. Available online: https://jon.oberheide.org/files/cansecwest14-android-third-party.pdf.

42. M. Nauman, S. Khan, and X. Zhang. Apex: Extending Android permission model and enforcement with user-defined runtime constraints. In *Proceedings of the Fifth ACM Symposium on Information, Computer and Communications Security (ASIACCS'10)*, Beijing, China. ACM, New York, 2010.

43. S. Neuner, V. van der Veen, M. Lindorfer, M. Huber, G. Merzdovnik, M. Mulazzani, and E. R. Weippl. Enter sandbox: Android sandbox comparison. In *Proceedings of the Third Workshop on Mobile Security Technologies (MoST)*, San Jose, CA, 2014.

44. D. Octeau, P. McDaniel, S. Jha, A. Bartel, E. Bodden, J. Klein, and Y. L. Traon. Effective inter-component communication mapping in Android: An essential step towards holistic security analysis. In *Proceedings of the 22nd USENIX Security Symposium (USENIX Security 13)*, pp. 543–558. USENIX, Washington, DC, 2013.

45. M. Ongtang, S. McLaughlin, W. Enck, and P. McDaniel. Semantically rich application-centric security in Android. In *Computer Security Applications Annual Conference* (ACSAC'09), Hawaii, 2009.

46. M. Ongtang, K. Butler, and P. McDaniel. Porscha: Policy oriented secure content handling in Android. In *Proceedings of the 26th Annual Computer Security Applications Conference (ACSAC'10)*, Austin, TX, pp. 221–230. ACM, New York, 2010.

47. R. Pandita, X. Xiao, W. Yang, W. Enck, and T. Xie. WHYPER: Towards automating risk assessment of mobile applications. In *Proceedings of the 22Nd USENIX Conference on Security (SEC'13)*, Washington, DC, pp. 527–542. USENIX Association, Berkeley, CA, 2013.

48. D. Papamartzivanos, D. Damopoulos, and G. Kambourakis. A cloud-based architecture to crowdsource mobile app privacy leaks. In *Proceedings of the 18th Panhellenic Conference on Informatics (PCI'14)*, Athens, Greece, pp. 59:1–59:6. ACM, New York, 2014.

49. Y. Park, C. Lee, J. Kim, S.-J. Cho, and J. Choi. An Android security extension to protect personal information against illegal accesses and privilege escalation attacks. *Journal of Internet Services and Information Security (JISIS)*, 2(3/4):29–42, 11, 2012.

50. Y. Park, C.H. Lee, C. Lee, J. Lim, S. Han, M. Park, and S.-J. Cho. RGBDroid: A novel response-based approach to Android privilege escalation attacks. In Presented as part of the *Fifth USENIX Workshop on Large-Scale Exploits and Emergent Threats.*, San Jose, CA. USENIX, Berkeley, CA, 2012.

51. C. Qian, X. Luo, Y. Shao, and A. T. S. Chan. On tracking information flows through JNI in Android applications. In *2014 44th Annual IEEE/IFIP International Conference on Dependable Systems and Networks (DSN)*, Atlanta, GA, pp. 180–191. June 2014.

52. J. Qiu, B. Yadegari, B. Johannesmeyer, S. Debray, and X. Su. A framework for understanding dynamic anti-analysis defenses. In *Proceedings of the Fourth Program Protection and Reverse Engineering Workshop (PPREW-4).*, New Orleans, LO. ACM, New York, 2014.

53. L. Qiu, Z. Zhang, Z. Shen, and G. Sun. Apptrace: Dynamic trace on Android devices. In *IEEE International Conference in Communications (ICC)*, Kuala Lumpur, Malaysia, 2015.

54. Z. Qu, V. Rastogi, X. Zhang, Y. Chen, T. Zhu, and Z. Chen. Autocog: Measuring the description-to-permission fidelity in Android applications. In *Proceedings of the 2014 ACM SIGSAC Conference on Computer and Communications Security (CCS'14)*, Arizona, pp. 1354–1365. ACM, New York, 2014.

55. M. Rangwala, P. Zhang, X. Zou, and F. Li. A taxonomy of privilege escalation attacks in Android applications. *International Journal on Security Network*, 9(1):40–55, February 2014.

56. S. Rasthofer, S. Arzt, and E. Bodden. A machine-learning approach for classifying and categorizing Android sources and sinks. In *Proceedings of the Network and Distributed System Security Symposium (NDSS 2014)*. San Diego, CA, February, 2014.

57. S. Rasthofer, S. Arzt, E. Lovat, and E. Bodden. Droidforce: Enforcing complex, data-centric, system-wide policies in Android. In *2014 Ninth International Conference on Availability, Reliability and Security (ARES)*, Fribourg, Switzerland, pp. 40–49. September 2014.

58. A. Reina, A. Fattori, and L. Cavallaro. A system call-centric analysis and stimulation technique to automatically reconstruct Android malware behaviors. In *Proceedings of the Sixth European Workshop on System Security (EUROSEC)*. Prague, Czech Republic, April 2013.

59. F. Roesner, T. Kohno, A. Moshchuk, B. Parno, H. J. Wang, and C. Cowan. User-driven access control: Rethinking permission granting in modern operating systems. In *2012 IEEE Symposium on Security and Privacy (SP)*, San Jose, CA, pp. 224–238. May 2012.

60. S. Rosen, Z. Qian, and Z. Morely Mao. AppProfiler: A flexible method of exposing privacy-related behavior in Android applications to end users. In *Proceedings of the Third ACM Conference on Data and Application Security and Privacy (CODASPY'13)*, San Antonio, TX, pp. 221–232. ACM, New York, 2013.

61. B. Sanz, I. Santos, C. Laorden, X. Ugarte-Pedrero, and P. G. Bringas. On the automatic categorisation of Android applications. In *2012 IEEE Consumer Communications and Networking Conference (CCNC)*, Las Vegas, NV, pp. 149–153. January 2012.

62. B. Sanz, I. Santos, C. Laorden, X. Ugarte-Pedrero, P. Bringas, and G. lvarez. Puma: Permission usage to detect malware in Android. In *International Joint Conference CISIS12-ICEUTE12-SOCO12 Special Sessions*, Ostrava, Czech, Advances in Intelligent Systems and Computing, Vol. 189, pp. 289–298. Springer, Berlin Heidelberg, 2013.

63. B. Sanz, I. Santos, C. Laorden, X. Ugarte-Pedrero, J. Nieves, P. G. Bringas, and G. Lvarez Maran. Mama: Manifest analysis for malware detection in Android. *Cybernetics and Systems*, 44(6–7):469–488, 2013.

64. G. Sarwar, O. Mehani, R. Boreli, and M. Ali Kaafar. On the effectiveness of dynamic taint analysis for protecting against private information leaks on Android-based devices. In *10th International Conference on Security and Cryptography (SECRYPT 2013)*, Setubal, Portugal. ACM SIGSAC, SciTePress, July 2013.

65. R. Sato, D. Chiba, and S. Goto. Detecting Android malware by analyzing manifest flles. *Proceedings of the Asia-Pacific Advanced Network*, 36:23–31, 2013.

66. D. Sbirlca, M. G. Burke, S. Guarnieri, M. Pistoia, and V. Sarkar. Automatic detection of inter-application permission leaks in Android applications. *IBM Journal of Research and Development*, 57(6):10:1–10:12, November 2013.

67. D. Schreckling, J. Kstler, and M. Schaff. Kynoid: Real-time enforcement of fine-grained, uscr-defined, and data-centric security policies for Android. *Information Security Technical Report*, 17(3):71–80, February 2013.

68. D. Schreckling, J. Posegga, and D. Hausknecht. Constroid: Data-centric access control for Android. In *Proceedings of the 27th Annual ACM Symposium on Applied Computing (SAC'12)*, Riva (Trento), Italy, pp. 1478–1485. ACM, New York, 2012.

69. J. Park and R. Sandhu. 2004. The UCONABC usage control model. *ACM Transactions on Information System and Security*, 7(1):128–174, February 2004.

70. A. Shabtai, Y. Fledel, and Y. Elovici. Securing Android-powered mobile devices using SElinux. *Security Privacy, IEEE*, 8(3):36–44, May 2010.

71. R. K. Shahzad and N. Lavesson. Veto-based malware detection. In *2012 Seventh International Conference on Availability, Reliability and Security (ARES)*, Prague, Czech, pp. 47–54. August 2012.

72. S. Shekhar, M. Dietz, and D. S. Wallach. Adsplit: Separating smartphone advertising from applications. In *Proceedings of the 21st USENIX Conference on Security Symposium (Security'12)*, pp. 28–28. USENIX Association, Berkeley, CA, 2012.

73. F. Song and T. Touili. Model-checking for Android malware detection. In *Proceedings of Programming Languages and Systems*, Singapore, Singapore, Lecture Notes in Computer Science, Vol. 8858, pp. 216–235. Springer International Publishing, 2014.

74. K. Tam, S. J. Khan, A. Fattori, and L. Cavallaro. Copperdroid: Automatic reconstruction of Android malware behaviors. In *22nd Annual Network and Distributed System Security Symposium (NDSS 2015)*. San Diego, CA, 2015.

75. X. Wei, L. Gomez, I. Neamtiu, and M. Faloutsos. Permission evolution in the Android ecosystem. In *Proceedings of the 28th Annual Computer Security Applications Conference (ACSAC'12)*, Florida, pp. 31–40. ACM, New York, 2012.

76. D.-J. Wu, C.-H. Mao, T.-E. Wei, H.-M. Lee, and K.-P. Wu. DroidMat: Android malware detection through manifest and API calls tracing. In *Proceedings of the 2012 Seventh Asia Joint Conference on Information Security (ASIAJCIS'12)*, pp. 62–69. IEEE Computer Society, Washington, DC, 2012.

77. L. Xing, X. Pan, R. Wang, K. Yuan, and X. Wang. Upgrading your Android, elevating my malware: Privilege escalation through mobile OS updating. In *Proceedings of the 2014 IEEE Symposium on Security and Privacy (SP'14)*, pp. 393–408. IEEE Computer Society, Washington, DC, 2014.

78. R. Xu, H. Saïdi, and R. Anderson. Aurasium: Practical policy enforcement for Android applications. In *21st USENIX Security Symposium (USENIX Security 12)*, pp. 539–552. USENIX, Bellevue, WA, 2012.

79. W. Xu, F. Zhang, and S. Zhu. Permlyzer: Analyzing permission usage in Android applications. In *2013 IEEE 24th International Symposium on Software Reliability Engineering* (*ISSRE*), California, pp. 400–410. November 2013.

80. J. Yiming, G.-J. Ahn, Z. Ziming, and H. Hongxin. Riskmon: Continuous and automated risk assessment of mobile applications. In *Proceeding of the Fourth ACM Conference on Data and Application Security and Privacy* (*CODASPY*). San Antonio, TX, 2011.

81. Y. Zhou, X. Zhang, X. Jiang, and V. W. Freeh. Taming information-stealing smartphone applications (on Android). In *Proceedings of the Fourth International Conference on Trust and Trustworthy Computing* (*TRUST'11*), Pittsburgh, PA, pp. 93–107. Springer-Verlag, Berlin, Heidelberg, 2011.

82. Y. Zhauniarovich, G. Russello, M. Conti, B. Crispo, and E. Fernandes. Moses: Supporting and enforcing security profiles on smartphones. *IEEE Transactions on Dependable and Secure Computing*, 11(3):211–223, May 2014.

83. M. Zheng, M. Sun, and J. C. S. Lui. Droid analytics: A signature based analytic system to collect, extract, analyze and associate Android malware. In *Proceedings of the 2013 12th IEEE International Conference on Trust, Security and Privacy in Computing and Communications* (*TRUSTCOM'13*), Melbourne, Australia, pp. 163–171. IEEE Computer Society, Washington, DC, 2013.

84. Y. Zhou and X. Jiang. Detecting passive content leaks and pollution in Android applications. *Proceedings of the 20th Annual Symposium on Network and Distributed System Security*. San Diego, CA, 2013.

85. J. Zhu, Z. Guan, Y. Yang, L. Yu, H. Sun, and Z. Chen. Permission-based abnormal application detection for Android. In *Information and Communications Security*, Hong Kong, China, Lecture Notes in Computer Science, Vol. 7618, pp. 228–239. Springer, Berlin, Heidelberg, 2012.

86. A. P. Felt, E. Ha, S. Egelman, A. Haney, E. Chin, and D. Wagner. Android permissions: User attention, comprehension, and behavior. In *Proceedings of the Eighth Symposium on Usable Privacy and Security* (*SOUPS '12*), Washington, DC, 14pp. ACM, New York, Article 3, 2012.

87. StageFright, Android media playback engine, https://source.android.com/devices/media/.

88. J. J. Drake, Google Stagefright media playback engine multiple remote code execution vulnerabilities, http://www.securityfocus.com/bid/76052.

89. S. Fahl, M. Harbach, H. Perl, M. Koetter, and M. Smith. 2013. Rethinking SSL development in an appified world. In *Proceedings of the 2013 ACM SIGSAC Conference on Computer & Communications Security* (*CCS '13*), Berlin, Germany, pp. 49–60. ACM, New York, 2013.

90. D. R. Thomas, A. R. Beresford, T. Coudray, T. Sutcliffe and A. Taylor, The lifetime of Android API vulnerabilities: Case study on the JavaScript-to-Java interface. In *Proceedings of the Security Protocols Workshop*. Cambridge, UK, 2015.

Chapter 5

An Empirical Study on Android for Saving Nonshared Data on Public Storage*,†

Xiangyu Liu, Zhe Zhou, Wenrui Diao, Zhou Li, and Kehuan Zhang

Contents

---

* All vulnerabilities described in this chapter have been reported to corresponding companies. Alipay has fixed this vulnerability in the latest version. We have gotten IRB approval before all experiments related to human subjects.

† This is an extended version of conference paper: X. Liu, Z. Zhou, W. Diao, Z. Li, K. Zhang. An empirical study on Android for saving non-shared data on public storage. In *Proceedings of the 30th IFIP International Information Security and Privacy Conference (IFIP SEC2015)*. Hamburg, Germany. May 26–28, 2015.

Abstract

With millions of apps that can be downloaded from official or third-party markets, Android has become one of the most popular mobile platforms today. These apps help people in all kinds of ways and thus have access to lots of user's data that in general fall into three categories: sensitive data, data to be shared with other apps, and nonsensitive data not to be shared with others. For the first and second types of data, Android has provided very good storage models: an app's private sensitive data are saved to its private folder that can only be accessed by the app itself, and the data to be shared are saved to public storage (either the external SD card or the emulated SD card area on internal flash memory). But for the last type, that is, an app's private (nonshared) and nonsensitive data, we found that there is a big problem in Android's current storage model that essentially encourages an app to save its nonsensitive private data to shared public storage that can be accessed freely by all other apps.

At first glance, it seems no problem to do so, as those data are nonsensitive after all, but it makes an implicit assumption that app developers could identify and differentiate all nonsensitive data from sensitive data correctly. In this chapter, we will demonstrate that this assumption is invalid with a thorough survey on information leaks of some most popular apps that had followed Android's recommended storage model. Our studies showed that highly sensitive information from apps with billions of users can be easily hacked by exploiting the mentioned problematic storage model. Although our empirical studies are based on a limited set of apps, the identified problems are never isolated or accidental bugs of those apps being investigated. On the contrary, the problem is rooted in the vulnerable storage model recommended by Android (even including the latest Android 4.4 at the time of writing). To mitigate the threat, we proposed a dynamic program analysis framework that works at the system call layer to capture all suspicious writings to the SD card, as well as an automatic tool that can trace such operations back to their origins in apps and then apply a patch to fix such problems.

5.1 Introduction

The last decade has seen the immense evolution of smartphone technologies. Today's smartphones carry much more functionality than plain phones, including email checking, social networking, gaming, and entertainment. These emerging functionalities are largely supported by the vast number of mobile applications (*apps*). As reported by [2], the number of Android apps on Google Play is hitting 137 million.

When apps are performing their tasks, it is inevitable for them to touch our private information, like emails, contacts, and various accounts. Generally, when people download and install an app, they trust it and hope it will protect their private information carefully. To support the protection of private data, Android system allocated for each app a private folder that can only be accessed by the app owner.

Since each app now has its own private folder, it seems natural and absolutely right to save all private data that should not be shared with others to this secure folder, but unfortunately, according to our survey to be presented in Section 5.4, this is not the case. We found that, for private data, many apps will first try to decide what are sensitive and what are not, then only store the identified sensitive private data to the secure private folder, and save the other nonsensitive private data to a "private" folder on shared public storage, even though the data should never be shared with other apps.

Although there are some reasons and benefits by differentiating sensitive private data from nonsensitive private data and save the latter to shared public storage, which will be discussed in Section 5.7, we believe that such practice is wrong and could lead to serious attacks. *App's private data, which are not supposed to be shared with other apps, should never be saved to shared public storage,* no matter if they are sensitive or not, since most app developers are not experts in security and privacy, thus could make wrong decisions on what are sensitive and what are not. According to our survey in Section 5.4, we indeed have found some very popular apps leak information that are obvious sensitive (like full name, phone numbers, etc.), which shows even the companies with high reputation are not so credible under such a wrong storage model. What's worse, even if all developers have made the right decisions on each piece of information generated by their own apps, it would still be problematic, because nonsensitive information from multiple apps/sources could become sensitive and lead to serious privacy attacks. In Section 5.4, we will give examples of information that is not sensitive at first glance, but has a huge privacy impact when combined with other public information. Some examples described in Section 5.5 demonstrate how such nonobvious sensitive information could be used in serious privacy attacks.

Surprisingly, the storage model is still recommended in official Android documents, and the latest Android 4.4 emphasized a misleading concept called "app-private directory." The following statement is quoted from official Android developer documents*: *"If you are handling files that are not intended for other apps to use, you should use a* **private storage directory** *on the* **external storage** *by calling getExternalFilesDir()."* According to the same page, the **external storage** is defined as a place to *"store public data on the shared external storage."* So this guideline is actually asking apps to save app-private data to shared public storage. More relevant details on storage security in latest Android 4.4 will be given in Section 5.3.

---

* http://developer.android.com/guide/topics/data/data-storage.html.

Although it has been a hot topic to sniff and infer sensitive information from Android devices and there are a number of studies demonstrating how to launch attacks successfully, the work presented in this chapter is different. Here, we want to report a serious problem with the private data storage model that has been followed by many apps for years, while the previous works either trick the user to grant high-profile permissions [5] or exploit the system flaws in side-channel attacks [21]. It is also worth noting that the attacks proposed in this chapter are new and different from attacks that have been reported, even if the target is the same app. For example, our second attack against `WhatsApp` is based on nonobvious sensitive data (photo names), while the attack reported [17] took `WhatsApp` database as the target, and our attack succeeds in its latest version, even after the reported vulnerability was fixed.

Since the reported problem goes beyond the vulnerability of any specific app, current data protection mechanisms, like data encryption, do not work well. To address this problem, developers may be required to update their codes by saving all private data to a securely protected private folder instead of the one located on shared public storage. However, expecting all the developers to obey the security rules and fix this security issue is impractical. Another more effective solution requires some updates to Android system by enforcing fine-grained access control on "app-private" data. In order to mitigate this problem as much as possible, and avoid bringing unnecessary trouble to users, we propose to augment existing security framework on public storage by instrumenting the original APIs and automatically checking the ownership of files and directories.

Our contributions: We summarize our contributions as follows:

- We revisit the Android data storage model and first study whether 17 popular Android apps store the data correctly. Our study reveals that 13 apps (billions of installations on smartphones in total*) leave user's private information on shared storage, which leads to serious privacy risks. Then, a large-scale analysis was conducted and the result shows such a problem widely existed among apps, which means it is not a bug that happens on several apps but a wrong storage model provided to developers.
- We design three concrete attacks based on the sensitive information collected from some popular apps, including inferring user's location (`WeChat`), identifying the owners of phone numbers acquired from `WhatsApp`, and inferring the chatting buddy (`KakaoTalk`). These attacks are new and can achieve high accuracy with low profile.
- We proposed a mitigation method and developed supporting tools that can patch vulnerable Apps automatically through some dynamical code analysis techniques.

Roadmap: The rest of this chapter is organized as follows. We begin with the adversary model in Section 5.2, followed by an overview of the Android storage model in Section 5.3. Survey results on private information leakage of Android apps will be given in Section 5.4. Then we present three concrete attacks in Section 5.5 and defense technologies in Section 5.6. Section 5.7 will be devoted to discuss the fundamental reasons that lead to our discovered problem as well as some limitations of our work. We will compare our work with prior related researches in Section 5.8 and will finally conclude this chapter with Section 5.9.

---

* The detailed statistical data are shown in Table 5.3 and Figure 5.3, which actually contain installations on other platforms except Android, but Android has a big advantage in the size of these users.

5.2 Adversary Model

The adversary studied in our chapter is interested in privacy attacks by only exploiting the app-private data located on shared public storage due to the wrong model introduced in this chapter. In order to get app-private data and perform privacy attacks, it is assumed that a malicious app with permissions of access to external storage (i.e., the READ_EXTERNAL_STORAGE or WRITE_EXTERNAL_STORAGE permission, READ and WRITE for short), and the Internet (i.e., the INTERNET permission) has been installed successfully on an Android phone updated to the latest version (Android 4.4). The malicious app will read certain app-private folders selectively on shared public storage, searching for information that can be used for a privacy attack, and then upload such information to a malicious server where the intensive data analysis and concrete attacks will be done. Note that there is no need to scan the whole shared storage or upload all app-private data, because it is reasonable to assume that the adversary has studied many apps in the market beforehand and already knows where to read and how to extract useful information from app-private data.

Based on the assumptions mentioned earlier, it is hard to detect the attack. First, it requires only two very common permissions. We did a statistical analysis on the permissions requested by 34,369 apps, the top 10 permissions are shown in the Figure 5.1. About 94% of apps request INTERNET permission, and about 85% of apps request the permission to access external storage, ranking No. 1 and No. 3 respectively. Second, it only uploads data when Wi-Fi is available and filters out nonuseful app-private data to minimize data size. Third, all network- and CPU-intensive operations are done on a remote malicious server instead of the mobile phone, which can eliminate the alarm in CPU and power usage. Finally, the malicious app will not take any other malicious behavior to help the attack, like exploiting other system vulnerabilities or bypassing existing security mechanisms, so it is hard to be tagged as malicious by current antivirus software. The victim of such an attack could be the owner of the compromised Android phone. However, depending on on the apps installed and the information leaked through app-private data, attacks can be extended to user's family members, friends, and colleagues.

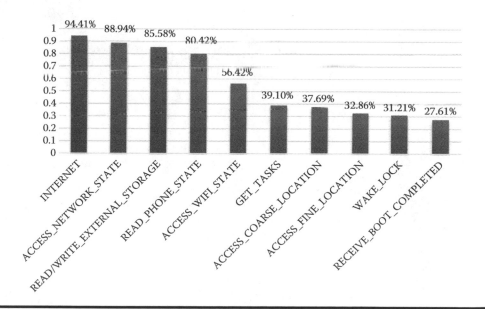

Figure 5.1   Top 10 permissions requested by 34,369 apps.

5.3 Overview of Android Storage Model

In Android, there are several different storage options that, according to their access control mechanisms, can be divided into three categories: *system*, *app-specific*, and *public*, as shown in Figure 5.2. The system storage is the directory where the whole Android OS is located and protected by Linux access control mechanism. The app-specific storages are places that are under control of a specific app (i.e., the owner of these folders) and can only be read and written by that app. Android system provides three types of app-specific storage: *internal storage,** shared preferences*, and *SQLite databases*. The last storage option is shared public storage that is used to share data among apps, like downloaded documents, video, photos, etc.

As shown in Figure 5.2, for system and app-specific storage, Android relies on the discretionary access control mechanism provided by underlying Linux Ext2/4 file system to enforce access control. Android creates an account for each app (UID) and system process that would be the owner of a specific folder and controls who can access that folder. However, there is no fine-grained access control on shared public storage that is only protected with READ and WRITE permissions. As a result, an app can read or write any folder at any time once it has acquired corresponding permissions. The Android system even sets the permissions to rwxrwx---- for the shared public folders on built-in flash memory that were originally protected by Ext2/4 with fine-grained access control mechanism, which we believe is to emulate the behavior of FAT file system and be consistent with the one used on SD card.

Misleading of new features on Android 4.4: There are several security enhancements on the shared public storage in the latest Android 4.4, which could give people a false sense that Android 4.4 is immune from the attacks introduced here. First, it emphasized the concept "app-private directory," which are not intended for other apps to use, and app reads or writes files in its own private directory does not require READ or WRITE permissions [38]. At first glance, many people, even including some security experts, may think that the Android system will save the app-private data in secure

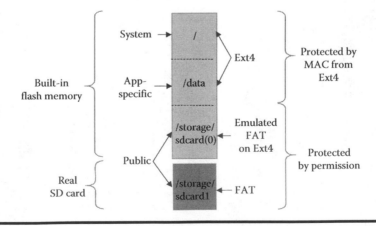

Figure 5.2 Overview of Android storage options and security.

---

* The name "internal storage" is actually very confusing. It should be understood as "being accessed only by app internal code" instead of the physical location of the storage (i.e., built-in flash memory) because Android also uses a term "external storage" that is shared public storage and could be either on build-in flash-memory or SD card.

places, and the problem discussed in this chapter is gone. However, that is totally wrong! The Android system actually creates a folder (named "./android/data/") on shared public storage to save all app-private data. As mentioned earlier, there is no fine-grained access control on shared public storage; any app with appropriate permission thus can access those app-private data and apply privacy attacks.

Another misleading feature introduced in latest Android 4.4 is the permission WRITE_MEDIA_ STORAGE. This permission controls the write operation on a **real** SD card and will only be granted to system process. It can be taken as a nonpublic permission, which means normal apps could not write to a **real** SD card. However, there is no dedicated permission to control the read operation on the SD card, which makes the latest Android version still vulnerable to the attack on the wrong app-private storage model. This has been confirmed by our home-brewed apps and third-party apps downloaded from Google Play market.

5.4 Survey on Information Leaks from Shared Storage

It is true that some app-private data are not sensitive. However, the model of letting apps write their private data to shared public storage relies on a strong assumption that app developers can make the right decision to tell sensitive data from nonsensitive ones. In this section, we will present a survey of the information leaks through the app-private data saved on shared public storage, which will show that such an assumption is problematic. The survey includes two parts: the first is a detailed examination of information in "shared" app-private data for 17 popular apps (in Section 5.4.1), and the other is a more general and large-scale study on the "shared" app-private data on public storage (in Section 5.4.2).

5.4.1 Investigation of Popular Apps

What apps have been surveyed: According to our adversary model, the attackers are interested in privacy attacks over "shared" app-private data, so we have selected 17 most popular apps on Google Play from three categories: "social networking," "instant messaging," and "online payment," which are believed to be more likely to touch users' sensitive information. The apps, including their user numbers, are shown in Figure 5.3, while the versions of these apps installed on our mobile phones are listed in Table 5.1.

How to check the app-private data: These 17 apps are downloaded and installed on all three Samsung Galaxy S3 mobile phones. Then we manually simulate three different users on three phones respectively, including account registration, adding friends, sending messages, etc. Finally, we check the shared public storage, search sensitive data for each app, and classify the collected data. Following are the details of user privacy data extraction for some apps.

1. **Viber** is an instant messaging app with 300 million users [7]. The text file .userdata saved under ~/viber/.viber/ reveals a lot of user's information, including the user's real name and phone number. Besides, it also contains the path pointing to the user's profile photo, for example, ~/viber/media/User Photos/xxx.jpg.
2. **WhatsApp** is an instant messaging app with 450 million users [7]. The user's profile photo is stored under the directory ~/WhatsApp/.shared/, with file name tmpt. The profile photos of user's friends are saved under the directory ~/WhatsApp/Profile

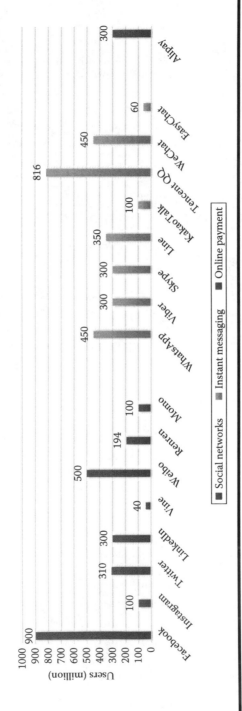

Figure 5.3   The number of users installed the 17 popular apps. (Details are shown in the Appendix with citations.)

Table 5.1   Versions and Users of the 17 Apps Installed on Our Android Phones

Category	App Name	Version	Users (Millions)
	Facebook	13.0.0.13.14	900
	Instagram	6.2.2	100
	Twitter	5.18.1	310
Social networks	LinkedIn	3.3.5	300
	Vine	2.1.0	40
	Weibo	4.4.1	500
	Renren	7.3.0	194
	Momo	4.9	100
	WhatsApp	2.11.186	450
	Viber	4.3.3.67	300
	Skype	4.9.0.45564	300
Instant messaging	Line	4.5.4	350
	KakaoTalk	4.5.3	100
	Tencent QQ	4.7.0	816
	WeChat	5.2.1	450
	EasyChat	2.1.0	60
Online payment	Alipay	8.0.3.0320	300

pictures/ after the profiles are viewed, and they are named by profile owners' phone numbers without any obfuscation.

3. **LinkedIn** is a business-oriented social networking app with 300 million users [20]. This app cache the photos into the directory ~/Android/data/com.Linkedin.android/cache/li_images/. Though the images stored under the directory can be many, the user's profile photo is unique in file size and modified time. In general, the user's profile photo is about 30 kB, while others are about 3 kB (if the user does not see the HD profile photo).

4. **KakaoTalk** is an instant messaging app with 100 million users [7]. If user A chats with user B, the app will create a content folder has the same name in both the Android phones under the path ~/Android/data/com.kakao.talk/contents/. The files on the two phones also have the same name (10 digital numbers), size and stored in the same path, like ~/Android/data/com.kakao.talk/contents/14 digital numbers/xx/.

5. **Tencent QQ** is an instant messaging app in China with 816 million users [25]. Under the path ~/tencent/com/tencent/mobi-leqq/, it is very easy for us to acquire a user's QQ account from the log file named as "com.tencent.mobileqq.xxx.log."

6. **Weibo** is a Chinese microblogging service with 500 million users [40]. There is a file named by the user's UID under the path `~/sina/weibo/page`, and we can acquire the user's username and even her email address if she used it to register her account.

7. **Alipay** is a online payment service in China with 300 million users [34]. We could find a user's phone number from two cache files, one is the `_meta` file in the directory `~/com.eg.android.Alipa-yGphone/cache/`, the file content reveals the user's phone number directly, and also points out the other file in the same directory that discloses the user's phone number.

8. **Renren** is a social networking app with 194 million users [31]. A folder named by the user' UID is stored in the directory `~/Andro-id/data/com.renren.mobile.android/audio/cache/JasonFileCache`. Even user's visit histories are also stored in this folder, which contains the name, UID of user's friends. The audio files are named as the format `UID+hash value`, which also disclose the user's UID. We can find the user's personal home page by the URL *http://www.renren.com/UID* in a browser.

9. **Momo** is a location-based services instant messaging app with 100 million users [18]. We can find a folder named as the user's account in the directory `~/immomo/users/`. Through searching the account number, we can not only get her Momo profile information, but also infer her location and even track her.

10. **EasyChat** is an instant messaging app with 60 million users [9]. The file `pjsip_log.txt` under the directory `~/Yixin/log/` shows us all the call records information in plaintext. For example, we can extract information: "From: 8525984xxxx, To: 1471434xxxx, Start: 2014-03-02 14:39:11, End: 2014-03-02 14:39:31."

11. **Audio files**. Almost all the popular instant messaging apps, like WhatsApp, Line, WeChat, Tencent QQ, and KakaoTalk, store the audio files into shared storage directly without any process.

How the information is leaked: By studying the 17 sample apps, we found 10 of them leak various user-related information through app-private data on shared public storage, including full name, account name and emails, etc., as shown in Table 5.2. Such information is leaked in different forms that will be discussed in the following text.

Leak through text file: Some apps store the user's profile into a text file. For example, Viber directly saves user's real name, phone number, and path of profile photo into a plain-text file `~/viber/.viber/.userdata` without any encryption (~ represents the root directory of public storage). User's username, email address* are stored by Weibo in a file named by the user's UID. Some apps keep text logs that also reveal quite rich information. EasyChat keeps call records in a file `~/Yixin/log/pjsip_log.txt`, so caller's number, callee's number, and call duration can be easily recovered by simply parsing each record.

Leak through photo: The social network apps usually cache user profile photos in shared public storage, like LinkedIn in our studies. The photo itself is nonobvious sensitive information, since we can easily get millions of such photos with Google. However, our study shows user's LinkedIn profile photos can be linked to her identity with high possibility, as shown in Figure 5.4.

Leak through file name: We found several apps organize data related to a person or friend into a dedicated file named with sensitive or nonobvious sensitive information. For example,

---

* If the user uses email to register Weibo account. The user also can use phone number to register an account.

Table 5.2 Sensitive Data Acquired from Smartphone A

Sensitive Information	App Name	Content/Remarks
User identity	Viber	Name
	Renren	UID
	Weibo	UID
	Linkedin	Profile photo
Phone number	Viber	8525984xxxx
	Alipay	1521944xxxx
	EasyChat	8525984xxxx
Email	Weibo	lixxxxxx@163.com
	Renren[a]	lixxxxxx@gmail.com
Account	Tencent QQ	8387xxxxx
	Renren	2388xxxxx
	Momo	3120xxxx
	Weibo	2648xxxxxx
	WeChat	QQ account/Phone number
Connection	EasyChat	Call records
	LinkedIn	Profile photos
	KakaoTalk	Names of content directories
	WhatsApp	Phone numbers of friends: 1213572xxxx ...
	Renren	Accounts of friends: 8295xxxxx ...

[a] The email found by `grep` command in Table 5.2 is from a "log.txt" file left by Renren old version (5.9.4).

WhatsApp stores user's friends photos in the path ~/`WhatsApp/Profile pictures/` with that friend's phone number as file name. They seem meaningless but could have significant privacy implications when combined with other public information, for example, the data from social networks. A sample attack will be given in Section 5.5 to show that how this nonobvious sensitive information could be used in privacy attacks.

Leak through folder name: Some apps use account name as folder name directly, like Renren, Momo, and KaokaoTalk will create a folder with the same name in two users' phones if they chatted with each other. We also find that a file sent to each other will have the same file name and saved in the same path. Such a naming convention reveals the connections among people and even can be leveraged to infer a user's chat history.

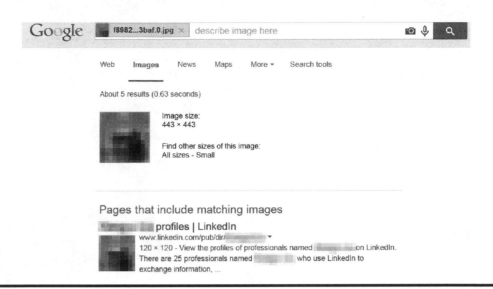

Figure 5.4   The result of searching a user's LinkedIn profile photo.

Leak through other file: Files left by uninstalled apps or put by the user may also disclose user's information, like email addresses. We leverage the command `/system/bin/sh -c grep -r @xxx.com` path to match and extract email addresses from files (only three types, .txt, .doc(x), .pdf) in shared storage.

What information has been leaked: As shown in Table 5.2, there is indeed some important sensitive information leaked through the app-private data on shared public storage. To better understand the privacy implication of such leak, it is better to use personal identifiable information (PII) [26], a well-known definition for private data, to classify and evaluate the leaked information. We defined two categories of sensitive data, *obvious sensitive data* and *nonobvious sensitive data*, by refining the concept of PII as follows:

- *Obvious sensitive data*: It contains identifiers in PII related to user's real-world identity, including full name, phone numbers, addresses, date of birth, social security number, driver's license ID, and credit card numbers.
- *Nonobvious sensitive data*: It contains identifiers in PII related to user's virtual world identity and also friends' information. The virtual world identifiers include email addresses, account name, profile photos, etc. We also consider the friends' information as shown in previous works [28]; they can be used to uniquely identify the user.

According to the definition, we divided the surveyed apps into three categories with privacy protection level from high to low, based on the type of sensitive data revealed. Table 5.3 shows the results of our evaluation.

How to infer user's identity: Without doubt, user's identity information is a key issue in any privacy research. We present five approaches to get a user's identity information by leveraging the sensitive data acquired, shown as follows:

1. A user's name and profile photo can be acquired from Viber.
2. A user's identity can be acquired by searching her Renren UID.

Table 5.3 Privacy Protection Level of the 17 Apps

Privacy Level	App Name	Remarks
★★★	Facebook, Twitter, Instagram, Skype	No comments
★★	Line, Vine, WeChat	Audio files without encryption
★	WhatsApp, LinkedIn, Viber, KakaoTalk, Tencent QQ, Alipay, Renren, Weibo, Momo, EasyChat	Detailed problems are shown in Section 5.4.1

3. A user's identity can be acquired by searching her Weibo Username/UID.
4. A user's identity can be acquired by searching the user's LinkedIn profile photo on Google image search.
5. We could find someone on Facebook with a high probability by the email addresses extracted from shared public storage and also the usernames acquired from other apps, since people prefer to use the same username and email address among their various social networking apps [1].

Apparently, leaking obvious sensitive data should be prohibited and requires immediate actions from the app developers and Android developing team. The damage of leaking nonobvious sensitive data is less clear. For instance, attackers holding a user's photo still have no way to learn her real-world identity unless the user is famous or acquainted with attackers. However, our research shows it is possible to infer obvious sensitive data from nonobvious ones (in Section 5.5) and the latter should also be well protected.

5.4.2 Investigation of Apps on a Large Scale

The earlier analysis shows that all 17 popular apps have saved their app-private data to shared public storage, and 10 of them have leaked lots of privacy-related information. In order to have a better understanding of the scale and seriousness of this issue, it is necessary for us to investigate more apps. However, this is a difficult task in general, because accurate evaluation relies on intensive use of these apps, including account registration, sending messages, etc. Too many deep simulation of people's daily use operations will make dynamic analysis tools not effective and, especially, will not scale to the large number of Android apps. Therefore, we take static analysis instead; the basic idea is scanning the app code for specific patterns that indicates the written app-private data on shared public storage. Again, we only focus on the three categories that are most likely to touch user's sensitive information, so we selected 1,648 different apps from our app repository (including 34,369 apps downloaded previously).

The previous static analysis of apps, like [3,15], convert the app code from the DEX format to a JAR or Java source code, and then leverage WALA [13] or Soot [39] to complete the further analysis. Such analysis methods are intuitive and easy to operate, but in general, some information will be lost when converting app code to JAR or Java source code. Therefore, we leveraged Apktool [19] to decompile APK to smali code that contains all the original APK information and even can be repackaged to a new app. Then we did static analysis directly on the smali code.

Table 5.4   Methods of Getting Path of Shared Public Storage

Category	Methods
Call API	`getExternalStorageDirectory()`
	`getExternalStoragePublicDirectory()`
	`getExternalFilesDir()`
	`getExternalFilesDirs()`
	`getExternalCacheDir()`
	`getExternalCacheDirs()`
Hardcode path	`"/sdcard", "/sdcard0", "/sdcard1"`

Determining whether app-private data is really sensitive or not requires more knowledge of the app itself and even combining with other app's information; therefore, it is difficult to automate. We think an app is highly risky if it intends to create "sensitive" app-private folders or files on shared public storage. Specifically, "sensitive" app-private folder or file needs contain at least one of the following keywords, including `log`, `cache`, `files`, `file`, `data`, `temp`, `tmp`, `account`, `meta`, `uid`, `history`, which are learned from the apps by heuristic. For each keyword, we define our patterns similar to "whole word match" in Linux, but with more restrictions. For example, the patterns of keyword `log` contain folders (`/user_log`, `/log`, ...) and files (`user_log.txt`, `user.log`, ...). What's more, we build a control flow graph (CFG) of an app's smali code to confirm whether the "sensitive" data are truly written on the shared public storage, since the data may be written in internal storage or other secure places. In general, there are several ways to get the path of shared public storage, we summarize them as shown in Table 5.4. Also, we consider developer's use of general methods to create files (e.g., `FileOutputStream`) and folders (e.g., `mkdir`), and ignore the situations that developers use Linux system commands to create files (like `touch`).

We marked each function f in the CFG based on the following three criteria:

1. Whether it contains at least one of the methods described in Table 5.4
2. Whether it contains at least one of the patterns we defined earlier
3. Whether it uses `FileOutputStream` to create files or `mkdir` to make folders

We implemented our detecting method (shown in Algorithm 5.1) on the marked CFG after the earlier processes; the depth parameter was set as 3 according to our experience. The results show that 489 apps from the 1648 apps analyzed intend to write some app-private data to shared storage. However, such a static scanning method has its limitations. So we randomly chose 30 apps from the 489 suspicious apps, and manually operated these apps and did our evaluation, just like the work performed on the popular apps. The result shows that 27 apps truly wrote app-private data in shared public storage, which indicates that such a privacy leakage problem revealed in this chapter widely exists and that it is very serious among apps. Also, the false positive of our method is low.

Algorithm 5.1 Detecting vulnerable apps

Input: Class_set *C*, Keyword_Patterns_set *KS*, Path_API_set *PA*, Write_API_set *WA*
Output: bool sensitive

```
 1: for class c in C do
 2:     for function f in c do
 3:         condition.clear();
 4:         DFS(f, depth);
 5:         if condition == Union(KS, PA, WA) then
 6:             return true;
 7:         end if
 8:     end for
 9: end for
10: return false;
11:
12: procedure DFS(function f, int depth)
13:     if depth == 0 then
14:         return;
15:     end if
16:     for all element e in f.mark do
17:         condition(e)=true;
18:     end for
19:     for all callee ce of f do
20:         DFS(ce, depth-1);
21:     end for
22: end procedure
```

5.5 Attacks Based on Nonobvious Sensitive Data

In this section, we present three example attacks based on the nonobvious sensitive information extracted from app-private data saved on shared public storage. We begin with a brief introduction to the design of a malicious Android app called Smuggle App-Private Data (SAPD), followed by a detailed description of three concrete attacks: attacking victim's location, identifying the owners of the phone numbers acquired from the WhatsApp friends list, and even getting chatting buddy and user's chat history. Again, we want to emphasize that the three attacks are different from previous attacks that have been reported.

5.5.1 Attack Preparation

Since it is assumed that attackers have already studied the vulnerable apps and know where and how to get useful information from app-private data, the natural next step is to develop a malicious app that could dig out and upload such information to a server under their control. As a result, the malicious app needs permission to access (more accurately, to read) shared public storage and Internet. As discussed in Section 5.2, such permissions are common and not alarming.

The weakest part of this malicious app might be the potential outstanding network traffic footprint, especially for users with a limited 3G plan. We implemented two optimizations in our app prototype SAPD to get around this limitation. First, try to minimize the uploaded data. For example, for information leaked through file name or folder name, it is only necessary to upload the list of file names instead of the whole file. Another optimization is to upload data only when a Wi-Fi network is available. It can be achieved by simply using the `WiFiManager` class provided by Android SDK, with the cost of requesting an extra permission `ACCESS_WIFI_STATE`. However, SAPD achieved the same goal by reading public available files in `procfs` from Linux kernel of Android. More specifically, Android put the parameters of address resolution protocol in file `/proc/net/arp` and other wireless activities in `/proc/net/wireless`, so by reading the files mentioned earlier, SAPD is able to know whether a WiFi network is connected or not.

After having extracted the useful information and identified appropriate data uploading timing, the last thing that SAPD will do is just sending the data out to a malicious server where all subsequent attacking steps will take place. To minimize the possibilities of being caught due to suspicious CPU usage or abnormal battery consumptions, SAPD only reads and uploads filtered useful data, and it will never perform any kind of intensive computations or Internet communications.

5.5.2 Inferring Your Location

In this section, we describe our attack on inferring user's location from data in shared storage. Location of a phone user is considered as sensitive from the very beginning, and there are already a lot of research works on inference attacks and protections. In recent years, location-based social discovery (LBSD) has become popular and is widely adopted by mobile apps. The locations of users are updated and viewable by friends or even strangers. For the latter case, to protect a user's privacy, only the distance between the user and the viewer is revealed. However, recent works show such protection is very weak. Chen et al. analyzed LBSD on a large scale and also show it is possible to reidentify users with high precision [8]. A recent work by Li et al. [23] demonstrates that the geolocation of a user can be revealed. Our attack also aims to identify a user's location from the LBSD network but we make improvements by combining the profile information extracted from the victim's phone. We show the attack would be more accurate and could be a realistic threat. As a showcase, we demonstrate our attack on `WeChat` app.

The LBSD module in `WeChat` is called "People Nearby," through which the user can view information of other users within a certain distance. The public information includes nickname, profile photo, posts (called *What's Up*), region (city level), and gender. As described in Section 5.4.1, the user's phone number or QQ user is left by WeChat on shared storage. This information will be collected by SAPD we built and sent to one of our servers (denoted as S_1). These servers are installed with emulated Android environment for running WeChat app. S_1 will query the server of Tencent (the company operating WeChat) for profile information. In the meantime, the attacker needs to instruct another server (denoted as S_2) to run WeChat using fake geolocations, check People Nearby, and download all the profile information and the corresponding distances from discovered nearby users. The profile information from S_1 is then compared with the grabbed profile information (downloads from S_2) in another server (denoted as S_3) following the steps shown in Figure 5.6. If a match happens, S_2 will continue to query People Nearby two more times using different geolocation (faked) to get two new distances. Through the three-point positioning method, the target's accurate location can be inferred. Since the target user may keep moving, the three servers need to synchronize on user's information. Figure 5.5 shows the processes of our attack, and we elaborate the technical details in each stage as follows:

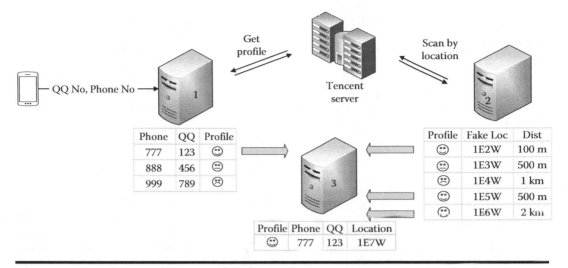

Figure 5.5 The schematic diagram of location attack.

5.5.2.1 Getting Users' WeChat Account

Though WeChat user ID is not stored on shared storage, QQ user ID and phone number are stored instead. They are bound to WeChat account and have to be unique for each user. Therefore, our app collects these two data and sends them to our server S_1 for profile querying.

5.5.2.2 Getting Users' Profile Information

Next, the attacker uses QQ user ID or phone number to query Tencent server for the user's profile information. The returned profile consists of five fields: nickname, profile photo, posts (*What's Up*), region, and gender. Our task is to assign the location information for this profile. Unfortunately, this field is not provided directly from the Tencent server and updated according to the user's geo location. What we do here is run WeChat on another server S_2, frequently refresh People Nearby to get as many profiles and distances until a profile match happens, and use the corresponding distances to infer the location. The challenge here is to extract the profile and distances from WeChat as there is no interface exposed from WeChat to export this information. After we decompile its code, we found the app invokes an Android API `setText` from `android.widget.TextView` to render the text on screen whenever a profile is viewed. We, therefore, instrument this API and dump all the texts related to profiles into logs. This helps us to extract the *Nickname, What's Up, and Region*. In general, we can distinguish a person from others based on these three features. However, it is also possible (with very low possibility) that different profiles have the same values of these three fields. If such a situation happens, we extract the size of profile photo and consider it in the matching process. We do not use the photo content since the comparing is more time consuming and size is enough.

5.5.2.3 Comparing Process

The comparison processes in S_3 are shown in Figure 5.6. The first step we need to compare the *Nickname* from S_1 and S_2, if they are the same, we will continue to compare their *What's Up*

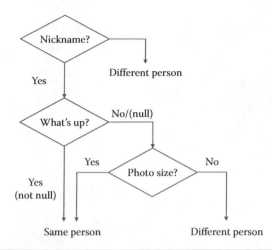

Figure 5.6 The processes of profile information comparison.

information, if the *What's Up* are not blank and the same, we think the two users are the same person. If the *What's Up* information is different or blank, we will continue to compare their profile photos size for another round of check. We add this additional step since the user could post new *What's Up* information while our profile information stored on S_1 has not been updated to the latest status. Here, we use "Shark for Root" [14], a network traffic capturer to intercept the packets and extract the photo size information from pcap file.

If a match happens, we get another two distances between the target and our faked coordinates to calculate the target's location. We use an app called "Fake GPS location" [22] to fake the server's geolocation to different places. To make WeChat work more stable when this app fakes GPS, "Allow mock locations" option has to be turned off. This is achieved through setting the access right of the directory /system as rw and then moving "com.lexa.fakegps-1.apk" from /data/app to /system/app. We also enable the "GPS satellites" option in Location services to let apps use GPS to pinpoint the location. Meanwhile, the "Google's location service" option was disabled to limit apps using data from Wi-Fi and mobile networks determining the approximate location.

We first set the default location of "Fake GPS location" to a spot within our university. For densely populated places, we added several more anchor points. Since People Nearby only displays a limited number of users (about 100), using more anchor points and merging the list of discovered people will increase the chance of hitting the target user. Next, we use a script to automatically refresh People Nearby. For each point, to load all the people's information that appeared in People Nearby, the script clicks on the links to people's profile one-by-one through triggering event KEYCODE_DPAD_DOWN. This process has to request data from Tencent. To avoid raising alarm from Tencent, the script sleeps a while before changing to a new anchor point. If someone we have known appeared in the list of People Nearby, another WeChat will be launched to get new two distances by mocking its GPS to another two locations near the place. With the help of a program that implements the three point positioning algorithm, we can easily calculate her location.

5.5.2.4 Attack Evaluation

We evaluate our attack on 20 participants. Each participant has installed WeChat with People Nearby turned on (so their profiles will be open to view). Our attack successfully revealed the live locations for most of the participants and are verified by them. To point out, some of the inferred locations are not exact where the user stays, but they are all within the acceptable range.

5.5.3 Inferring Your Friends' Identities

Besides inferring the identity of the phone owner, the identities of her friends are also attractive to the adversary, which could lead to more accurate profiling of the phone owner or attacks against her friends. However, achieving this goal is not straightforward for the attacker. Let us take a look at WhatsApp; only the profile photos and phone numbers of the user's friends are accessible and it is difficult for the adversary to directly learn who they are. On the other hand, studies have shown that Internet users tend to use default privacy settings of social network and expose their information (including full name) to visitors, even strangers [6]. Moreover, people tend to use the same photo for different social networks [10]. These two observations inspire us to match the profile photos between WhatsApp and social networks to infer the identities. Figure 5.7 illustrates this attack and we elaborate the following steps.

Getting photos from WhatsApp: After an adversary obtains the photos in the victim's phone, she could opt to transmit all of them to the server for later analysis. However, this will incur a lot of network overhead and is less stealthy. Instead, we leverage a special feature of the WhatsApp server on adding friends to reduce the volume of data required for transmission. Specifically, a registered WhatsApp user (say U_A) can get the profile photo of any other user (say U_B) by adding phone number of U_B into U_A contact list, and there is no need for permission from U_B. Therefore, our malicious app SAPD collects the phone numbers and compiles them into a standard contact file (with file extension .vcf) and sends to our server. After that, the attacker loads the file into her smartphone's contact list and then launches WhatsApp to download the profile photos. To notice, downloading a profile photo is only triggered when the account owner clicks the user' profile photo. Therefore, we write a monkeyrunner script to automate this process by simulating the clicks. Compared to directly sending photos, this approach can reduce about 3 MB network traffic (taking 50 photos as an example) and hence is very helpful to make SAPD more stealthy.

Getting photos from social networks: Since an Internet user is likely to use default privacy settings on social networks, her user name, full name (if required), friends, etc., are usually open to public. In particular, her friends' photos are usually grouped together and stored into her public gallery (e.g., user X's friends' photos are stored under https://www.facebook.com/X/friends). We validate this setting in popular social network sites, including Facebook, LinkedIn, RenRen, and Weibo, and we believe this setting should apply in most cases. In order to grab these photos, the user's corresponding

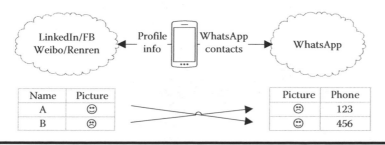

Figure 5.7 The diagram of attack user's WhatsApp friends.

social network user ID is required and is usually not directly available for the adversary. However, a large number of users also installed the mobile clients of these social networks and our study already shows they tend to leave user ID in shared storage within the reach of attackers (LinkedIn, RenRen, and Weibo). In case the user ID is not directly revealed, we found user's email can be used in replace (Facebook). As shown in Section 5.4.1, emails can be grabbed through scanning shared storage. After retrieving user's user ID or email, we use a crawler built by ourselves to query the websites and download all the friends' photo and profile information (e.g., full name and address) into the attacker's server.

Comparing photos: After we have the photos from WhatsApp and social networks, the next step naturally is to find matches. There are already a bunch of image matching techniques proposed and some advanced ones are able to recognize faces and match the photos of one person in different contexts. Using the advanced ones would improve the chance of successful matching but it is also time consuming. In fact, a simple algorithm, like sampling regions and calculating similarity, is enough for our scenario as users prefer to use the same profile photo in different social networks. Therefore, we adopt a simple matching algorithm described in [32].

Attack evaluation: We evaluate this attack by running our developed app on the participant's phone and also collect her friends' profiles using the server we set up. The app is able to collect 50 profiles containing phone numbers and photos from her WhatsApp contact. By searching the photos, we collect 29 social network profiles (12 from LinkedIn, 19 from Facebook, 11 from Renren, and 5 from Weibo, some are the same people), and they are all correctly matched as verified by the participant. This result indicates the phone user's friends' identities could be inferred with high chance through our attack.

5.5.4 Inferring Your Chatting Buddy

Instant messaging apps usually keep chatting logs under shared storage. In some cases, the logs are stored in plain text (e.g., Easy Chat) and the list of the user's chatting buddies will be directly exposed to the adversary. Encrypting the logs seems to fix the problem, but our attack on KakaoTalk shows the problem is still not solved.

In fact, KakaoTalk creates a tag when conversation happens and then dumps the attachments (e.g., photos and audio files) under a folder named using the tag. As shown in Figure 5.8, the tag between smartphone A and smartphone B is 87204252525678 (a 14-digit number). Though it is difficult to recover user IDs from the tag, it is still feasible to infer the chatting buddies.

Assuming the malicious app SAPD is installed on a number of phones, it will keep collecting the tags and transmitting them to the attacker's server. When tags from two different phones match, it is clear that the owners of the phones have chatted some time. The chance of matching is relatively small considering the large user base of KakaoTalk; however, attacker can leverage social engineering

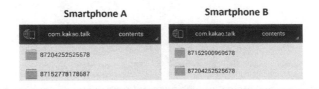

Figure 5.8 The contents folders of KakaoTalk on two smartphones.

tricks (e.g., sending spams) to propagate SAPD from the infected user only to her friends to increase the success rate. After the entities of conversation are inferred, the attacker could further recover the time point for each chat by leveraging the creation time of attachment files.

5.6 Defense

Our survey of popular Android apps shows leaks of sensitive information from app-private data stored on public storage do exist. Given the number of all apps in the wild, the issues we found are very likely to be the tip of the iceberg. Against this problem, several guidances and solutions are proposed but all of them require huge efforts from both app developers and users. Instead, we propose a new framework that extends the existing Android system and truly isolates app-private data from other apps.

5.6.1 *Existing Solutions*

Since there is no finer-grained protection for public storage, the data need to be carefully examined before saving to it. Presently, there is no tool as far as we know that is capable of automating this checking process, and the developers have to manually scrutinize their code and keep sensitive data protected. The CERT Oracle Coding Standard for Java [24] states two coding guidances regarding storing sensitive information:

1. *Encrypting data*: Before an application saves sensitive information to public storage, it needs to be encrypted. Following this rule would mitigate most data leakage problems (information may still be leaked through side-channels, like file size), but unfortunately this rule is not correctly enforced in practice. Though Android provides cryptographic API, many developers make critical mistakes when using them, including using ECB mode and nonrandom IV for CBC encryption [11], and sensitive information from the encrypted files could be partially or even fully revealed to attackers. Ironically, even the developers of most popular apps have made some critical mistakes that would directly reveal the encryption keys [17]. What is worse, our research shows even if developers had used encryption routines correctly, they do not always identify the sensitive objects accurately. As shown in Section 5.4.1, user's account number and phone number are used as filenames, which can be abused to infer a wide spectrum of users' sensitive information.

2. *Saving to internal storage*: Android uses Ext2/4 for internal storage and an app can only access its own data stored on internal storage. Therefore, it is encouraged to keep sensitive data there. Specifically, data can be saved to a file created in the application data directory with permission set to `MODE_PRIVATE`. However, developers would like to only identify sensitive data and throw all the other data to public storage, which will be elaborated upon in Section 5.7.

Expecting all the developers to obey security rules or fix security issues is impractical. On the other hand, users could protect themselves: by converting the format of public storage from FAT to Ext2/4, the built-in access control mechanism of Ext2/4 will be inherited, and storage of different apps will be well separated. Yet, this migration might be only suitable for advanced users: the Android system has to be rooted and additional apps have to be installed to support Ext2/4 format public storage; Windows users have to install additional software to browse and read files on it; the existing files have to be backed up and restored when formatting. None of these steps are easy for common users.

5.6.2 Mitigations at Android System Side

The problem will not be fixed in the near future if following these paths described earlier. On the contrary, modifying the Android system and pushing the upgrades to users' devices would be an easier way to mitigate the security issues. For this purpose, we propose to augment the existing security framework on public storage by instrumenting the API checkPermission() [37]; the framework is described as follows:

Architecture: We design a new module named *ownership checker* as shown in Figure 5.9, which works on the Android middleware layer and can achieve mandatory access control for app-private data. Specifically, when the targets are public resources, like shared music directory, the access is permitted (certainly, the app needs to hold necessary permissions), while the targets are located in app's private folder, the access is only permitted when the calling app matches the owner, if not, the *ownership checker* will return PERMISSION_DENIED even if the app has requested the READ or WRITE permission. To enforce such a rule, we create a system file owner_checker.xml storing the mapping between apps and resources, similar to access control lists of Ext4 file system. And the Android system code, which manages the checkPermission(), is modified to read the mapping before actual file operations. An exception will be thrown to the caller if a mismatch happens.

Ownership inference: The ownership mapping between apps and resources needs to be established before actual access happens, which, however, is not a trivial task. We have to deal with the situation that the public storage has already kept apps' data before our module is installed, while the owner of the data is not tracked. To fix the missing links, we exploit a caveat regarding naming convention: an app usually saves data to a folder whose name is similar to its package name. For example, Viber saves data to ~/viber/ and its package name is com.viber.voip (app's UID and package name can be acquired from packages.xml under /data/system). Therefore, we initialize the mappings by scanning all the resources. For a given resource, we assign the owner app if the resource location and app package name share a nontrivial portion. To notice, this initialization step could not infer all the ownerships when an app stores the data in a folder whose name is irrelevant.

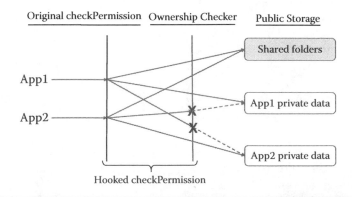

Figure 5.9   Proposed defense framework, adding ownership checker to restrict apps accessing other apps' private data.

While the access will be blocked for this case, we provide an interface for users to add the ownerships. When a failed access happens or the resource has no ownership associated, a dialog is prompted to user to ask if the access is permitted. If the permission is granted, a new mapping will be added. To reduce the hassles to users, the user-driven access control model [30] can be integrated to automatically assign ownership based on the user's implicit actions.

5.6.3 Mitigation at Android Application Side

Defense at the Android system layer requires modification of current infrastructure, which takes time and may have other side effects such as compatibility problems, so here we also proposed a mitigation solution from the app side by performing automatic code analysis and patching to redirect all vulnerable file access operations from shared external storage to secure private folders.

5.6.3.1 Overview

The overall structure and working flow of our proposed method is illustrated in Figure 5.10. There are three key components: vulnerable operation detection, problem origition locating, and code patching. The central part to detect vulnerable operations is a dynamic code analysis engine that is driven by Monkey, a testing automation tool for Android. It monitors every system call related to file operations and checks whether it is vulnerable. If true, the system will try to trace from that system call back to problem origins. Finally, the system patches the code by redirecting the target file from shared external storage to internal private folders.

5.6.3.2 Detect Vulnerable Operations

Although static code analysis method has been used in Section 5.4.2 to perform a large-scale survey of writing nonshared data to external storage, this method is not accurate enough and could have false negatives. So we developed a dynamic analysis method based on a key observation that Android fundamentally is built on a Linux kernel; thus, all file operations are achieved via Linux system calls. By monitoring those system calls and checking their parameters, it is easy to extract file name and path and pick out vulnerable operations.

Our dynamic analysis tool is built on Strace, ADB, and Monkey. Strace is a powerful tool on Linux systems that can be used to get system call information. We downloaded and modified its source code and ported it to the Android platform. Since the goal is to check vulnerable file operations on shared storage, the modified Strace program only monitors file-related APIs, like openat. In fact, all files need to be opened before any reading and writing operations, and the file path is

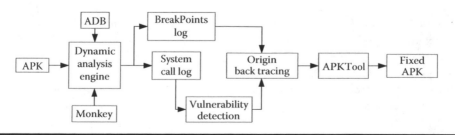

Figure 5.10 Proposed defense solution at application side.

passed as parameters to system calls, so by checking such parameters, our tool can decide whether the current operation is vulnerable or not. ADB is a standard Android debugging tool that is used to run, control, and interact with apps, while Monkey is the testing automation tool for the Android platform that can simulate user behaviors. When combined together, these three tools then can execute a given app, feed it with various inputs, and check and record the parameters.

5.6.3.3 Trace Back to Problem Origins

Once a vulnerable operation is identified, it is necessary to locate the exact statement that triggers this operation before diving to fix the problem. It seems straightforward to get the problem origins in an app from identified vulnerable system calls by looking at the stack since records of all function caller–callee relationships are stored there. Unfortunately, it is a complex job due to limitations of the current version of Strace. Strace indeed has a feature to print stack trace starting from version 4.9; however, such functionality is not well-supported on ARM (and thus most Android) platforms. As a result, we could not get a complete stack trace that can let us trace back to the original function in a given app.

To bypass this limitation, we proposed a hybrid approach using (ADB) breakpoints and Strace. First, ADB is used to set breakpoints at each method, and whenever it is triggered, will print a log with method name and a time stamp. Next, Strace is used to monitor system calls, which will print a time stamp for any vulnerable writing operations to a log file. Finally, log files from ADB and Strace are compared and aligned, so any vulnerable operation can be traced back to a specific method. The earlier process can be repeated within that method if necessary until the exact Dalvik virtual machine instruction that triggers the vulnerable system is identified.

5.6.3.4 Patch the Code

Once the origin of this problem is identified, the last step is to patch the code automatically. Our approach uses APKTool [19] to disassemble APK files into smali code, then find the origin instruction and modify its parameters. Later, the modified smali code is converted back to APK format again.

5.7 Discussion

Evolvement of Android storage model: In the early stage, Android phones were equipped with limited onboard flash memory (only 256 MB for the first Android phone, HTC Dream). In the meantime, its storage can be expanded through large volume SD card (could be several GB) that is usually shipped together. This storage model forces Android app developers to save most of the data to a FAT formatted SD card, while the sensitive information, like passwords, is saved to more secured Ext2/4 formatted flash memory. The size of on-board flash memory is gradually enlarged; however, the app developers still prefer to use the SD card, even forcing Android to simulate FAT format on a compartment of flash memory. While keeping compatibility for existing apps reduces the workload for developers, it brings severe security and privacy risks, as shown in our attacks. Notably, there have been continuous efforts on improving the security of storage model by Google, including introducing READ_EXTERNAL_STORAGE permission in Android 4.1 and WRITE_MEDIA_STORAGE in Android 4.4. Unfortunately, these enhancements do not solve the fundamental issues with the storage model. We proposed a new framework that enforces finer-grained access control mechanisms on the secured part of storage system (SD card and part of flash memory simulating SD card). This approach does not require changes from existing apps or intensive efforts from users and could be easily applied.

Limitations of app study: We launched a study on existing apps to understand the scale of the problem (see Section 5.4.2). We built a tool running static analysis on the app's code and used a set of heuristics to determine if the app saves sensitive information to an unprotected shared storage. This simple tool identifies a large number of potentially vulnerable apps and shows reasonable accuracy from our sampling result. However, it inevitably suffers from false negatives (e.g., the file name does not contain the keywords we used) and false positives (e.g., the information saved is not sensitive). We leave the task of building a more accurate detector as future work.

5.8 Related Work

5.8.1 Harvesting Users' Sensitive Data from Mobile Devices

Since mobile devices now hold a lot of people's private information, they have become popular targets for attackers. Attacks like stealing users' chat history [17] have proved their feasibility in the real world. These attacks usually depend on certain vulnerabilities identified from the victim's apps. Instead, our attacks exploit a general design flaw of the Android system—the lack of finer-grained control on shared storage, and we demonstrated the effectiveness of attacks on a number of popular apps from different categories.

In addition to stealing user's sensitive information directly, a lot of research focused on inferring user's private information, especially the location. Srivatsa and Hicks [36] showed that a set of users' location traces can be deanonymized through correlating their contacts with their social network friends' list. Using users' public comments, Zheng et al. showed that it is feasible to infer users' locations and activities [43]. Also, a recent work by Zhou et al. [44] targets to infer information of users from more perspectives, including identities, locations, and health information. While their works exploit side channels of the Android system, our work takes advantage of the data on shared storage and corresponding attacks are easier to carry out.

5.8.2 Inferring Users' Information from Social Network

Social networks allow users to share information with their friends. It is another hub of user-sensitive information and its privacy issues have been well studied in several previous papers [4,16,27,41]. Mislove et al. [27] found that users with common attributes are more likely to be friends and often form dense communities, and they proposed a method of inferring user attributes that are inspired by previous approaches to detecting communities in social networks. Goga et al. [16] proposed an attack to link the user's identity to her accounts on different social networks. This attack exploits only innocuous activity that inherently comes with posted content. Also, user's group memberships [41] and the "friend finder" function [4] also could leak user's identity (deanonymization attack). They studied how to use network characteristics to infer the social situations of users in an online society.

These works focused on social networks themselves, that is to say, all analyzed data come from one or several social networks. However, our attack combines the sensitive data from users' phones (shared storage) and public data from social networks, which is more powerful.

5.8.3 Protecting Users' Privacy on Mobile Devices

To defend against existing or potential attacks tampering with user's privacy, a bunch of defense mechanisms have been proposed. Roesner et al. [30] proposed user-driven access control to manage the access to private resources while minimizing user's actions. Ongtang et al. [29] proposed a finer-grained access control model (named Saint) over installed apps. It addressed some limitations of Android security through install-time permission-granting policies and run-time interapplication communication policies. Mobiflage [35], developed by Skillen and Mannan, was leveraged to enable plausibly deniable encryption on mobile devices by hiding encrypted volumes within random data on a device external storage. Users are needed to comply with certain requirements, but Mobiflage also has high risks to be weakened even if users follow all the guidelines that make it not practical. To protect users' location, several methods were proposed, like [33] that restricts the information disclosed from locations.

Besides, efforts have also been focused on code analysis to block the information leakage. Enck et al. developed TaintDroid [12] to prevent users' private data from being abused by third-party apps. DroidScope [42] is another dynamic Android malware analysis platform. Based on the idea of virtualization, it reconstructed both the OS-level and Java-level semantics simultaneously and seamlessly.

5.9 Conclusion

It is known that the shared public storage on Android is insecure due to its coarse-grained access model. Therefore, it is highly recommended that saving sensitive data there should be avoided. In this chapter, we carry out a large-scale study on existing apps to determine whether app developers follow this rule and the result turns out to be gloomy: a significant number of apps save sensitive data into the insecure storage, some of the apps are even top ranked in the Android market. By exploiting these leaked data, it is possible to infer a lot of information about the users, severely violating users' privacy. The security of users' data under the current model heavily relies upon app developers, but requiring all app developers to identify the sensitive objects and correctly handle them is impractical. Instead, we propose a new defense framework by extending the existing Android systems that achieve both compatibility and finer-grained access control. To summarize, our study reveals unneglectable security issues in shared storage and further actions need to be taken to fix the problems.

References

1. B. Acohido. 'Like' it or not, sharing tools spur privacy concerns. http://usatoday30.usatoday.com/tech/news/2011-07-05-social-media-privacy-concerns_n.htm.
2. AppBrain. Top Android apps and games on Google Play—Appbrain.com. http://www.appbrain.com/stats/number-of-android-apps.
3. K. W. Y. Au, Y. F. Zhou, Z. Huang, and D. Lie. Pscout: Analyzing the Android permission specification. In *Proceedings of the 2012 ACM Conference on Computer and Communications Security*, pp. 217–228. ACM, 2012.
4. M. Balduzzi, C. Platzer, T. Holz, E. Kirda, D. Balzarotti, and C. Kruegel. Abusing social networks for automated user profiling. In *Recent Advances in Intrusion Detection* (*RAID*), pp. 422–441. Springer, 2010.

5. S. Bugiel, L. Davi, A. Dmitrienko, T. Fischer, A. R. Sadeghi, and B. Shastry. Towards taming privilege-escalation attacks on Android. In *19th Annual Network & Distributed System Security Symposium (NDSS)*, vol. 17, pp. 18–25, 2012.

6. Enid Burns. Social media privacy a concern for only a quarter of Internet users. http://www.redorbit.com/news/technology/1113003885/social-media-internet-security-facebook-privacy-settings-111413/.

7. Stephanie Ellen Chan. The 10 most popular mobile messaging apps in the world. http://readwrite.com/2014/03/06/10-biggest-popular-mobile-messaging-apps-world-whatsapp.

8. T. Chen, M. A. Kaafar, and R. Boreli. The where and when of finding new friends: Analysis of a location-based social discovery network. In *Seventh International AAAI Conference on Weblogs and Social Media (AAAI)*, 2013.

9. China Daily. China telecom profit increases 17 percent in 2013. http://www.tmcnet.com/usubmit/2014/03/19/7733866.htm.

10. J. Duffy. Get organized: Update your profile picture. http://www.pcmag.com/article2/0,2817,2398859,00.asp.

11. M. Egele, D. Brumley, Y. Fratantonio, and C. Kruegel. An empirical study of cryptographic misuse in Android applications. In *Proceedings of the 2013 ACM SIGSAC Conference on Computer & Communications Security*. ACM, 2013.

12. W. Enck, P. Gilbert, B. G. Chun, L. P Cox, J. Jung, P. McDaniel, and A. Sheth. Taintdroid: An information-flow tracking system for realtime privacy monitoring on smartphones. In *OSDI*, vol. 10, pp. 1–6, 2010.

13. S. J. Fink et al. TJ Watson libraries for analysis (wala).

14. Wireshark Foundation. Shark for root. https://play.google.com/store/apps/details?id=lv.n3o.shark.

15. C. Gibler, J. Crussell, J. Erickson, and H. Chen. Androidleaks: Automatically detecting potential privacy leaks in Android applications on a large scale. In *Proceedings of the Fifth International Conference on Trust and Trustworthy Computing*, TRUST'12, pp. 291–307, 2012. Springer-Verlag.

16. O. Goga, H. Lei, S. H. Krishnan Parthasarathi, G. Friedland, R. Sommer, and R. Teixeira. Exploiting innocuous activity for correlating users across sites. In *Proceedings of the 22nd International Conference on World Wide Web (WWW)*, pp. 447–458. International World Wide Web Conferences Steering Committee, 2013.

17. A. Hern. WhatsApp user chats on Android liable to theft due to file system flaw. http://www.theguardian.com/technology/2014/mar/12/whatsapp-android-users-chats-theft.

18. J. Horwitz. Flirty Chinese chat app Momo passes 100 million registered users. http://www.techinasia.com/flirty-chinese-chat-app-momo-passes-100-million-users/.

19. iBotPeaches. Apktool—A tool for reverse engineering Android APK files. https://ibotpeaches.github.io/Apktool/, 2015.

20. M. Isaac. LinkedIn tops 300 million users worldwide. http://recode.net/2014/04/18/linkedin-tops-300-million-users-worldwide/.

21. S. Jana and V. Shmatikov. Memento: Learning secrets from process footprints. In *2012 IEEE Symposium on Security and Privacy (SP)*, pp. 143–157. IEEE, 2012.

22. Lexa. Fake GPS location. https://play.google.com/store/apps/details?id=com.lexa.fakegps.

23. M. Li, H. Zhu, Z. Gao, S. Chen, L. Yu, S. Hu, and K. Ren. All your location are belong to us: Breaking mobile social networks for automated user location tracking. In *ACM International Symposium on Mobile Ad Hoc Networking and Computing (MobiHoc)*, 2014.

24. F. Long, D. Mohindra, R. C. Seacord, D. F. Sutherland, and D. Svoboda. *The CERT Oracle Secure Coding Standard for Java*. Addison-Wesley Professional, 2011.

25. We Are Social Ltd. We are social. http://wearesocial.net/blog/2014/02/facebook-buys-whatsapp-wechat-closes/.

26. E. McCallister. *Guide to Protecting the Confidentiality of Personally Identifiable Information*. Diane Publishing, 2010.

27. A. Mislove, B. Viswanath, K. P. Gummadi, and P. Druschel. You are who you know: Inferring user profiles in online social networks. In *Proceedings of the Third ACM International Conference on Web Search and Data Mining*, pp. 251–260. ACM, 2010.

28. A. Narayanan and V. Shmatikov. De-anonymizing social networks. In *2009 30th IEEE Symposium on Security and Privacy*, pp. 173–187. IEEE, 2009.

29. M. Ongtang, S. McLaughlin, W. Enck, and P. McDaniel. Semantically rich application-centric security in Android. *Security and Communication Networks*, 5(6):658–673, 2012.

30. F. Roesner, T. Kohno, A. Moshchuk, B. Parno, H. J. Wang, and C. Cowan. User-driven access control: Rethinking permission granting in modern operating systems. In *2012 IEEE Symposium on Security and Privacy (SP)*, pp. 224–238. IEEE, 2012.

31. Sabrina. Renren active users up to 194 million in q2 2013. http://www.chinainternetwatch.com/3335/renren-active-users-up-to-194-million-in-q2-2013/.

32. R. Santos. How do I compare two images to see if they are equal? http://www.lac.inpe.br/JIPCookbook/6050-howto-compareimages.jsp.

33. R. Shokri, G. Theodorakopoulos, C. Troncoso, J. P. Hubaux, and J. Y. Le Boudec. Protecting location privacy: Optimal strategy against localization attacks. In *Proceedings of the 2012 ACM Conference on Computer and Communications Security*, CCS'12, pp. 617–627. 2012.

34. Sida. Alipay is the largest mobile payments platform in the world. http://www.chinainternetwatch.com/6183/alipay-the-largest-mobile-payments-plat/form-in-the-world/#ixzz2yqgsnbkd.

35. A. Skillen and M. Mannan. On implementing deniable storage encryption for mobile devices. In *20th Annual Network & Distributed System Security Symposium*, 2013.

36. M. Srivatsa and M. Hicks. Deanonymizing mobility traces: Using social network as a side-channel. In *Proceedings of the 2012 ACM Conference on Computer and Communications Security*, pp. 628–637. ACM, 2012.

37. Android Team. checkpermission. http://developer.android.com/reference/android/content/pm/PackageManager.html#checkPermission(java.lang.String, java.lang.String).

38. Android Team. Using the external storage. http://developer.android.com/guide/topics/data/data-storage.html#filesExternal.

39. R. Vallée-Rai, P. Co, E. Gagnon, L. Hendren, P. Lam, and V. Sundaresan. Soot-a Java bytecode optimization framework. In *Proceedings of the 1999 Conference of the Centre for Advanced Studies on Collaborative Research*, page 13. IBM Press, 1999.

40. W. Wee. Sina Weibo passes 500 million users. http://www.techinasia.com/sina-weibo-500-million-users-but-not-monetizing-mobile/.

41. G. Wondracek, T. Holz, E. Kirda, and C. Kruegel. A practical attack to de-anonymize social network users. In *2010 IEEE Symposium on Security and Privacy (SP)*, pp. 223–238. IEEE, 2010.

42. L. K. Yan and H. Yin. Droidscope: Seamlessly reconstructing the OS and Dalvik semantic views for dynamic Android malware analysis. In *USENIX Security Symposium*, pp. 569–584, 2012.

43. V. W. Zheng, Y. Zheng, X. Xie, and Q. Yang. Collaborative location and activity recommendations with GPS history data. In *Proceedings of the 19th International Conference on World Wide Web (WWW)*, pp. 1029–1038. ACM, 2010.

44. X. Zhou, S. Demetriou, D. He, M. Naveed, X. Pan, X. Wang, C. A. Gunter, and K. Nahrstedt. Identity, location, disease and more: Inferring your secrets from Android public resources. In *Proceedings of the 20th ACM Conference on Computer and Communications Security*. ACM, 2013.

MOBILE OPERATING SYSTEM VULNERABILITIES

Chapter 6

Security Analysis on Android and iOS

Jin Han

Contents

Abstract

Android and iOS, as the top two mobile operating systems in terms of user base, both have their own security weaknesses. In this chapter, by analyzing the security architectures of Android and iOS, distinct security challenges on both platforms will be revealed and discussed. Furthermore, a systematic comparison framework is presented that is used to compare the security and privacy protection on Android and iOS. With detailed proof-of-concept attacking code and malicious application examples, it is shown that security mechanisms utilized on Android and iOS are still far from perfect and should be further improved to mitigate existing vulnerabilities.

6.1 General Introduction

The security mechanisms on modern mobile platforms are very different from those on the traditional desktop environment—the security goal of mobile operating systems is to make the platforms inherently secure rather than to force users to rely upon third-party security solutions. Among all the modern mobile operating systems, Google's Android and Apple's iOS are the top two mobile operating systems in terms of user base [1]. Other mobile platforms, such as Windows Phone and BlackBerry, only added up to less than 4% of the total market by 2015 and, therefore, have much less impact to the overall mobile users' safety, compared to the top two players. In this chapter, security analysis is conducted for both Android and iOS.

Google's Android mobile operating system has many unique features—it is an open-source architecture; it has unique application framework and Dalvik virtual machine (VM); it has a different security model and also different distribution model compared to iOS. All these features lead to security challenges specific for Android, which are discussed and analyzed in Section 6.2. Among all the Android security challenges to be revealed in Section 6.2, some of them can be mitigated and have been actively discussed and investigated by researchers, while others cannot be easily solved with the current design of the Android platform.

iOS is Apple's mobile operating system, which is used on Apple's mobile devices, such as iPhone, iPad, and iPod touch. Different from Android, iOS follows a closed-source model, where source code of the underlying architecture and implementation details of its security mechanisms are not available to the public. Though it is debatable whether such obscurity provides better security, iOS has been generally believed to be one of the most secure commodity operating systems. In Section 6.3, a generic attack vector is provided that can be utilized to launch serious attacks on iOS devices that are not jailbroken. Malicious applications created via such an attack vector are able to bypass the application examine process from Apple and then execute attacking code within restrictions from iOS sandbox. A detailed analysis of these attacks and their mitigation mechanisms will both be presented in Section 6.3.

It is important to understand the security challenges on both Android and iOS. But it is even better if there is a way of comparing these two platforms and conclude which platform provides a better architecture for security and privacy protection of mobile users. In Section 6.4, the first security comparison framework for Android and iOS is introduced, where more detailed security vulnerabilities in mobile applications are found and discussed. Finally, implications of all the analysis results given in this chapter are discussed and summarized in Section 6.5.

6.2 Security Challenges on Android

Android is unique—it is currently the biggest mobile platform in terms of user base; it is an open-source architecture and it has a very special distribution model. In order to understand the security challenges on Android, one has to be familiar with its architecture and security features. In this section, a brief introduction to Android architecture is given, then a list of security challenges that Android is facing will be revealed and discussed. This list of challenges provided in this section is not (and are probably not possible to be) comprehensive due to the fast-evolving nature of mobile platforms.

6.2.1 Android Architecture Overview

In general, Android operating system consists of five different software components that form four software layers. The five software components (from bottom to top) include Linux kernel, libraries, Android runtime, application framework, and applications, as illustrated in Figure 6.1.

6.2.1.1 Linux Kernel

This is the bottom layer of Android operating system. Android is built from Linux kernel on top of which Google then made certain architectural changes. This layer provides basic system functionalities such as memory management and disk management. This layer also provides a level of abstraction on top of the device hardware, and it contains all the essential hardware drivers like camera, keypad, and display.

6.2.1.2 Libraries

On top of Linux kernel, there is a set of libraries that perform common tasks. These libraries include the open-source web browser engine WebKit, well-known C library libc, SQLite database that is used for storage and sharing of application data, and SSL libraries used for secured Internet communications. More importantly, this layer also includes a set of Java-based libraries that are specific to Android development. These Android libraries contain the following core libraries (and a detailed list can be found from the Android Developer site [3]):

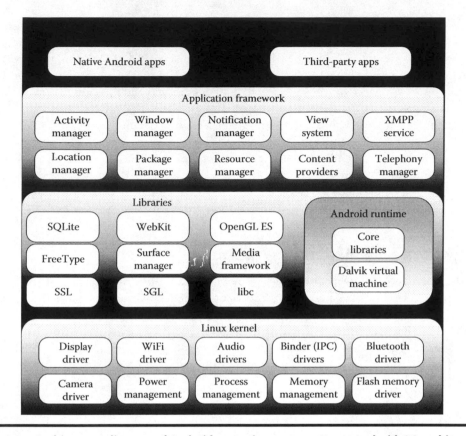

Figure 6.1 Architecture diagram of Android operating system. (From Android OS architecture, http://www.c4learn.com/android/android-os-architecture/.)

- *android.app*—Contains high-level classes encapsulating the overall Android application model
- *android.content*—Contains classes for accessing and publishing data on a device
- *android.database*—Contains classes to explore data returned through a content provider
- *android.opengl*—Provides an OpenGL ES static interface and utilities
- *android.os*—Provides basic OS services, message passing, and interprocess communication on the device
- *android.text*—Provides classes used to render or track text and text spans on the screen
- *android.view*—The fundamental building blocks of application user interfaces
- *android.widget*—Contains (mostly visual) UI elements to use in application screen
- *android.webkit*—Provides a set of classes to allow apps to have web-browsing capabilities

6.2.1.3 Android Runtime

The third component of the architecture lies also on the second layer, which is on top of the Linux kernel. Before Android 5.0, Android Runtime provides a key component called Dalvik VM that is a variation of Java VM specially designed and optimized for Android. The Dalvik VM enables every Android application to run in its own process, with its own instance of the Dalvik VM.

The Android Runtime also provides a set of core libraries that enable Android application developers to write Android applications using standard Java programming language. Starting from Android 4.4, "the Dalvik runtime is no longer maintained or available and its bytecode format is now used by Android RunTime (ART)" [4]. ART is the new version of Android Runtime, and starting from Android 5.0, ART is the only included runtime that has completely replaced Dalvik runtime.

6.2.1.4 Application Framework

The Application Framework layer provides many higher-level services to applications in the form of Java classes, which mainly include the following key services:

- *Activity manager*—Controls all aspects of the application lifecycle and activities
- *Content providers*—Allows applications to publish and share data with other applications
- *Resource manager*—Provides access to embedded resources such as strings, color settings, and UI layouts
- *Notifications manager*—Allows applications to display notifications to the user
- *View system*—An extensible set of views used to create application user interfaces

6.2.1.5 Applications

All the Android applications are at the top layer. These applications include both native applications that are provided by Google or OEMs, and also third-party applications that can be downloaded and installed from Google Play. Examples of such applications include browser apps such as Chrome, social media apps such as Facebook and Twitter, game apps such as Angry Birds, etc.

Given its unique architecture, Android is facing various security challenges that are specific for this platform. The first one is its long security fix cycle, which is the most naive security vulnerability on Android, but could be the most difficult one to mitigate due to Android's special distribution model.

6.2.2 Long Security Fix Cycle

In an effort to increase adoption of the Android OS, Google created the Open Handset Alliance to build cooperation between hardware manufacturers to facilitate implementation on their devices [5]. Through this cooperation, Google provides the base open-source operating system, then device manufacturers and telecommunications carriers modify this base to differentiate their offerings from other Android devices and provide added value to their customers [6].

Figure 6.2 shows the patch cycle from the initial discovery of a vulnerability until the patch eventually reaches the user's Android device. Patch cycle events A–C are typical of traditional software product (which are also close to iOS's patch cycle). However, for Android, D–G are needed and will be appended to the patch cycle. The additional states added to the cycle come from Google's cooperation with multiple manufacturers and carriers. Whenever a patch to Android becomes necessary, Google provides an update through their open-source forum and manufacturers then proceed to port the update to their customized version of the operating system.

As explained in Vidas et al. [6], it is not uncommon to observe at least 2 months (and sometimes much more) of delay between an Android update and an actual deployment of the update by the

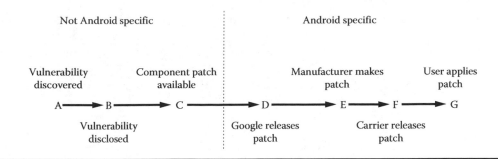

Figure 6.2   Lifecycle of an Android patch. (From Vidas, T. et al., All your droid are belong to us: A survey of current android attacks, in: *Proceedings of the Fifth USENIX Conference on Offensive Technologies,* **USENIX Association, Berkeley, CA, 2011.)**

major manufacturers. The Android's slow security patch cycles can cause increased vulnerability. Once a vulnerability in Webkit or Linux is discovered, it is generally patched and released quickly by the open-source community, but the corresponding Android patch may not be available to users for months. The big gap in the Android patch cycle leaves a long attack window for Android devices. In addition, due to Android's Linux foundation, lower-level attacks are simpler when compared to less ubiquitous OSs, as attackers do not need to learn a new kernel (or newer version of an existing kernel).

In order to mitigate this "long security fix cycle" vulnerability, Android must provide timely updates and collaborate with hardware manufacturers to minimize the attack window. However, it is inevitable that this vulnerability will continue to exist unless big changes have been made with the current distribution model of the Android operating system.

6.2.3 Problematic Permission Model

On Android, an application needs to explicitly declare the permissions it requires. A user will be presented with these permissions when installing an application so that the user can choose not to proceed if he or she is unwilling to grant the corresponding permissions. A complete list of Android application permissions can be found at [7].

Android permission model has several vulnerabilities from different perspectives. First of all, a user can only choose not to install an application, but is not given the flexibility of still installing the app while denying part of the permissions requested. To solve this security problem, a number of platform-level extensions have been proposed including Apex [8], MockDroid [9], TISSA [10], AppFence [11], and Aurasium [12] that extend the current Android permission framework to provide fine-grained controls of system resources accessed by untrusted third-party applications. Shekhar et al. [13] and Pearce et al. [14] further provide specific techniques to separate permission usage of advertising libraries and hosting applications to prevent permissions from being abused by advertisers.

From another perspective, common users may not understand what are the application permissions presented during installation. The user studies conducted by Felt et al. [15] show that only 17% of the users will pay attention to permissions during installation, and only 3% of the users could correctly understand those permissions. This indicates that current Android permission warnings do not help most users make correct security decisions. This weakness can only be mitigated with better user education and better permission or risk explanation, but is very hard to completely eliminate.

The application developers, on the other hand, are also not clear about which permissions are necessary to be requested during app development, and thus, try to request more permissions than needed [16]. The additionally requested permissions can then be utilized by permission redelegation attacks [17] launched from other third-party apps. Felt et al. [17] demonstrated that a less privileged Android application is able to perform a privileged task through another third-party application with higher privilege. A similar privilege escalation attack that is implemented with return-oriented programming is introduced by Davi et al. [18]. The defense mechanism for such privilege escalation attack is proposed by Bugiel et al. [19].

6.2.4 Easy Reverse Engineering and Repackaging

Android apps are written in Java and then compiled into Dalvik bytecode files. However, since hardware-independent Dalvik bytecode files contain lots of information of the original Java sources [20], they are easier to be reverse engineered compared to native apps such as Windows executables or iOS apps. As a consequence, app code modification and repackaging can be easily done for Android apps, which lead to further security issues, such as malware injections [21,22]. Though both static analysis methods [23,24] and dynamic auto-detection methods [25] have been proposed, repackaging and malware injection is still one of the main threats on the Android platform. The root cause is mainly due to the openness of the Android platform, which allows third-party apps that are not from the official app store (i.e., Google Play) to be installed on the devices.

6.2.5 Easy User Tracking

Android devices provide several different options of identifiers, which can all be used to identify a unique device. These identifiers include IMEI number, MAC address, or Android system ID [26,27]. Due to the problematic permission model on Android, third-party apps or ads libraries can easily obtain sensitive data such as GPS information of a device. Combining with the device identifier, it will be very easy for an external third-party to track an Android user remotely. To mitigate user tracking problem, techniques such as AppFence [11] that feeds fake data (e.g., IMEI, phone numbers) to third-party apps would help. However, such solutions always come with a trade-off between security and usability. For example, if fake GPS data are always provided to third-party apps, apps such as Google Maps and Yelp will be impossible to be used at all. Thus, in order to have both privacy and usability, large amounts of manual efforts are still needed in terms of user customization and configuration.

6.3 Security Challenges on iOS

Different from Android, iOS is a closed-source platform, which does not suffer from most of the security challenges introduced in the last section. iOS is controlled and only distributed by Apple. Therefore, iOS security patches have much shorter release cycles compared to Android. iOS app permissions are requested during application runtime when needed, which makes it possible to grant and revoke each permission dynamically. Also, it can be easier for the users to understand as the context of the permission request is usually very obvious. iOS executables are ARM binaries, which are much harder to be reverse engineered or repacked, compared to Android dex files.

In addition, iOS prevents third-party apps from obtaining device identifiers, such as IMEI number, which mitigates the user tracking problem.

There are also two important security features that make iOS platform secure—the vetting process and the application sandbox. Any third-party applications developed for iOS devices are required to go through Apple's application vetting process and can only appear on the official iTunes App Store upon approval. When an application is downloaded from the store and installed on an iOS device, it is given a limited set of privileges, which are enforced by iOS application sandbox. Although details of the vetting process and the sandbox are kept as black box by Apple, it was generally believed that these iOS security mechanisms are effective in defending against malware [28], and thus, iOS has been generally considered as one of the most secure commodity operating systems [29].

In this section, a generic attack vector to the iOS platform will be revealed, which enables third-party applications to launch attacks on nonjailbroken iOS devices. Following this generic attack mechanism, multiple proof-of-concept attacks can be constructed, such as cracking device PIN and taking snapshots without user's awareness. All the applications embedded with the attack codes are able to pass Apple's vetting process and work as intended on nonjailbroken devices. These proof-of-concept attacks have shown that Apple's vetting process and iOS sandbox have weaknesses that can be exploited by third-party applications. In order to defend against the proposed attacks, corresponding mitigation strategies for both vetting and sandbox mechanisms will be discussed at the end of this section.

6.3.1 iOS Platform Overview

iOS is Apple's mobile operating system, which is used on Apple's mobile devices, such as iPhone, iPad, and iPod touch. iOS follows a closed-source model, where source code of the underlying architecture and implementation details of its security mechanisms are not available to the public. Though it is debatable whether such obscurity provides better security, iOS has been generally believed to be one of the most secure commodity operating systems [29]. iOS adopts various security measures [30], such as secure boot chain, application code signing, remote wipe, and address space layout randomization.

iOS is designed in a way that both the operating system and the applications running on it are highly controlled by Apple. Unlike other mobile platforms, third-party applications on iOS are given a more restricted set of privileges [31]. In addition, any third-party application developed for iOS must go through Apple's application vetting process before it is published on the official iTunes App Store. While some users and developers favor such restrictions for better security, others prefer to have more controls over the device for additional functionalities, such as allowing to install pirated software and allowing applications to change the themes of the device. To attain such extended privileges, an iOS device needs to be jailbroken. Jailbreaking is a process of installing modified kernel patches that allow a user to have root-level access of the device so that any unsigned third-party applications can run on it. Although jailbreaking is legal [32,33], it violates Apple's end user license agreement and voids the warranties of the purchased mobile devices. Jailbreaking is also known to expose to potential security attacks, such as iKee [34].

6.3.1.1 Application Vetting Process

Without jailbreaking a device, the only way of installing a third-party application on iOS is via the official iTunes App Store. Any third-party application that is submitted to iTunes Store needs to be reviewed by Apple before it is published on the store. This review process is known as Apple's application vetting process. The vetting covers several aspects, including detection of malware, detection of copyright violations, and quality inspection of submitted applications. The App Review Guidelines [35] provide the rules and examples for user interface designs, functionalities, and contents (e.g., no pornography or religious-offensive contents) that are required for the submitted applications to pass the vetting process. Although the vetting process is kept secret by Apple, it is generally regarded as highly effective as no harmful malware had been reported on iTunes App Store [28] until the recent Xcode attack [36]. Only grayware (which stealthily collects sensitive user data) had been reported on iTunes Store and was removed from the store upon reporting [28,37].

6.3.1.2 Application Sandbox

On iOS, Apple utilizes another security measure—application sandbox—to restrict privileges of third-party applications running on a device. The sandbox is implemented as a set of fine-grained access controls, enforced at the kernel level. When an application is installed, it is given a home directory where it has unfettered read/write access to the files under this directory. However, with the sandbox restrictions, files and folders of other applications cannot be accessed. In order to access user data or control system hardware (e.g., camera or vibration), applications also need to call respective application programming interfaces (APIs) provided on iOS. The APIs are hooked by the sandbox so that validations of these API invocations can be performed dynamically. The sandbox mechanism serves as the last line of security defense that limits malicious applications from accessing system services or exploiting resources of other applications.

6.3.1.3 iOS Frameworks and APIs

To facilitate development of third-party applications, a collection of frameworks are provided in Cocoa Touch [38], which is an abstract layer of iOS operating system. These frameworks are mainly implemented in Objective-C language and consist of both public frameworks and private frameworks. Public frameworks are application libraries officially provided by Apple for third-party developers while private frameworks are intended only for Apple's internal developers. Each framework provides a set of APIs with which third-party applications can access required system resources and services. Similar to frameworks, APIs can also be categorized into public APIs and private APIs.

Public APIs allow third-party applications to access user information and control hardware of iOS devices, such as camera, Bluetooth, and WiFi. To prevent third-party developers from abusing public APIs, Apple adopts the permission request policy. When an application tries to access certain sensitive user data, an alert window will be shown to the user. The user can then decide to allow or deny the permission request. On iOS 5, permission requests are only used when applications attempt to access location information or send push notifications. Starting from iOS 6, permission requests have been extended to include access to users' contacts, photos, reminders, and calendar events.

Private APIs are meant to be used by Apple's internal developers. Private APIs may exist in both public and private frameworks. Though not officially documented, private APIs include various functions that could be used by a third-party application to escalate its restricted privileges. Thus, Apple explicitly forbids third-party developers from using private APIs and rejects applications once the use of private APIs is detected during the vetting process. On the other hand, private APIs can still be used by third-party applications that are designed to run on jailbroken devices.

Such applications are available through Cydia [39], which is an unofficial application market built for jailbroken iOS devices.

6.3.2 Generic iOS Attack Vector

On iOS, third-party applications are treated "equally" [31] in the sense that they are given an equal set of privileges once installed. Though this set of privileges is restricted, third-party iOS applications can still access certain user data and even transmit these data out of the device for malicious purposes [40]. However, as Apple does not provide an official list of user data that can be accessed by third-party applications, it is up to security researchers to reveal such a list [40–42]. Security researchers manually explore all the possible means of accessing sensitive user data and embed data access codes in their proof-of-concept applications [41,42], which were submitted to Apple's vetting process to test whether they are able to get approved and appear on the official iTunes App Store. Such experiments also provide certain insights on the vetting process, which has been kept as black box by Apple.

As revealed by existing research [40–42], the most typical iOS attack vector is to utilize private APIs in third-party applications. As mentioned previously, iOS private APIs exist in both private frameworks and part of public frameworks. When used by third-party applications, private APIs may provide additional privileges to the applications and thus are explicitly forbidden and rejected by the vetting process. However, if the invocation of private APIs could evade detection from the vetting process, then it can be used to construct iOS attacks that perform various malicious functionalities. In this section, two ways of dynamically invoking private APIs are presented, which enable the malicious applications to pass the vetting process without being detected. Such dynamic loading mechanisms guarantee the success of the first stage in the attack vector. The second attack stage is to identify useful private APIs that are not restricted by iOS application sandbox. Utilizing the useful private APIs identified, researchers [41,42] managed to implement multiple serious attacks that cover a wide range of privileged functionalities. These attacks can be embedded in any third-party applications, and they work effectively on nonjailbroken iOS devices.

Although the generic attack vector includes two stages, these two stages are not isolated—what private API needs to be utilized decides the way of its dynamic invocation. Thus, in the following, SMS-sending and PIN-cracking attacks are presented as two examples to explain the underlying mechanisms of the entire attack vector. Then other attacks that can be implemented with this attack vector are also introduced and the implications of these attacks are discussed.

6.3.2.1 Attacks via Dynamically Loaded Frameworks

When implementing a third-party iOS application that uses private APIs, the normal process is to link the corresponding framework statically (in the application's Xcode [43] project) and import the framework headers in the application's source code. For example, if a developer wants to send SMS programmatically in his application, CoreTelephony.framework needs to be linked, and CTMessageCenter.h needs to be imported in the application code. After preparing those preconditions, the SMS-sending private API can then be called as follows:

```
[[CTMessageCenter sharedMessageCenter]
sendSMSWithText:@"A testing SMS" serviceCenter:niltoAddress:@"+19876543210"];
```

In the code mentioned earlier, the static method sharedMessageCenter returns an instance of CTMessageCenter class and then invokes the private API call "sendSMSWithText:service-Center:toAddress:," which performs the SMS-sending functionality on iOS 5. Third-party application can utilize this method to send premium rate SMS, and the sent SMS will not even appear in the SMS outbox (more precisely, it does not appear in the default iOS Message application). Thus, a user would be totally unaware of such malicious behavior until the user receives his next phone bill.

However, this standard way of invoking private APIs can be easily detected by the vetting process, even though only the executable binary of the compiled application is submitted for vetting. One way of detecting this API call is to simply use string matching (e.g., "grep") on the binary, as the name of the function call appears in the binary's objc methname segment (and also other segments). Moreover, the framework name and class name also appear in the binary as imported symbols. In this example SMS-sending code, although CoreTelephony is a public framework, CTMessageCenter.h is a private header (i.e., CTMessageCenter is a private class); thus, importing it in the source code can be detected by performing static analysis on the application's binary file. In order to pass Apple's vetting process, the application cannot link the framework statically.

To avoid being detected, the framework has to be loaded dynamically and the required classes and methods need to be located dynamically. In our attacks, we utilize Objective-C runtime classes and methods to achieve this goal. The example SMS attack code that illustrates the dynamic loading mechanism is given as follows:

```
1: NSBundle *b = [NSBundle
bundleWithPath:@"/System/Library/Frameworks/CoreTelephony.framework"];
2: [b load];
3: Class c = NSClassFromString(@"CTMessageCenter");
4: id mc = [c performSelector:NSSelectorFromString(@"sharedMessageCenter")];
5: // call "sendSMSWithText:serviceCenter:toAddress:" dynamically by
    utilizing NSInvocation
...
```

In this code, the first two lines are used to load the CoreTelephony framework dynamically, without linking this framework in the application's source code. The path of this library is fixed on every iOS device, which is under the /System/Library/Frameworks/ folder. Note that not only public frameworks can be loaded dynamically but private frameworks (which is under /System/Library/PrivateFrameworks/) can also be loaded dynamically using the same method. According to existing research [41], Apple's sandbox does not check the parameter of [NSBundle load] to forbid accessing these frameworks under /System/Library folder.

NSClassFromString at the third line is a function that can locate the corresponding class in memory by passing it the class name, which is similar to the "Class.forName()" method in Java reflection. At the fourth line, the sharedMessageCenter method is called via "performSelector:". At last, in order to call a method with more than two parameters (which is "sendSMSWith-Text:serviceCenter:toAddress:" in this case), the NSInvocation class is utilized.

Although this code dynamically invokes the private API call, it may need certain obfuscation in order to avoid the detection from static analysis during the vetting process. The last step of generating the actual attack code is to obfuscate all the strings appearing in the previous example code. There are various ways of obfuscating strings in the source code. One simple technique is to create a constant string that includes all 52 letters (both upper and lower cases), 10 digits and common symbols. Then all the strings in the code mentioned can be generated dynamically at runtime by selecting corresponding positions from this constant string. Some of the proof-of-concept applications utilize this method to obfuscate strings in the attack codes, and some others adopt a complex obfuscation mechanism, which involves bitwise operations and certain memory stack operations that are more difficult to be detected. However, according to existing research [41], obfuscation may not be necessary when constructing such attacks, as the vetting process does not seem to check all text segments in the binary.

6.3.2.2 Attacks via Private C Functions

Information about private Objective-C classes and methods in the Cocoa Touch frameworks can be obtained from the iOS runtime headers [44], which are generated using runtime introspection tool such as RuntimeBrowser [45]. An example of directly utilizing these Objective-C private APIs has been introduced in the previous subsection. However, Objective-C private classes and methods are not the only private APIs that can be used in third-party applications.

When reverse engineering the binary files of each framework, researchers [41] found that there are a number of C functions in these frameworks that can be invoked by third-party applications, which do not appear in the iOS runtime headers [44] and cannot be found with RuntimeBrowser [45]. In order to invoke these C functions, one will need to dynamically load the framework binary and locate the function at runtime. The following code segment is part of the PIN-cracking attack from [41], which illustrates how the dynamic invocation of private C functions can be realized.

```
void *b = dlopen("/System/Library/PrivateFrameworks/MobileKeyBag.framework/
              MobileKeyBag", 1);
int (*f)(id, id, id) = dlsym(b, "MKBKeyBagChangeSystemSecret");
...
int r = f(oldpwd, newpwd, pubdict);
...
```

In this code segment, dlopen() is used to load the binary file of the private framework MobileKeyBag, which returns an opaque handle for this dynamic library. Utilizing this handle and dlsym(), it is then possible to locate the address where the given symbol MKBKeyBagChangeSystemSecret is loaded into memory. This address is casted into a function pointer so that it can be directly invoked later on in any malicious third-party applications.

Although the code segment may look simple, it is actually not easy to identify which C functions can be invoked to serve for malicious apps, especially when just the framework binary is given. Even after the C functions are identified and located, it takes further tedious work to figure out the correct parameter types and values to pass to the C functions. And in many cases, even all parameters are correct, these functions may be restricted by iOS sandbox and thus will not function correctly within third-party applications. To speed up the manual reverse engineering process when analyzing the given framework binaries, usually static analysis tools need to be built and utilized [41] to disassemble the framework binaries and obtain assembly instructions that are relatively easy to read.

After this MKBKeyBagChangeSystemSecret function is successfully invoked, the rest of the attack code is straightforward—one can simply use brute force to crack the password. Four-digit PIN is widely used to lock iOS devices and has a password space of 10^4. According to experiments in [41], it takes 18.2 minutes on the average (of 16 trials on two iPhone 5 devices) to check the whole PIN space (10^4). This gives an average speed of 9.2 PINs per second.

The cracking speed can be further sped up, by building a PIN dictionary so that common PINs are checked first. If the given PIN is in birthday format (mmdd/ddmm), it takes only about 40 seconds to crack the PIN on average. Since such a PIN-cracking attack uses the low-level C functions, it will not trigger the "wrong password" event on the iOS device that is implemented at higher level (Objective-C functions) in the framework code. Thus, there is no limit on the number of attempts for the brute force attacks when cracking the device PIN. It is the same procedure to crack four-digit PIN and complex password using this method, but the latter will take much longer time than PIN due to its large password space.

6.3.2.3 Other Attacks Implemented with the Generic Attack Vector

The SMS-sending attack and the PIN-cracking attack previously presented explain how the entire attack vector is constructed. The former uses private Objective-C functions, while the latter uses private C functions. With the same dynamic invocation mechanisms that are able to bypass the vetting process, other attacks can also be implemented as long as new sensitive private APIs can be identified, which are overlooked by the iOS sandbox.

Previous research [41,42] has manually analyzed the 180+ public and private iOS frameworks and managed to identify seven sets of sensitive APIs that are not restricted by iOS sandbox. Utilizing these APIs and the dynamic invocation mechanisms, seven attacks can be implemented on iOS, which are listed in Table 6.1. Most of the attacks in Table 6.1 work on both iOS 5 and iOS 6 (which is the default iOS version on iPhone 5). The last two attacks (SMS-sending and email-sending) currently only work on iOS 5, but not iOS 6. The APIs of sending SMS and emails on iOS 6 have been substantially changed to prevent such attacks. Starting from iOS 7, Apple has fixed all of these sensitive private APIs that have been published by security researchers [46,47]. However, this does not mean that there are no other sensitive private APIs that can be exploited.

The severity of most of existing attacks would be significantly increased when the attack code is embedded in an application that can keep running in the background. Take the snapshot attack as an example. By calling the private API [UIWindow createScreenIOSurface], an application can capture the current screen content of the device. When continuously running at the background, this application can take snapshots of the device periodically and send these snapshots back to the developer's server for further analysis. Such snapshot-taking attack may reveal user's email content, photos, and even bank account information; thus, it should be avoided on any mobile devices.

Similar to the *snapshot-taking* attack, the *call-blocking* and *PIN-cracking* attacks also become more serious when they are used in an application that can continuously run at the background. Besides the seven attacks presented here, this generic attack vector can be used to construct other attacks as long as there are security sensitive functions on iOS that are not restricted by iOS sandbox. As each iOS version will include new functionalities to the platform, each iOS update may introduce new attacks from malicious third-party applications based on this attack vector.

6.3.3 Attack Mitigation for the Generic Attack Vector

Existing attacks [41,42] have shown that Apple's current vetting and sandbox mechanisms have weaknesses that can be exploited by third-party applications to escalate their privileges and perform serious attacks on iOS users. To mitigate such attacks, the vetting process should be improved to identify dynamic invocations. In addition, the iOS sandbox should also be enhanced to further defend against attacks utilizing private APIs.

Table 6.1 Seven Attacks That Can Be Implemented on iOS

#	Attack Name	Description	iOS 5	iOS 6	iPhone	iPad
1	PIN-cracking	Crack and retrieve the PIN of the device	✓	✓	✓	✓
2	Call-blocking	Block all incoming calls or the calls from specified numbers	✓	✓	✓	
3	Snapshot-taking	Continuously take snapshots for current screen (even the app is at background)	✓	✓		✓
4	Secret-filming	Open camera secretly and take photos or videos without the user's awareness	✓	✓	✓	✓
5	Tweet-posting	Post tweets on Twitter without user's interaction	✓	✓	✓	✓
6	SMS-sending	Send SMS to specified numbers without the user's awareness	✓		✓	
7	Email-sending	Send emails using user's system email accounts without the user's awareness	✓		✓	✓

Source: Han, J. et al., Launching generic attacks on iOS with approved third-party applications, in: *Proceedings of the 11th International Conference on Applied Cryptography and Network Security (ACNS 2013)*, Banff, AB, Canada, June 25–28, 2013.

6.3.3.1 Improving Application Vetting Process

Static analysis can be used to determine all the API calls that are not invoked with reflection (i.e., dynamic invocations), and it can provide the list of frameworks that are statically linked in the application. Thus, an automated static analysis is able to detect the standard way of invoking private APIs, as what is probably being used by Apple in its current vetting process. However, the existing static analysis needs to be improved to detect suspicious applications based on certain code signatures. For example, one suspicious code signature could be applications containing any dlopen() or [NSBundle load] invocations whose parameters are not constant strings (which match the cases of existing attacks [41,42]). However, as static analysis alone is not sufficient to determine whether a suspicious application is indeed a malware or not, manual examination and dynamic analysis should be utilized to examine such suspicious applications.

In many cases, manual examination may not be able to find malicious behaviors of the examined applications, because the malicious functions may not be performed for every execution. Instead, they can be designed in the way that such functions are only triggered when certain conditions have been satisfied. Examples of such conditions include time triggers or button triggers. When a malicious application uses such trigger strategy, the manual inspection may not find any suspicious behaviors during the vetting process. Such malicious applications can only be detected by utilizing fuzz testing [48] (or in the extreme case, using symbolic execution [49]), where different inputs are used to satisfy every condition of the application code. Furthermore, in order to determine whether sensitive user data are transferred out of the device, dynamic taint analysis [50] is an effective approach to serve this purpose. However, since it is expensive to apply fuzz testing and dynamic taint analysis on every application, the vetting process may choose to run such examinations only on selected suspicious applications.

6.3.3.2 Enhancement on iOS Sandbox

6.3.3.2.1 Dynamic Parameter Inspection

From the perspective of iOS sandbox, a straightforward defense to attacks that utilize the dynamic loading functions (such as [NSBundle load] and dlopen()) is to forbid third-party applications to invoke these functions. However, it is not practical to completely forbid the invocation of dynamic loading functions, since frameworks, libraries, and many other resources need to be dynamically loaded for benign purposes at runtime. Even Apple's official code, including both framework code and application code (which is automatically generated by Xcode), utilizes dynamic loading functions extensively to load resources at runtime. On the other hand, since sensitive APIs can be hooked by utilizing the application sandbox, the parameters of these APIs can be checked at runtime. Thus, it is useful if Apple's sandbox is modified in the way that the parameter values passed to dynamic loading functions are examined, and accessing files under a specific folder is forbidden.

One way of implementing this approach is to forbid third-party applications to dynamically load any frameworks under "/System/Library/" folder. However, a sophisticated attacker may be able to completely reverse engineer a given framework binary, locate all the code regions in the binary that are needed for launching his attack, and then copy only the needed code regions from the binary and insert them into his application code. In this way, he does not need to dynamically load framework binaries in his malicious applications. Therefore, this parameter-inspection approach is not able to completely defend against these attacks, though it can increase the attack complexity.

6.3.3.2.2 Privileged IPC Verification

Another technique of enhancing the sandbox is to dynamically check the privilege of the identity that makes sensitive API calls. For example, a third-party application should not have the privilege to invoke MKBKeyBagChangeSystemSecret API, which is used in the PIN-cracking attack. Such private APIs should only be invoked by processes or services with the system privilege. However, directly restricting the access to private APIs may not effectively prevent the attacks. By analyzing the implementation of several private APIs (in assembly code), existing research [41,42] found that the private APIs eventually use interprocess communication (IPC) methods, which communicate with the system service process, to complete the functionalities of the private APIs. For example, MKBKeyBagChangeSystemSecret API uses perform command() method to communicate with the system service (with service bundle id = "com.apple.mobile.keybagd"). This means that instead of invoking private APIs, an application can also use such an IPC method to directly send command to the system service process to perform the same functionality.

In order to defend against such attacks, for each privileged system service, the recipient of the command (which is the service process itself) needs to check the sender of the command to verify whether the sender has the valid privilege to make such IPC. To enable this IPC verification, the system service process needs to maintain a list of privileged IPC commands that are checked dynamically when an IPC is received. Compared to the parameter-inspection approach, privileged IPC verification provides better defense against the PIN-cracking, call-blocking, and snapshot-taking attacks as the corresponding privileged functionalities should not be used by any third-party applications. However, this approach alone is not sufficient to mitigate the other four attacks listed in Table 6.1. For these four attacks, the corresponding functionalities should be provided to applications due to usability reasons, but at the same time, it needs to be ensured that user interactions are involved when these functionalities are performed.

6.3.3.2.3 Service Delegation Enhancement

From iOS 6, Apple starts using the XPC Service, which allows processes to communicate with each other asynchronously so that it can be used for privilege separation. Originally on iOS 5, the SMS and email APIs are implemented as "view controller" classes that are created and used within a third-party application process. Therefore, applications can manipulate these view controller classes to send out SMSs and emails programmatically without users' interaction. However, starting from iOS 6, the SMS and email functionalities are now delegated to another system process utilizing XPC Service, which is completely out of the process space of third-party applications. Thus, a third-party application on iOS 6 is no longer able to send SMSs or emails programmatically without user's interaction. Also, starting from iOS 7, the service delegation mechanism has also been implemented for the Twitter service, which stops the tweet-posting attack.

The secret-filming attack, however, cannot be easily mitigated using such service delegation. Instead of using a unified user interface, iOS enables third-party applications to create their own customized user interfaces for taking photos or videos. If the same service delegation mechanism is applied, then the camera interface will be identical across different applications as it is provided by system service. Thus, more precisely, service delegation is able to defend against camera device abuse, but its implementation may greatly impact user experience.

6.3.3.2.4 System Notifiers for Sensitive Functionalities

In order to mitigate the threat of secret filming, while preserving the functionality and filming, which of using camera in third-party applications on iOS, one possible solution is to add a half-transparent system notifier on the screen (e.g., at the upper-right corner) whenever the camera device is being used. This notifier can be shown using the XPC mechanism so that the notifier is handled by a system daemon process, which is outside of the control of third-party applications. In this way, whenever the camera is being used (either taking photos or taking videos), the system notifier is shown on the screen to alert the user. A similar notifier for location tracking has been implemented and used on iOS for several versions.

6.3.4 More Security Challenges on iOS Platform

Different from the Android platform, iOS is relatively more secure, as less attacks have been found on iOS. There have been no major iOS attacks until the recent XcodeGhost attack [36,51]. Even for XcodeGhost, it hardly counted an iOS attack, but rather a cloud or compiler attack—basically, a malicious version of Xcode was uploaded to the Chinese cloud file sharing service "Baidu," and then downloaded by some iOS developers in China. Chinese developers unknowingly compiled iOS apps using the modified Xcode IDE and distributed those infected apps through the App

Store. Thus, to prevent such an XcodeGhost attack, there is nothing that Apple can do, other than removing those apps that are possibly compiled by the modified Xcode. The security enhancement should be at the developer side, who should not download and use the modified version of Xcode.

Before XcodeGhost, most iOS malwares, such as iKee [52] and Dutch 5 ransom [53] worms only work on jailbroken iOS devices where an SSH server is installed with the default root password unchanged. Other iOS malwares known to the public, such as iSAM created by Damopoulos et al. [54] (which focuses more on malware propagation methods), also exploit vulnerabilities existing only on jailbroken iOS devices. When iOS devices have been jailbroken, it is much more difficult to protect the platform, especially if the password of the root user is not changed. Because it will be similar to directly give the root access to attackers over Internet.

Certain attack mechanisms have also been discovered for nonjailbroken iOS devices. Miller [55] discovered that iOS code signing could be bypassed, allowing attackers to allocate a writeable and executable memory buffer. A malicious application can then exploit this vulnerability to generate and execute attack code at runtime. Wang et al. [42] put forward another novel approach to evade the app review process by making the apps remotely exploitable and introducing malicious control flows by rearranging signed code. These vulnerabilities discovered by Miller and Wang have been fixed or mitigated by Apple in recent iOS releases.

Another main iOS challenge comes from the nonofficial distributions of the iOS apps. As mentioned earlier, Cydia [39] is an application store that distributes applications that only run on jailbroken iOS devices. But even for nonjailbroken iOS devices, there are also nonofficial distribution methods. In particular, "enterprise" or "developer" applications are distributed under the enterprise/developer certificates. Since their "enterprise" applications are not distributed via the App Store, they are not regulated by the Apple's vetting process [56]. According to existing research [56], many of these "enterprise" applications tend to use private APIs that are crucial and security sensitive. Thus, they could pose more security threats than normal third-party applications distributed via the official App Store.

6.4 Systematic Security Comparison Framework for Mobile Platforms

As can be seen from the previous two sections, each platform has its own specific security challenges. However, which platform has a better architecture for security and privacy protection for mobile users remains a heated debate. Some claim that Android is better since it makes the complete permission list visible to users and it takes an open-source approach [57]. Some argue that iOS is better because (1) Apple screens all applications before releasing them to the iTunes App Store (aka Apple's vetting process), (2) Apple has complete control of its hardware so that OS patches and security fixes are more smoothly applied on all devices, and (3) the open-source nature of Android makes it an easier target of attacks than iOS [58]. Others [59,60] suggest that the two platforms achieve comparable security but in different ways. These different voices clearly raise the need for establishing a baseline for security comparison among different mobile platforms.

6.4.1 Overview of Android and iOS Security Features

As mentioned earlier, mobile security is very different from desktop PC security as various security mechanisms are adopted and enabled as default on current mobile operating systems. The security features used by Android and iOS are listed in Table 6.2.

Table 6.2 Security Model Comparison: Android versus iOS

Security Feature	Android	iOS
Privilege notification	Yes	Partial
Approval/vetting process	Partial	Yes
Digital signing	Yes	Yes
Binary encryption	Since v4.1	Yes
Sandboxing	Yes	Yes
Data encryption	Yes	Yes
Damage control	Yes	Yes
Address space layout randomization	Since v4.0	Since v4.3

■ *Privilege notification*: On Android, an application has to explicitly declare the privileges it requires. A user will be presented with these privileges when installing an application so that he can choose not to proceed if he is unwilling to grant the corresponding privileges. A complete list of privileges used by Android apps can be seen at [7], which consists of more than 130 items. On iOS, however, all third-party applications are treated "equally" in the sense that they are given the same set of privileges as default. Until iOS 5, the only privileges that require users' explicit acknowledgment are for accessing location information and for sending push notifications. However, since iOS 6, more privileges are added into the "privacy" section in iOS user settings, so that iOS users have more control over applications' privileges. By iOS 9, the privacy setting on iOS includes location, contacts, calendar, reminders, photos, Bluetooth sharing, microphone, camera, health, HomeKit, and motion and fitness. As can be seen from the previous list, the set of privileges that can be controlled by users on iOS are still far less than on Android.

■ *Approval/vetting process*: Approval from Apple is needed before an application is distributed via the iTunes Store. Apple screens each uploaded application to check whether it contains malicious code or violates Apple's privacy policy before releasing it on the iTunes Store. This vetting process is not well documented, and there have been cases where malicious applications passed the vetting process but had to be removed later from the iTunes Store [61]. On the Android platform, Bouncer [62] was implemented by Android team that provides automated scanning of Google Play for potentially malicious software. However, Bouncer does not require developers to go through an application approval process. It simply performs a set of analysis on the applications that are already released on Google Play. If an application is found to be malicious by Bouncer, it will then be removed from the online store.

■ *Signing and encryption*: On both platforms, every application is digitally signed with a certificate. The difference is that Android applications are signed by developers, and iOS applications are signed by Apple. Furthermore, signing on Android is mainly used in the sandbox for resource sharing—applications that are signed by the same private key can be assigned with the same process ID on Android. In addition to signing, iOS application binaries are also partially encrypted to mitigate unauthorized distribution. Each application downloaded from the iTunes Store has to be decrypted first in memory before launching. Starting from Android 4.1, all paid applications on Google Play are encrypted with a device-specific key before they are delivered and stored on an Android device.

■ *Other features*: iOS uses a sandboxing policy and Android uses UNIX UIDs to separate each individual application. Both platforms provide the service of encrypting users' confidential data, which could also be remotely erased once the device is lost. In addition, both platforms have kill switches in the hands of Google/Apple that can be used to remove malicious applications from the users' phones remotely. This feature limits the potential damage of a malicious application by preventing it from spreading widely. Finally, starting from Android 4.0 and iOS 4.3, both platforms provide address space layout randomization to help protect the systems and applications from exploitation due to memory vulnerabilities.

The general comparison shows that both platforms employ a number of common defense mechanisms, but also have their own distinct features. Android's privilege notification has some security advantage, but it pushes the most important security checking work to its end users who might not have expertise in security and may not even read or understand those privileges listed during application installation [15]. On iOS, the approval process provides a certain degree of defense against malicious applications. However, its capability is limited and can be bypassed sometimes [61]. Thus, a systematic comparison of the applications on these two platforms is needed to fully understand the effectiveness of these two different security architectures.

However, as introduced in previous sections, applications on Android are mainly written in Java and run on Dalvik [63] VMs. On the other hand, iOS applications are written in Objective-C and/or Swift and run natively as ARM [64] binaries. As the architectures of these two platforms are completely different, a reasonable approach is needed, which can perform a fair comparison on security and privacy provided by Android and iOS. One possible choice is to compare the *cross-platform applications* on these two platforms [31]. Two applications (one on Android and the other on iOS) are considered to be two versions of the same *cross-platform application* if they have the same set of functionality. For example, both Android and iOS have a Facebook application that provides the same functionality.

When given a cross-platform application, it will be interesting to examine the difference in the usage of their security sensitive APIs. A security-sensitive API (SS-API) is a public API provided for third-party applications by the underlying mobile operating systems that may have access to private user data or control over certain device components (e.g., Bluetooth and camera). When comparing thousands of such cross-platform applications for their SS-API usage, then the results could reflect the effectiveness of the security features on Android and iOS. However, in order to analyze the similarities and differences of the SS-API usage, the first challenge is to compare the application privileges and develop an SS-API mapping between Android and iOS.

6.4.2 Comparing Application Privileges

To compare the security architecture of Android and iOS, one of the most important comparison perspectives is to find out the similarity and difference on restricting the privileges for the third-party applications running on these platforms. However, it is not clear how such privileges can be compared as they might be of different granularity on the two platforms. To make things more complicated, although Google provides a comprehensive list of application permissions for Android [7], there is no official documentation specifying what privileges are allowed for third-party applications on iOS—this is possibly due to the close-source nature of iOS platform.

As the APIs on these two platforms are constantly evolving, Android 4.0 and iOS 5.0 are chosen to be used in the rest of this section when comparing the application privileges on both platforms. Given the 122 application permissions supported on Android 4.0 [7], huge manual works had been done by Han et al. [31] to find out what is the exact privilege obtained in each permission by examining the functionality of all APIs related to this permission according to the mapping of Android permission to API provided by [16]. With further investigation from both online advisories and offline iOS documentations on Xcode, whether each privilege available on Android is supported on iOS, and how it is supported on the iOS platform has been revealed in [31]. The overview of the analysis result is given in Table 6.3.

Although the term "permission" used on Android platform is concise, it also implies that there is access control in the architecture, which iOS barely has. Thus, in the rest of this chapter, SS-API type is used to refer to a group of SS-APIs that require the same privilege to access certain private data or sensitive service. The name and scope of the SS-API types follow the official Android permission list [7]. As shown in ·Table 6.3, among all the Android SS-API types, three of them (PERSIS-TENT_ACTIVITY, RESTART_PACKAGES and SET_PREFERRED_APPLICATIONS) have deprecated, and four of them (such as BRICK) do not really exist in Android as there are no API calls, content providers, or intents in Android related to these SS-API types [16]. The rest of the SS-API types are then divided into three groups, as explained in the following text.

Table 6.3 A Classification of Android Application Privileges

Group of Privileges	# of SS-API Types	Example SS-API Types
Does not actually exist in Android		
Already deprecated in Android or no Android API corresponds to it	7	SET_PREFERRED_APPLICATIONS BRICK
Reserved by Android system		
Only for OEMs, not granted to third-party apps, that is, these privileges can only be used by apps signed with system keys	42	DELETE_CACHE_FILES WRITE_SECURE_SETTINGS
Not supported on iOS		
Either iOS does not have such device, for example, removable storage, or iOS does not allow third-party apps to have such privilege	51	CHANGE_NETWORK_STATE MODIFY_AUDIO_SETTINGS
Both supported by iOS and Android		
Third-party apps have these privileges on iOS as default	20	BLUETOOTH READ_CONTACTS RECORD_AUDIO

6.4.2.1 Privileges Reserved for Android System Applications

The openness concept of Android and its online documentation may have given a misleading understanding to users and developers that a third-party Android application can obtain any privilege. However, this is not true—many SS-APIs are only provided for original equipment manufacturers (OEMs) and are not granted to third-party applications. Examples of these API types include DELETE_CACHE_FILES, INSTALL_LOCATION_PROVIDER, and FACTORY_TEST.

Since there is no official documentation specifying which privileges are reserved for OEMs on Android, this list of SS-API types can only be identified by analyzing the protection level tags in the frameworks/base/core/res/AndroidManifest.xml file, as API types reserved for system applications are labeled as android:protectionLevel="signatureOrSystem" or android:protectionLevel= "signature" in this firmware configuration file. In order to validate this list, a testing application can be developed that tries to access all SS-APIs on Android. With such auto test and validation, 42 SS-API types are found to be reserved for system applications on Android, which are not granted to third-party applications unless users explicitly give them the root privilege.

6.4.2.2 Privileges Not Supported on iOS

Among the rest of the SS-API types that can be used by Android third-party applications, more than 2/3 of these SS-API types are not supported on iOS. The reasons are either because iOS does not have corresponding functionality/device or iOS just does not allow third-party applications to have such privileges. For example, MOUNT_FORMAT_FILESYSTEMS permission on Android allows third-party apps to format file systems for removable storage. However, there is no removable storage for iPhone, iPad, or iPod Touch; thus, there would be no such SS-API type on iOS.

Some other SS-API types, including KILL_BACKGROUND_PROCESSES, PROCESS_ OUTGOING_CALLS, RECEIVE_SMS, are not allowed to third-party apps on iOS. Some of these privileges are not allowed due to security reasons, but some others are not. Although it is not officially documented, APIs for changing global settings that would affect the user experience (UX) are usually disallowed by Apple, and that is one of the reasons why there are still many people who jailbreak their iPhones. Examples of such SS-API types include MODIFY_AUDIO_SETTINGS, SET_TIME_ZONE, SET_WALLPAPER, WRITE_SETTINGS, etc. Although this would limit the capability of third-party applications, it is still reasonable from the UX perspective. For example, it could be a disaster if you are waiting for an important call, but a third-party application mutes the sound globally without your awareness.

6.4.2.3 Privileges Supported by Both Android and iOS

The last group of privileges in Table 6.3 contains the SS-API types supported on both Android and iOS. A comprehensive list of these SS-API types is given in Table 6.4. Note that although there are only 20 SS-API types both supported on Android and iOS, these SS-APIs cover the access rights to the most common resources/services, including user calendar, contacts, Bluetooth, Wi-Fi state, camera, and vibrator.

6.4.3 Systematic Security Comparison Framework

With SS-API mappings introduced earlier, it is now possible to perform a systematic comparison for the usage of security sensitive APIs in cross-platform apps on Android and iOS. The comparison framework was proposed in [31] and contains several components, as illustrated in Figure 6.3.

Table 6.4 Security-Sensitive–Application Programming Interface Types Supported on Both Android and iOS

SS-API Type	Abbreviation	Description and Explanation
ACCESS_LOCATION	LOC	Allows to access the location info (This type corresponds to both ACCESS_COARSE_LOCATION and ACCESS_FINE_LOCATION in [7].)
ACCESS_NETWORK_INFO	ANI	Allows to access information about networks (This SS-API type corresponds to both ACCESS_NETWORK_STATE and ACCESS_WIFI_STATE in [7].)
BATTERY_STATS	BAT	Allows to collect battery statistics
BLUETOOTH	BLU	Allows to connect to Bluetooth devices
BLUETOOTH_ADMIN	BTA	Allows to discover and pair Bluetooth devices
CALL_PHONE	PHO	Allows to initiate a phone call
CAMERA	CAM	Allows to access the camera device
CHANGE_WIFI_MULTICAST_STATE	CWS	Allows applications to enter Wi-Fi multicast mode
FLASHLIGHT	FLA	Allows access to the flashlight
INTERNET	INT	Allows to open network sockets
READ_CALENDAR	CAL	Allows to read the user's calendar data
READ_CONTACTS	CON	Allows to read the user's contacts data
READ_DEVICE_ID	RDI	Allows to read the device ID
RECORD_AUDIO	RAU	Allows an application to record audio
SEND_SMS	SMS	Allows to send SMS messages
USE_SIP	SIP	Allows an application to use SIP service
VIBRATE	VIB	Allows the access to the vibrator
WAKE_LOCK	WAK	Allows to disable auto-lock or screen-dimming
WRITE_CALENDAR	CAL	Allows to write the user's calendar data
WRITE_CONTACTS	CON	Allows to write the user's contacts data

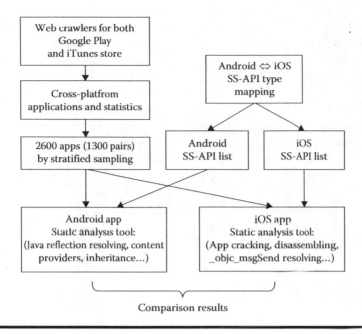

Figure 6.3   Overview of the cross-application security comparison framework.

The first step in the framework is try to find out what applications exist on both Android and iOS. This is not a trivial step, which involves crawling all the information for millions of applications provided by Google Play and iTunes App Store, and then use both machine learning classifiers and manual refinement to identify the available cross-platform applications.

In the second step, the executable files of all the cross-platform applications are passed to static analysis tools built for both Android applications (Dalvik bytecode) and iOS applications (Objective-C executables) in order to identify what SS-APIs are used in these applications. Finally, the Android and iOS versions of the same cross-platform application are compared for their SS-API usage.

6.4.4 Comparison Results and Analysis

The systematic security comparison were performed with 1300 pairs of selected cross-platform applications [31]. The results show that 73% of the applications on iOS access additional SS-APIs compared to their replicas on Android. The additional SS-APIs invoked are mostly for accessing sensitive resources such as device ID, camera, user contacts, and calendar, which may cause privacy breaches or security risks without being noticed. In the following, detailed comparison results are given from two different perspectives: SS-API usage difference for applications and the SS-API usage difference for third-party libraries.

6.4.4.1 Comparisons on SS-API Types in Cross-Platform Apps

The first comparison focuses on the 20 SS-API types that are both supported on Android and iOS. It is interesting to see how differently these SS-API types are used on the two platforms for cross-platform applications. The results show that the total amount of SS-API types that are used by 1300 Android applications is 4582, which indicates that each Android application uses 3.5 SS-API types on average. In comparison, the corresponding 1300 iOS applications access a total amount of 7739 SS-API types, which has on average 5.9 types per iOS application. 948 (73%) of the applications on iOS access additional SS-API types compared to its Android version.

Among the 20 different SS-API types, some of them are accessed almost equally by the applications on both platforms. For example, INTERNET is required by 1247 Android applications and 1253 iOS applications. However, some other SS-API types are used much more often by iOS applications compared to Android applications. For example, READ_DEVICE_ID type is required by 510 Android applications and 925 iOS applications. Another example is that the CAMERA type is required by 172 Android applications and 601 iOS applications.

To obtain a detailed understanding of the SS-API usage difference results, typical applications in each SS-API type are selected to conduct further investigation. The results show that famous applications such as Twitter and XECurrency do not access READ_DEVICE_ID APIs on Android. However, on their corresponding iOS version, five locations in Twitter's code and six locations in XECurrency's code are observed to read the device ID. Another typical instance is the famous free game app, Words With Friends. Compared to its Android version, the additional SS-API types accessed by its iOS version include (but are not limited to): BATTERY_STATS, CALL_PHONE, CAMERA, and FLASHLIGHT. The result of the popular game application Angry Birds also shows that compared to its Android version, Angry Birds on iOS additionally reads the user contacts data, as API call ABAddressBookGetPersonWithRecordID and ABAddressBookCopyArrayOfAllPeople are observed in the code.

Overall, the findings in the comparisons on the 20 SS-API types both supported on Android and iOS show that iOS third-party applications turn to access more often to some devices (such as camera and vibration) and are more likely to access sensitive data such as device ID, user contacts, and calendar. Further analysis shows that some of these APIs are actually invoked by the third-party libraries used in these applications (such as IMAdView and MobclixRichMediaWebAdView classes in the Words With Friends application). Thus, it is also important to analyze the SS-API usage of the third-party libraries on both platforms.

6.4.4.2 SS-API Usage of Third-Party Libraries

The eight most commonly used advertising and analytic libraries on Android (and their SS-API usage) are listed in Table 6.5, and the eight most common libraries on iOS are listed in Table 6.6.

As shown in Tables 6.5 and 6.6, the results clearly indicate that the most commonly used third-party libraries, especially advertisement and analytic libraries on iOS, access much more SS-APIs compared to the libraries on Android. A likely explanation of this phenomenon could be because on iOS the SS-APIs can be accessed more stealthily compared to on Android, where applications need to list out the types of SS-APIs they need to access during installation. The privileges to use these SS-APIs on iOS are granted to third-party applications as default without users' awareness, which gives certain freedom for advertisement and analytic libraries to access user data and sensitive resources.

However, evidence found in open-source applications show that third-party libraries only contribute a portion of the difference of the SS-API usage for cross-platform applications; the other part of the difference is caused by the application's own code.

Table 6.5 Most Common Advertising/Analytic on Android

Library Name	App Adopt Ratio (%)	SS-API Types (Abbr. Given in Table 6.4)
com/google/ads	21.7	ANI, INT
com/flurry/android	19.1	LOC, INT
com/google/android/apps/analytics	12.5	ANI
com/tapjoy	7.9	INT, RDI
com/millennialmedia/android	7.3	ANI, INT, RDI
com/admob/android/ads	4.4	LOC, INT
com/adwhirl	3.8	LOC, INT
com/mobclix/android/sdk	3.2	LOC, ANI, INT, RDI

Table 6.6 Most Common Advertising/Analytic Libraries on iOS

Library Name	App Adopt Ratio (%)	SS-API Types (Abbr. Given in Table 6.4)
Flurry	19.9	LOC, INT, RDI
GoogleAds	15.9	ANI, INT, RDI, SMS, VIB, WAK
Google Analytics	9.8	INT
Millennial Media	9.3	LOC, ANI, CAM, INT, CON, RDI, VIB
TapJoy	9.1	ANI, INT, RDI
AdMob	7.2	LOC, INT, CON, RDI
AdWhirl	6.9	LOC, ANI, INT, RDI
Mobclix	3.7	LOC, ANI, BAT, CAM, FLA, INT, CAL, CON, RDI, SMS, VIB

6.4.4.3 SS-API Differences Inside Application Code

There are at least two factors that have strong correlations with the SS-API usage differences between iOS and Android applications.

The most natural reason that may be expected is the implementation difference between the two versions of cross-platform applications. For example, ACCESS_NETWORK_INFO_APIs are only used by the iOS version of WordPress, but not by its Android version. In its iOS version, several API calls in WPReachability class are invoked, which are used to test the reachability to the WordPress hosts. However, for the Android version of WordPress, there is no code for testing any reachability. For example, when posting a blog to the server, the code of Android WordPress simply checks the return value of the posting function to see whether the connection is successful or failed. But on iOS, many Objective-C classes in the WordPress code will actively check the reachability beforehand and notify the users if the network is not reachable. Such implementation difference leads to the SS-API usage difference that WordPress on iOS uses the additional ACCESS_NETWORK_INFO_APIs compared to its Android version. Similar evidence can be found in the source code of MobileOrg application.

Such coding difference is also the main reason causing the difference in using the CAMERA SS-APIs. Taking the popular applications such as eBuddyMessenger and SmackIt, in their iOS versions, the user profile photo in the setting can either be chosen from the pictures stored on device or by directly taking the photo with the device's camera. However, their Android versions do not provide such photo-taking option. Note that such implementation difference does not only exist in the application's own code, but also for the same third-party libraries on two platforms. For example, CAMERA SS-APIs are used by OpenFeint library on iOS, but not by its Android version, which is caused by the same reason mentioned earlier.

Evidence has also been found that even the functionality of the two versions of cross-platform application are the same, some SS-APIs are intentionally avoided to be used on Android. The open-source application WordPress can be used to explain this phenomenon.

Compared to its Android version, WordPress on iOS uses the additional READ_DEVICE_ID APIs. In the WordPress iOS code, runStats method of WordPressAppDelegate reads the UUID, OS version, app version, language, device model, and then sends them to http://api.wordpress.org/iphoneapp/update-check/1.0/ to check whether this application needs to be updated. On the Android platform, the code of WordPress performs the same functionality—in the wpAndroid class, uploadStats method tries to retrieve the same set of data and sends these data back to the WordPress server to check for update. However, there is one major difference for the WordPress code on Android compared to the code on iOS. In its iOS code, the UUID is retrieved by directly calling UIDevice.uniqueIdentifier, which returns the device unique ID. In contrast, for its Android version, the UUID used is a random ID that is unique, but not the real device ID. It is a unique ID that is randomly generated and stored as the first record in WordPress's own SQLite database on the Android device. Thus, the different way of obtaining UUID is the reason that WordPress on iOS uses the additional READ_DEVICE_ID SS-API type.

The reason that developers do not try to avoid using the device ID on iOS is probably because of the same reasons mentioned earlier—on Android, an application needs to show the list of SS-API types it needs to access to the user during installation, while on iOS, no such notification is given to the user. This could also be the main reason that causes the difference in accessing SS-API types such as READ_CONTACTS and READ_CALENDAR because before iOS 6 there are no privilege notifications when the third-party apps are trying to access user contacts, calendar, photos, or reminders.

6.5 Summary

In this chapter, we have listed and discussed security challenges from both Android and iOS. On Android, due to its special distribution model, security patches have much longer release cycles compared to other mobile platforms. Its current permission model also has many problems—a user can only choose not to install an application, but is not given the flexibility of still installing the app while denying part of the permissions requested; most users may even not understand what are the application permissions presented during installation, and the application developer on Android is also not clear about which permissions are necessary to be requested during app development.

Android applications are also easy to be reverse engineered and repacked, which leads to further security issues, such as malware injections. Android also makes it easy to track its users, which leads to further privacy issues.

On the other hand, iOS does not suffer from most of the security challenges existing on Android. However, it has its own security weaknesses—for instances, its application vetting process can be easily bypassed—and its application sandboxing mechanisms are also not reliable. In this chapter, we have shown that malicious applications with serious attacks can be constructed utilizing vulnerabilities exist in iOS, which are accepted by the official iTunes App Store. Though part of the suggested mitigation mechanisms have been adopted in the new versions of iOS platforms, it is still possible to construct more attacks via the same attack vector.

In the last part of this chapter, a systematic security comparison for Android and iOS is presented. The comparison results show that applications on iOS tend to use more security sensitive APIs compared to their counterparts on Android and are more likely to access sensitive resources that may cause privacy breaches or security risks without being noticed. Further investigation revealed a strong correlation between such difference and the lack of application privilege list on the current iOS platform. Such results may imply that Apple's vetting process is not as effective as Android's explicit privilege list mechanism in restricting the privilege usage by third-party application developers. And therefore, applications on iOS could potentially be using more privileges compared to applications on Android, at least for applications from both official online markets.

The analysis results of this chapter can benefit all readers from different backgrounds. From the perspective of ordinary mobile users, people should be aware that both iOS and Android security are not perfect. There could be malicious applications existing on the official app stores of both platforms that have not been detected by existing security tools. Users should be aware that the third-party applications installed on their smartphones can introduce more security and privacy risks than commonly expected. It is also suggested to further verify and check the permissions of each application (in the settings) of both platforms and then either use security tools to limit the permission usage on Android or disable unnecessary permissions that granted to those applications on iOS.

For security researchers, the results from this chapter show that there are still many research gaps in the mobile security field. For example, there is still not enough research done on how different mobile systems can be compared in terms of their security and privacy. What are the ways of converting a benign application into a malicious one and trick users to install it is also an interesting but open topic. More importantly, different from traditional phones, the most severe problem of modern smartphones is probably the fact that the local users of these systems have much less control capability compared to the remote vendors. If any research projects can solve this ultimate problem, it will probably lead to a revolution of the existing mobile OS market, and after the legacy smartphone platforms have vanished, one or several new platforms will be created, which would provide much better security and privacy protection of their users.

References

1. IDC Research, Inc. 2015. Smartphone OS market share. http://www.idc.com/prodserv/smartphone-os-market-share.jsp, accessed May 8, 2016.
2. Android OS architecture. http://www.c4learn.com/android/android-os-architecture/, accessed May 8, 2016.
3. Android developer reference. 2016. http://developer.android.com/reference/packages.html, accessed May 8, 2016.

4. Debugging ART garbage collection. 2015. https://source.android.com/devices/tech/dalvik/gc-debug.html, accessed May 8, 2016.

5. Open handset alliance. http://www.openhandsetalliance.com/, accessed May 8, 2016.

6. T. Vidas, D. Votipka, and N. Christin. All your droid are belong to us: A survey of current Android attacks. In *Proceedings of the Fifth USENIX Conference on Offensive Technologies*. USENIX Association, Berkeley, CA, 2011.

7. Android Manifest.permission. 2016. http://developer.android.com/reference/android/Manifest.permission.html, accessed May 8, 2016.

8. M. Nauman, S. Khan, and X. Zhang. Apex: Extending Android permission model and enforcement with UserDefined runtime constraints. In *Proceedings of the Fifth ACM Symposium on Information, Computer and Communications Security*, pp. 328–332. ACM, New York, NY, 2010.

9. A. R. Beresford, A. Rice, N. Skehin, and R. Sohan. MockDroid: Trading privacy for application functionality on smartphones. In *Proceedings of the 12th International Workshop on Mobile Computing System and Applications*, pp. 49–54. ACM, New York, NY, 2011.

10. Y. Zhou, X. Zhang, X. Jiang, and V. W. Freeh. Taming information-stealing smartphone applications (on Android). In J. M. McCune, B. Balacheff, A. Perrig, A.-R. Sadeghi, and A. Sasse (Eds.), *Proceedings of the Fourth International Conference on Trust and Trustworthy Computing*, pp. 93–107. Springer-Verlag, Berlin, Heidelberg, 2011.

11. P. Hornyack, S. Han, J. Jung, S. Schechter, and D. Wetherall. These aren't the droids you're looking for: Retrofitting Android to protect data from imperious applications. In *Proceedings of the 18th ACM Conference on Computer and Communications Security*, pp. 639–652. ACM, New York, NY, 2011.

12. R. Xu, H. Saidi, and R. Anderson. Aurasium: Practical policy enforcement for Android applications. In *Proceedings of the 21st USENIX Security Symposium*, pp.27. USENIX Association, Berkeley, CA, 2012.

13. S. Shekhar, M. Dietz, and D. S. Wallach. Adsplit: Separating smartphone advertising from applications. In *Proceedings of the 21st USENIX Security Symposium*, pp.28. USENIX Association, Berkeley, CA, 2012.

14. P. Pearce, A. P. Felt, G. Nunez, and D. Wagner. Android: Privilege separation for applications and advertisers in Android. In *Proceedings of the Seventh ACM Symposium on Information, Computer and Communications Security*, pp. 71–72. ACM, New York, NY, 2012.

15. A. P. Felt, E. Ha, S. Egelman, A. Haney, E. Chin, and D. Wagner. Android permissions: User attention, comprehension, and behavior. In *Proceedings of the Symposium on Usable Privacy and Security* (*SOUPS*). ACM, New York, NY, 2012.

16. A. P. Felt, E. Chin, S. Hanna, D. Song, and D. Wagner. Android permissions demystified. In *Proceedings of the 18th ACM Conference on Computer and Communications Security*, pp. 627–638. ACM, New York, NY, 2011.

17. A. P. Felt, H. J. Wang, A. Moshchuk, S. Hanna, and E. Chin. Permission re-delegation: Attacks and defenses. In *Proceedings of the 20th USENIX Security Symposium* (*USENIX Security'11*), San Francisco, CA, 2011.

18. L. Davi, A. Dmitrienko, A.-R. Sadeghi, and M. Winandy. Privilege escalation attacks on Android. In M. Burmester, G. Tsudik, S. Magliveras, and I. Ilić (Eds.), *Proceedings of the 13th International Conference on Information Security*, pp. 346–360. Springer-Verlag, Berlin, Heidelberg, 2011.

19. S. Bugiel, L. Davi, A. Dmitrienko, T. Fischer, A.-R. Sadeghi, and B. Shastry. Towards taming privilege-escalation attacks on Android. In *Proceedings of the 19th Annual Network & Distributed System Security Symposium*, San Diego, CA, February 2012.

20. G. Nolan. *Decompiling Android*. Apress, New York, 2012.

21. W. Zhou, Y. Zhou, X. Jiang, and P. Ning. Detecting repackaged smartphone applications in third-party Android marketplaces. In *Proceedings of the Second ACM Conference on Data and Application Security and Privacy* (*CODASPY'12*), pp. 317–326. ACM, New York, NY, 2012.

22. S. Hanna, L. Huang, E. Wu, S. Li, C. Chen, and D. Song. Juxtapp: A scalable system for detecting code reuse among Android applications. In U. Flegel, E. Markatos, and W. Robertson (Eds.), *Proceedings of the Ninth Conference on Detection of Intrusions and Malware & Vulnerability Assessment* (*DIMVA'12*), pp. 62–81. Springer-Verlag, Berlin, Heidelberg, 2012.

23. L. Lu, Z. Li, Z. Wu, W. Lee, and G. Jiang. CHEX: Statically vetting Android apps for component hijacking vulnerabilities. In *Proceedings of the 2012 ACM Conference on Computer and Communications Security* (*CCS'12*), pp. 229–240. ACM, New York, 2012.

24. J. Crussell, C. Gibler, and H. Chen. Attack of the clones: Detecting cloned applications on Android markets. In *ESORICS 2012*, Lecture Notes in Computer Science, Vol. 7459, pp. 37–54. Springer, Berlin, Germany, 2012.

25. C. Zheng, S. Zhu, S. Dai, G. Gu, X. Gong, X. Han, and W. Zou. SmartDroid: An automatic system for revealing UI-based trigger conditions in Android applications. In *Proceedings of the Second ACM Workshop on Security and Privacy in Smartphones and Mobile Devices* (*SPSM'12*), pp. 93–104. ACM, New York, 2012.

26. T. Bray. Identifying app installations, 2011. http://android-developers.blogspot.sg/2011/03/identifying-app-installations.html, accessed May 8, 2016.

27. R. Stevens, C. Gibler, J. Crussell, J. Erickson, and H. Chen. Investigating user privacy in Android ad libraries. In *IEEE Mobile Security Technologies* (*MoST*), San Francisco, CA, 2012.

28. A. P. Felt, M. Finifter, E. Chin, S. Hanna, and D. Wagner. A survey of mobile malware in the wild. In *Proceedings of the ACM Workshop on Security and Privacy in Smartphones and Mobile Devices*, pp. 3–14. ACM, New York, NY, 2011.

29. macgasm.net: IT professionals rank iOS as most secure mobile OS, August 2012. http://www.macgasm.net/2012/08/17/it-professionals-rankios-as-most-secure-mobile-os/, accessed May 8, 2016.

30. iOS Security, a white paper by Apple. 2012. http://images.apple.com/iphone/business/docs/iOS_Security_May12.pdf, accessed May 8, 2016.

31. J. Han, Q. Yan, D. Gao, J. Zhou, and R. H. Deng. Comparing mobile privacy protection through cross-platform applications. In *Proceedings of the 20th Annual Network & Distributed System Security Symposium*, San Diego, CA, 2013.

32. D. Kravets. 2010. Jailbreaking iPhone legal, U.S. government says. *aBCNews*, http://abcnews.go.com/Technology/story?id=11254253, accessed May 8, 2016.

33. N. Statt. US government says it's now okay to jailbreak your tablet and smart TV. The Verge, 2015. http://www.theverge.com/2015/10/27/9622066/jailbreak-unlocked-tablet-smart-tvs-dmca-exemption-library-of-congress, accessed May 8, 2016.

34. G. Cluley. 2009. First iPhone worm discovered—Ikee changes wallpaper to Rick Astley photo. http://nakedsecurity.sophos.com/2009/11/08/iphone-worm-discovered-wallpaper-rick-astley-photo/, accessed May 8, 2016.

35. Apple. 2016. Official App Store review guidelines. https://developer.apple.com/app-store/review/guidelines/, accessed May 8, 2016.

36. J. Finkle. Apple cleaning up iOS App Store after first major attack. *Reuters*, September 21, 2015. http://www.reuters.com/article/us-apple-china-malware-idUSKCN0RK0ZB20150921, accessed May 8, 2016.

37. Malware for iOS? Not really. *TrendLabs*, June 2012. http://blog.trendmicro.com/trendlabs-security-intelligence/malware-for-ios-not-really/, accessed May 8, 2016.

38. Apple. 2016. Cocoa Touch. iOS technology overview. https://developer.apple.com/library/ios/documentation/Miscellaneous/Conceptual/iPhoneOSTechOverview/iPhoneOSTechnologies/iPhoneOSTechnologies.html, accessed May 8, 2016.

39. J. Freeman. 2016. Cydia, an alternative to Apple's App Store for jailbroken iOS devices. http://cydia.saurik.com/.

40. M. Egele, C. Kruegel, E. Kirda, and G. Vigna, PiOS: Detecting privacy leaks in iOS applications. In *Proceedings of the Network and Distributed System Security Symposium* (*NDSS*), San Diego, CA, February 2011.

41. J. Han, S. M. Kywe, Q. Yan, F. Bao, R. H. Deng, D. Gao, Y. Li, and J. Zhou. Launching generic attacks on iOS with approved third-party applications. In *Proceedings of the 11th International Conference on Applied Cryptography and Network Security* (*ACNS 2013*), pp. 272–289. Springer-Verlag, Berlin, Heidelberg, 2013.

42. T. Wang, K. Lu, L. Lu, S. Chung, and W. Lee. Jekyll on iOS: When benign apps become evil. In *Proceedings of the 22nd USENIX Security Symposium* (*USENIX Security*), pp. 559–572. USENIX Association, Berkeley, CA, 2013.

43. Apple Developer: Xcode, Apple's integrated development environment for creating apps for Mac and iOS. https://developer.apple.com/xcode/, accessed May 8, 2016.

44. N. Seriot. 2016. iOS Runtime headers. https://github.com/nst/iOS-Runtime-Headers, accessed May 8, 2016.

45. N. Seriot. 2016. Objective-C runtime browser, for Mac OS X and iOS. https://github.com/nst/RuntimeBrowser/, accessed May 8, 2016.

46. Apple. 2015. About the security content of iOS 7. https://support.apple.com/en-us/HT202816, accessed May 8, 2016.

47. Today Online. October, 2013. Local researchers help fix iOS security flaws. http://www.todayonline.com/tech/local-researchers-help-fix-ios-security-flaws, accessed May 8, 2016.

48. P. Godefroid, M. Y. Levin, and D. A. Molnar. Automated whitebox fuzz testing. In *Proceedings of the Network and Distributed System Security Symposium*, San Diego, CA, 2008.

49. S. Person, G. Yang, N. Rungta, and S. Khurshid. Directed incremental symbolic execution. In *Proceedings of the 32nd ACM SIGPLAN Conference on Programming Language Design and Implementation*, pp. 504–515. ACM, New York, NY, 2011.

50. M. G. Kang, S. McCamant, P. Poosankam, and D. Song. DTA++: Dynamic taint analysis with targeted control-flow propagation. In *Proceedings of the Network and Distributed System Security Symposium*, San Diego, CA, 2011.

51. J. Rossignol. What you need to know about iOS malware XcodeGhost. MacRumors, September 20, 2015. http://www.macrumors.com/2015/09/20/xcodeghost-chinese-malware-faq/, accessed May 8, 2016.

52. NakedSecurity: First iphone worm discovered—Ikee changes wallpaper to Rick Astley photo, November 2009. http://nakedsecurity.sophos.com/2009/11/08/iphone-worm-discovered-wallpaper-rick-astley-photo/, accessed May 8, 2016.

53. NakedSecurity: Hacked iPhones held hostage for 5 euros. November 3, 2009. http://nakedsecurity.sophos.com/2009/11/03/hacked-iphones-held-hostage-5-euros/, accessed May 8, 2016.

54. D. Damopoulos, G. Kambourakis, and S. Gritzalis. iSAM: An iPhone stealth air-borne malware. In J. Camenisch, S. Fischer-Hübner, Y. Murayama, A. Portmann, and C. Rieder, eds., *SEC 2011*, IFIP AICT, Vol. 354, pp. 17–28. Springer, Heidelberg, Germany, 2011.

55. C. Miller. Inside iOS code signing. In *Proceedings of the Symposium on SyScan*, Taiwan, 2011.

56. M. Zheng, H. Xue, Y. Zhang, T. Wei, and J. Lui. Enpublic Apps: Security threats using iOS enterprise and developer certificates. In *Proceedings of the 10th ACM Symposium on Information, Computer and Communications Security*, pp. 463–474. ACM, New York, NY, 2015.

57. K. Noyes. Why Android app security is better than for the iPhone. *PCWorld News*, August 2010. http://www.pcworld.com/businesscenter/article/202758/, accessed May 8, 2016.

58. Trend Micro: Android much less secure than iPhone. *Electronista News*, January 2011. http://www.macnn.com/articles/11/01/11/trend.micro.warns.android.inherently.vulnerable/, accessed May 8, 2016.

59. Android, iPhone security different but matched. *cNET News*, July 2010. http://www.cnet.com/news/experts-android-iphone-security-different-but-matched/, accessed May 8, 2016.

60. Smartphone Security Smackdown: iPhone vs. Android. *InformationWeek*, July 2011. http://www.informationweek.com/news/security/mobile/231000953, accessed May 8, 2016.

61. C. Sorrel. Apple approves, pulls flashlight app with hidden tethering mode. *Wired*, July 2010. http://www.wired.com/gadgetlab/2010/07/apple-approvespulls-flashlight-app-with-hidden-tethering-mode, accessed May 8, 2016.

62. H. Lockheimer. Android and security. Google Mobile Blog, February 2, 2012. http://googlemobile.blogspot.com/2012/02/android-and-security.html, accessed May 8, 2016.

63. D. Ehringer, The Dalvik virtual machine architecture. Technical Report, March 2010. http://davidehringer.com/software/android/The_Dalvik_Virtual_Machine.pdf, accessed May 8, 2016.

64. D. Seal. *ARM Architecture Reference Manual*. Pearson Education, 2001.

Chapter 7

Protecting Android Apps against Reverse Engineering

Wenjun Hu, Xiaobo Ma, and Xiapu Luo

Contents

Abstract

As the most popular mobile operating system, Android has received considerable attention from the security community. One major security concern is that Android apps are easily reversed and in turn modified and repackaged due to their inherent development and construction schemes, resulting in negative effects on the developers' reputation as well as the Android ecosystem. Hence, there is an acute demand for Android developers to take methods to protect the apps from reverse engineering. In this chapter, we will outline state-of-the-art weapons regarding antireverse engineering. Armed with these weapons, Android developers will make their apps much more secure and gain a significant tactical advantage in the reverse/antireverse engineering arms race.

7.1 Introduction

Android has become the most popular mobile operating system during these years. According to IDC, Android dominated the smartphone market with a share of 82.8% [15] in 2015 Q2. This popularity is accompanied by a large amount and wide variety of feature-rich Android apps.

However, the inherent mechanism regarding development and construction of Android apps makes them easy to be reversed, modified, and repackaged. To be specific, with the help of some publicly available tools such as smali/baksmali [9] and Apktool [14], crackers may easily reverse Android apps, thereby making these apps plagiarized or modified. Malicious code could then be injected into legitimate apps through reverse engineering for a variety of purposes such as fast spreading. Such malicious code injection is referred to as app repackaging, which poses a huge security risk. Zhou et al. [34] showed a worrisome fact that 5%–13% of the apps from third-party Android markets are repackaged. A recent survey [4] also showed that most of the top 100 paid and free Android apps are suffering from app repackaging.

App repackaging has brought a number of negative effects. For example, crackers could easily modify the client IDs of some advertisement components in an Android app, hence redirecting the revenue to their own pockets. App repackaging not only directly results in financial losses and impairing the reputation of the original Android developers but also impact the entire Android ecosystem. To prevent an Android app from being easily reversed and repackaged, it is highly desired that Android developers take measures to enhance the app protection.

Figure 7.1 outlines the protection methods available to Android developers. These methods mainly fall into two categories, that is, *antistatic analysis* and *antidynamic analysis*. To be more concrete, code obfuscation makes code obscure to understand and analyze by concealing its logic. Self-verification is capable of checking integrity to avoid app repackaging. Java reflection and dynamic loading make use of the dynamic characteristic of Java language to protect the apps. Native code provides a way to achieve stronger protection for bytecode protection. DEX modification can be used to hide methods during static analysis and recovered at run time. Antireverse engineering tools can make the related tools useless. Emulator detection is used to detect the dynamic

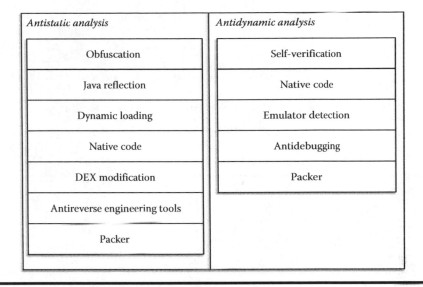

Figure 7.1 Protection methods for Android apps.

running environment. Antidebugging prevents dynamic debugging at the Java and native layers. Packer products always combine several methods for better protection and can be used by developers directly.

In this chapter, we will present the state-of-the-art weapons that raise the bar against reverse engineering. Armed with these weapons, Android developers will make their apps much more secure and achieve a significant tactical advantage in the reverse/antireverse engineering arms race.

7.2 Obfuscation

For efficient development management, the source code is generally written following a naming convention, which defines rules for choosing characters to represent identifiers, such as variables and methods. When an Android app is released to the public without extra processing, all the method and variable names will remain unchanged. Such straightforward publishment reduces the development cost but leaves hints for reverse engineering the app.

To prevent apps from being reversed, code obfuscation, which transforms readable code into obfuscated code, is introduced. It aims at making code obscure to understand and analyze by concealing its logic under the premise that code functionalities are preserved [6]. Christian and Clark introduced several methods to implement code obfuscation [5]. For example, lexical transformation targets at scrambling identifiers, which would be annoying to a reverse engineer. Control transformation could break up the flow-of-control of a procedure, and data transformation is capable of obfuscating data structures. In [25,33], several code obfuscation measures targeting at Android apps are presented. These measures include changing control flow graphs, data encryption, and renaming methods.

Google also realized the importance of code obfuscation and thus integrated ProGuard [23] into the Android build system as an optional (free) module. ProGuard achieves code obfuscation via lexical transformation, that is, semantically obscure renaming of classes, fields, and methods. Next, we will detail the steps of using ProGuard in an Android project.

```
 1 # This file is automatically generated by Android Tools.
 2 # Do not modify this file -- YOUR CHANGES WILL BE ERASED!
 3 #
 4 # This file must be checked in Version Control Systems.
 5 #
 6 # To customize properties used by the Ant build system edit
 7 # "ant.properties", and override values to adapt the script to your
 8 # project structure.
 9 #
10 # To enable ProGuard to shrink and obfuscate your code, uncomment this
11 # (available properties: sdk.dir, user.home):
12 #proguard.config=${sdk.dir}/tools/proguard/proguard-android.txt:proguard-project.txt
13
14 # Project target.
15 target=android-21
```

Figure 7.2   Content of `project.properties`.

7.2.1 Enable ProGuard

ProGuard is an optional module, and it is disabled by default. We will demonstrate how to enable it under Eclipse development environment. For other build systems such as Gradle and Ants, please refer to [11]. Upon the creation of an Android project, a file named `project.properties` is automatically generated in the root directory of the current project. The content of this file is shown in Figure 7.2, where one can uncomment line 12 to enable ProGuard.

7.2.2 Configure ProGuard

The `proguard.config` property in the file `project.properties` specifies the path of the configuration file, which in Android SDK revision 20 or higher is `$sdk.dir/tools/proguard/proguard-android-optimize.txt` and in revision prior to 20 is `$sdk.dir/tools/proguard/proguard-android.txt` [24]. The default configuration in `proguard-android.txt` for ProGuard only covers general cases. However, it is hard for ProGuard to analyze some cases correctly. Specifically, ProGuard might falsely remove some code that it considers unnecessary but the app actually needs. To resolve this problem, we manually set customized rules in `proguard-project.txt`. For example, if we encounter exceptions such as `ClassNotFoundException`, which happens when ProGuard strips away an entire class that our app calls, we can fix this exception by adding a line `-keep` in the configuration file.

7.2.3 Effectiveness of ProGuard

We use the code in Listing 7.1 to demonstrate the effectiveness of ProGuard. In the `MainActivity` class, we define a variable named `testStr` and a method named `testMethod` that is not invoked in the project. If we use Apktool to decompile the app built from the code without enabling ProGuard, we can find rich formation, including the source file name, the original variable and method name of the code, as described in Figure 7.3. However, if we decompile the app with ProGuard enabled (default configuration), all such information is stripped out, as depicted in Figure 7.4. We can see that there is no source file name, and the variable is renamed to a random alphabet character. Note that in this example there is no decompiled information about `testMethod` due to the removal of unused code by enabling ProGuard.

```
public class MainActivity extends Activity {
    public final String testStr = "hello world";
    @Override
    protected void onCreate(Bundle savedInstanceState) {
        super.onCreate(savedInstanceState);
        setContentView(R.layout.activity_main);
    }
    public int testMethod(int i, int j){
        return i + j;
    }
    ......
}
```

Listing 7.1   Code example for ProGuard.

```
1  .class public Lcom/example/androidtest/MainActivity;
2  .super Landroid/app/Activity;
3  .source "MainActivity.java"
4
5  # instance fields
6  .field public final testStr:Ljava/lang/String;
7
8  .method public testMethod(II)I
9      .locals 1
10     .param p1, "i"    # I
11     .param p2, "j"    # I
12
13     .prologue
14     .line 39
15     add-int v0, p1, p2
16
17     return v0
18  .end method
```

Figure 7.3   Decompiled smali code of demo app without enabling ProGuard.

```
1  .class public Lcom/example/androidtest/MainActivity;
2  .super Landroid/app/Activity;
3
4
5  # instance fields
6  .field public final a:Ljava/lang/String;
```

Figure 7.4   Decompiled smali code of demo app without enabling ProGuard.

According to the previous example, we observe that ProGuard can effectively obfuscate the code by removing unused code and renaming classes, fields, and methods with semantically obscure names. Although ProGuard cannot completely defeat reversing engineer, it would raise the bar for crackers. There is another commercial tool called DexGuard [7], which can provide more protection functionalities for Android apps than ProGuard. Besides renaming code entry names, DexGuard can encrypt strings, classes, and native libraries. It can also use Java reflection to hide access to sensitive APIs from reverse engineers. Figure 7.5 shows a part of decompiled code of an Android app protected by DexGuard. It's obvious that the class name and instance filed names are shown as weird characters. The powerful functionalities provided by DexGuard can increase the efforts needed to reverse Android apps.

```
1  .class public abstract Lo/ʰ;
2  .super Ljava/lang/Object;
3  .source ""
4
5  # interfaces
6  .implements Lo/^;
7
8  # instance fields
9  .field public ':Lo/ᴱ;
10 .field private ':Lo/^$if;
11 .field private ':I
12 .field public ':Landroid/content/Context;
13 .field public `:Landroid/content/Context;
14 .field public .:Lo/ɔ;
15 .field protected ,:Landroid/view/LayoutInflater;
```

Figure 7.5   Part of decompiled code of a DexGuard protected Android app.

7.2.4 Advantages and Disadvantages

With the help of ProGuard and DexGuard, developers could obfuscate their codes conveniently. Take ProGuard for example—a developer only needs to configure related files to enable obfuscation, which greatly ease developers' efforts in securing their apps. However, the obfuscation ability of ProGuard is limited, as it can only scramble those nonsystem identifiers. The unchanged system identifiers still expose rich information to crackers. Although DexGuard goes one step further, any determined and sophisticated cracker has a chance of discovering what the code is doing. Apvrille and Nigam recently presented possible methods to defeat ProGuard and DexGuard [3].

7.3 Self-Verification

To thwart app repackaging, the developers of the original Android apps should take self-verification into consideration. Self-verification aims at checking whether the app is modified or not at run time. We next describe several possible ways to implement self-verification in developing apps.

7.3.1 Verification of DEX

The `classes.dex` contains the classes compiled in the DEX format understandable by the Dalvik virtual machine. The DEX file includes the main functionalities of Android apps, and this file is always the target for app repackaging. Although the new run-time environment Android RunTime takes the place of DVM after Android 4.4, the `classes.dex` is still automatically generated after building the android application package (APK) and will be converted to OAT format during the booting stage. To verify if the DEX file is modified or not, we compare the DEX file's cyclic redundancy check (CRC) or hash to the predefined value. We can store the predefined value into an Android app's local resources or retrieve it from a remote server through network transmission. The code snippet in Listing 7.2 shows how to verify the `classes.dex` file.

7.3.2 Verification of APK

Under some conditions, there is no need to modify the `classes.dex` to achieve app repackaging. For example, Android developers publish free apps while pushing advertisements to make

```
public boolean isDexValidate(){
    String apkPath = this.getPackageCodePath();
    // Get predefined CRC value from local string resources
    Long dexCrc = Long.parseLong(this.getString(R.string.dex_crc));
    try {
        ZipFile zipfile = new ZipFile(apkPath);
        ZipEntry dexentry = zipfile.getEntry("classes.dex");
        if(dexentry.getCrc() != dexCrc)
            return false;
        else
            return true;
        } catch (IOException e) {
        }
}
```

Listing 7.2   An example of DEX verification.

revenue. For some advertisement SDKs, developers should specify proper ad client IDs in the `AndroidManifest.xml` for the providers of ad SDK to pay the revenue. Crackers then can easily modify the ad client ID to redirect the revenue to his or her own pocket. However, modification of `AndroidManifest.xml` has no impact on the classes.dex, and the verification of the DEX file will become invalid in this case. The main logic of verifying the whole APK is the same as the verification procedure of the DEX file, while the major difference is that we have no idea about the CRC or hash value of the full APK file content, meaning that we cannot embed the predefined CRC or hash value into local resources when developing apps. Maybe the only way we can work it out is through network communication.

7.3.3 Verification of Signature

Android OS requires the app to be signed by the developer's private key before installation. The developer's private key is self-certified and there is not a central certificate authority validating the signature. The app's signature is used to identify the author of an app (i.e., verify its legitimacy), as well as establish trust relationships between apps with the same signature [12]. As long as the signature is not leaked, crackers have to sign the app using their own signatures when they repackage the app. Hence, besides verification of `classes.dex` and APK, we can also determine if the app has been modified through verifying the signature. The predefined value of the signature can be obtained from local resources or remote servers. We demonstrate the workflow of verification of signature in Listing 7.3.

7.3.4 Advantages and Disadvantages

Self-verification ensures the integrity of an Android app and could further protect an Android app from being repackaged. Self-verification alone, however, can not effectively achieve antirepackaging. For example, crackers could first decompile the app and locate the verification functions; then they could modify the verification functions to constantly return true. Finally, crackers could circumvent self-verification and repackage the app. Therefore, it is necessary to combine self-verification and other protection methods to secure Android apps.

```
public boolean isSignatureValidate(String packageName) {
    PackageManager pm = this.getPackageManager();
    PackageInfo pi = null;
    int sig = 0;
    try {
        pi = pm.getPackageInfo(packageName, PackageManager.
            GET_SIGNATURES);
        Signature[] s = pi.signatures;
        sig = s[0].hashCode();
    } catch (Exception e1) {
        sig = 0;
    }
    int validateSig = Integer.parseInt(this.getString(R.string.
        validate_sig));
    if(sig == validateSig)
        return true;
    else
        return false;
}
```

Listing 7.3 An example of signature verification.

7.4 Java Reflection

As stated in Section 7.2, using Java reflection, DexGuard could hide access information to sensitive APIs from static reverse engineering. Java reflection can examine or modify the run-time behaviors of apps running in the virtual machine and could be exploited to further consolidate Android apps.

Listing 7.4 shows a simplified example of how to invoke method `MethodA` in class `Reflection` using Java reflection. In this example, the invoked method is hardcoded in the

```
public class Reflection {
    public void methodA(){
        System.out.println("Invoke methodA");
    }
}
public void reflectionInvoke() {
    Class[] consTypes = new Class[]{};
    Class reflectionCls = null;
    String className = "com.example.reflection.Reflection";
    String methodName = "methodA";
    try {
        reflectionCls = Class.forName(className);
        Constructor cons = reflectionCls.
        getConstructor(consTypes);
        Reflection reflectionIns = (Reflection) cons.
        newInstance(new Object[]{});
        Method method = reflectionCls.getDeclaredMethod(
                methodName, new Class[]{});
        method.invoke(reflectionIns, new Object[]{});
    } catch (Exception e){
    }
}
```

Listing 7.4 An example of Java reflection.

```
try
{
    Class localClass = Class.forName("com.example.reflection.Reflection");
    Reflection localReflection2 = (Reflection)localClass.getConstructor(arrayOfClass).newInstance(new Object[0]);
    localClass.getDeclaredMethod("methodA", new Class[0]).invoke(localReflection2, new Object[0]);
    return;
}
```

Figure 7.6   Decompiled code of the Java reflection demo.

```
24      static void Io0o0I0I(AdminService arg7) {  // has try-catch handlers
25          Class v0 = arg7.getClass();
26          String v1 = new String(CIOIIolc.Io0o0I0I(CIOIIolc.cOIcOOo(
27              Io0o0I0I.cOIcOOo("InVuJ0dpI2kyd2p7NjQ="),
28              CIOIIolc.cOIcOOo(ocOlcICo.cOIcOOo(1631, -37, -1).getBytes()))));
29          Method v0_1 = v0.getMethod(v1, Boolean.TYPE);
30          v0_1.invoke(arg7, Boolean.TRUE);
31          if(AdminService.cOIcOOo != null) {
32              AdminService.cOIcOOo.interrupt();
33          }
34
35          arg7.stopSelf();
36      }
```

Figure 7.7   Method hiding using DexGuard.

source code and, consequently, would be leaked to crackers. Unfortunately, invoked methods are critical information needed to perform reverse engineering.

Figure 7.6 shows the decompiled code, and it's easy to find which method to be invoked. To hide the method, one can further encrypt the method such as its name and the corresponding class name before invoking it and then decrypt at run time. Figure 7.7 demonstrates the decompiled code of OBad Android malware [29] that utilizes DexGuard for method hiding. From the figure, the variable v1 represents the method name that will be decrypted at run time. From this code snippet, it is really hard, if not impossible, to get the correct method name directly through static analysis.

7.5 Dynamic Loading

Android allows developers to perform dynamic class loading. Executable files including dex, jar, and apk could contain the dynamically loaded classes. Combined with Java reflection, code from external files can be executed. Therefore, one could extract the app's core logic (e.g., self-verification) into the external executable file and then utilize dynamic loading to execute the code. The external executable file could be stored as local resources or on remote servers.

We next present a typical dynamic loading example. The code snippets in Listing 7.5 demonstrate the routine of invoking method sayHello in class com.dexload.Example, during the process of DexClassLoader loading this class into the DVM. The first two parameters of DexClassLoader are the most important, where dexPath indicates the path of the dex, jar, or apk file, and getApplicationInfo().dataDir specifies the optimized directory. Normally, we set the optimized directory as the app's own data directory.

As stated earlier, to use DexClassLoader, one should set the optimized directory as the second parameter. This will bring some security concerns. For example, crackers are still capable of decompiling the dynamically loaded executable files since he could pull out the optimized executable files with a root access, even though the optimized directory is the app's own data directory. The risk of modification of the optimized executable files before running the app also remains.

```
DexClassLoader classLoader = new DexClassLoader(dexPath,
        getApplicationInfo().dataDir, null, getClassLoader());
try {
        Class<?> mLoadClass = classLoader.loadClass(
        "com.dexload.Example");
        Constructor<?> constructor = mLoadClass.getConstructor(
        new Class[] {});
        Object dexExample = constructor.newInstance(new
        Object[] {});
        Method sayHello = mLoadClass.getDeclaredMethod("sayHello",
        new Class[]{} );
        sayHello.setAccessible(true);
        sayHello.invoke(dexExample, new Object[]{});
} catch (Exception e){
}
```

Listing 7.5   Code example for dynamic loading.

Therefore, we need to go one step further to thwart these threats when utilizing dynamic loading. The following methods listed can be taken into consideration:

■ Verify the dynamically loaded executable files before executing to avoid modification attacks.
■ Encrypt the dynamically loaded executable files when saving them in the local storage and decrypt them at run time.
■ After Android revision 4.0, `dex` files can be dynamically loaded by invoking method `openDexFile` in class `dalvik.system.DexFile`. Hence, we could avoid the storage of the optimized output when using `DexClassLoader`. However, we must use Java reflection to invoke this method since it is not exported to developers. For more details, please refer to [31].

7.6 Native Code

Android apps are typically written in Java, with its elegant object-oriented design. However, at times, one needs to overcome the limitations of Java, such as memory management and performance, by programming directly at the native layer. Native developer kit (NDK) in Android allows developers to implement some parts of an app using native languages like C and C++. Compared with Java implementations, native implementations are much harder to reverse due to their dependence on the advanced RISC machine (ARM) instructions.

We could implement the key parts of an app at the native layer. Take the routines in Section 7.3 for example—all the verification schemes are conducted at the Java layer, and this is easy to bypass. Crackers can use tools such as Apktool to decompile the app and then directly modify the Dalvik instructions (e.g., making the `isDexValidate` method always return true). Finally, the modified app could be repackaged and resigned for installation.

Therefore, we could move the verification logic to the native layer to raise the bar of cracking. For native code development, please refer to the official documents [10]. For example, you can fetch your app's signature information in a method using Java language and then invoke this method at the native layer to check if the signature is changed or not. To bypass this signature verification, crackers should know some ARM reverse engineering knowledge. This native code implementation will then add one more protection vector for our Android app.

Additionally, packing methods for native code could be utilized to make an app much stronger against reverse engineering. For example, obfuscation and antidebugging can be applied to the native code. To be specific, inserting junk instructions and unordering instructions can be taken into consideration for obfuscation. For antidebugging, please refer to Section 7.10.2.

Making use of native code could increase the efforts needed for crackers to reverse the app, while developers will also take much more energy when developing native code compared with Java code. Since developers can have much more control over the lower layer resources when using native code, extra caution should be taken to avoid app crashes.

7.7 DEX Modification

In this section, we will introduce some techniques to hide the methods by modifying `classes.dex`. Axelle Apvrille presented an interesting idea about how to hide methods in `classes.dex` [2]. In the DVM, each method is represented by struct `DexMethod` in Listing 7.6.

The member `methodIdx` indexes into the `DexMethodId` list for the identity of this method (including the name and descriptor), represented as a difference from the index of the previous element in the list. The index of the first element in a list is represented directly. We then redirect `methodIdx` and `codeOff` of one method to those of another method to implement method hiding. We detail the following steps of method hiding:

Step 1: Modify the `DexMethod` struct of the method to hide.

- Set the `methodIdx` to the value 0x00, which means redirect this method to the previous method.
- Set the `codeOff` to the value of the previous method's `codeOff`.
- Update the next method's `methodIdx` based on the following formula: `next_method_idx = original_next_method_idx + original_method_idx`.

Figure 7.8 shows the effectiveness of the method redirecting [2].

Step 2: Fix the modified `classes.dex`

- Recalculate the modified `classes.dex`'s Adler32 and secure hash algorithm 1 (SHA1) values to update the checksum and signature members of the `classes.dex`'s header, respectively.

Step 3: Repackage and resign the app.

- Remove the folder of `META-INF` from the original APK file.
- Replace the APK's original `classes.dex` with the modified one.
- Resign the modified Android app.

```
struct DexMethod {
    u4 methodIdx; /* index to a method_id_item */
    u4 accessFlags;
    u4 codeOff;  /* file offset to a code_item */
};
```

Listing 7.6   Code example for dynamic loading.

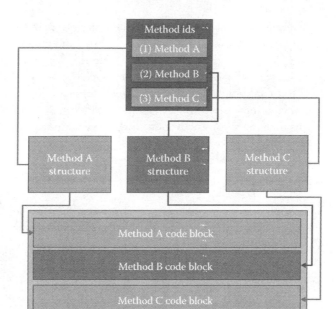

Figure 7.8 Redirect method B to method A resulting in hiding method B.

```
1098  # duplicate method ignored
1099  # .method public openNonAsset(Ljava/lang/String;)Ljava/io/InputStream;
1100  #    .locals 9
1101  #    .param p1, "paramString"    # Ljava/lang/String;
1102
1103  #    .prologue
1104  #    .line 92
1105  #    :try_start_0
1106  #    const-string v7, "android.content.res.AssetManager"
1107
1108  #    invoke-static {v7}, Ljava/lang/Class;->forName(Ljava/lang/String;)Ljava/lang/Class;
1109
1110  #    move-result-object v3
```

Figure 7.9 Decompiled code of the method hidden app.

After the previous process, the target method will be hidden when we decompile the app. Take the demo app created by Axelle for example—this app hides a method named `thisishidden` and redirects it to the method `openNonAsset`. Figure 7.9 shows a part of the decompiled results using the Apktool v2.0.0 for this demo app. We can infer that there are no decompiled results of the `thisishidden` method, while there exist two `openNonAssest` methods. Figure 7.9 shows the second `openNonAssest` is commented as duplicated by Apktool.

Since the hidden method still needs to be invoked at run time, typical ways to invoke the hidden method are presented as follows:

■ Read the `classes.dex` file into memory.
■ Recover the hidden method's `DexMethod` struct.
■ Load the recovered DEX with the proper class loader.
■ Use Java reflection to search and invoke the hidden method.

Please refer to [1] for complete implementation. The previous approach has some limitations. First, the characteristic of the decompiled code is evident enough to identify this approach is used to hide certain methods. Second, the value of `methodIdx` would be absolute if the method to hide stays as the first item, and this will bring problems when we modify the value of `methodIdx`. To overcome the second limitation, please ensure the method to hide is not the first one in the file `classes.dex`. This is not hard since the method items are in alphabetical order of method names.

Besides the approaches proposed earlier to hide methods, there exist other approaches. For example, we can remove a method's binary code (`code_item struct`) from `classes.dex` and recover it at run time. Zhang et al. also introduced some dynamic `dex` modification methods used by packers [32].

7.8 Antireverse Engineering Tools

During reverse engineering, most crackers will arm themselves with different kinds of reverse engineering tools to speed up the procedure. For Android apps reverse engineering, it is also the case. There are kinds of handy tools targeting at reverse engineering, including smali/baksmali, Apktool, dex2jar, JD-GUI, and JEB. However, most of the tools have their own bugs when processing reverse engineering. If our own apps can successfully trigger these bugs, we can slow down the race of reverse engineering. In this section, we will mainly discuss some antireverse engineering tools.

7.8.1 Bad Opcodes

Some decompile tools adopt linearly scanning schemes when parsing the opcodes. Therefore, bad opcodes will crash such tools. Tim Strazzere presented this technique in his report [28]. The basic idea is to modify the `classes.dex` to inject some bad opcodes. However, we must ensure the injected bad opcodes do not reside in the classes that will be executed at run time since the Dalvik virtual machine will verify all the related opcodes' validity before executing the app. To inject bad opcodes into an app, please follow the instructions:

1. Create a Java class that will never be invoked in the Android project. For example, we can construct a class named `Foo` with the method `noUseMethod` as the code snippets in Listing 7.7.
2. Export the Android project as an APK file and extract the `classes.dex`.
3. Open the extracted `classes.dex` with a hex editor and locate at the opcodes of the `noUseMethod`. For this method, its opcodes are listed in Listing 7.8.
4. Modify the opcodes to invalid ones such as `FFFF`, and then fix the `checksum` and `signature` members of the `classes.dex`'s header, respectively.

```
public class Foo {
    public static void noUseMethod(){
        String a = "hello world";
    }
}
```

Listing 7.7   Code example for bad opcodes.

```
1A00 4A15  // const-string v0, string@5450
0E00       // return-void
```

Listing 7.8   Opcodes in `noUseMethod`.

5. Replace with the original `classes.dex` with the modified one and then compress all the contents in the app root directory except the `META-INF` folder.
6. Rename the compressed file's suffix as apk and resign the app.

After the aforementioned steps, we finish injecting bad opcodes into our Android app. We then check the effectiveness of this measure against reverse engineering tools. If we use dex2jar to decompile the modified app, an exception will be thrown as shown in Figure 7.10, while the app can be successfully installed on the Android device as depicted in Figure 7.11. Also, there is not any problem to run the app. Using baksmali before version 1.3.3 to decompile the app suffers from the bad opcodes problem as shown in Figure 7.12, while baksmali after this version will treat unknown opcodes as an NOP. Apktool responses are the same as baksmali since it utilizes baksmali directly for code decompilation.

7.8.2 Bad Index

Strings in `classes.dex` are all stored in the string pool and then are accessed by the indexes to this string pool. However, if we set the string index to a nonexistence value, it will also cause some tools to crash. The steps to create a bad index are the same as the ones in Section 7.8.1

Figure 7.10   dex2jar failed to decompile the app injected with bad opcodes.

Figure 7.11   Install the app injected with bad opcodes successfully.

```
MindMacdeMacBook-Pro:Decode mindmac$ java -jar baksmali-1.3.3.jar Badcode.apk

UNEXPECTED TOP-LEVEL EXCEPTION:
org.jf.dexlib.Util.ExceptionWithContext: Unknown opcode: ff
        at org.jf.dexlib.Util.ExceptionWithContext.withContext(ExceptionWithCont
ext.java:54)
        at org.jf.dexlib.Code.InstructionIterator.IterateInstructions(Instructio
nIterator.java:92)
        at org.jf.dexlib.CodeItem.readItem(CodeItem.java:154)
        at org.jf.dexlib.Item.readFrom(Item.java:77)
        at org.jf.dexlib.OffsettedSection.readItems(OffsettedSection.java:48)
        at org.jf.dexlib.Section.readFrom(Section.java:143)
        at org.jf.dexlib.DexFile.<init>(DexFile.java:431)
        at org.jf.baksmali.main.main(main.java:269)
Caused by: java.lang.RuntimeException: Unknown opcode: ff
        at org.jf.dexlib.Code.InstructionIterator.IterateInstructions(Instructio
nIterator.java:56)
        ... 6 more
Error occured at code address 28
code_item @0x67ea4
```

Figure 7.12 baksmali before version 1.3.3 (including) suffers from the bad opcodes problem.

```
MindMacdeMacBook-Pro:Samples mindmac$ dex2jar.sh BadIndex.apk
this cmd is deprecated, use the d2j-dex2jar if possible
dex2jar version: translator-0.0.9.15
dex2jar BadIndex.apk -> BadIndex_dex2jar.jar
com.googlecode.dex2jar.DexException: while accept method:[Lcom/example/androidtest/Foo;.noUseMethod()V]
        at com.googlecode.dex2jar.reader.DexFileReader.acceptMethod(DexFileReader.java:694)
        at com.googlecode.dex2jar.reader.DexFileReader.acceptClass(DexFileReader.java:436)
        at com.googlecode.dex2jar.reader.DexFileReader.accept(DexFileReader.java:323)
        at com.googlecode.dex2jar.v3.Dom2jar.doTranslate(Dex2jar.java:85)
        at com.googlecode.dex2jar.v3.Dex2jar.to(Dex2jar.java:261)
        at com.googlecode.dex2jar.v3.Dex2jar.to(Dex2jar.java:252)
        at com.googlecode.dex2jar.v3.Main.doData(Main.java:63)
        at com.googlecode.dex2jar.v3.Main.doData(Main.java:35)
        at com.googlecode.dex2jar.v3.Main.doFile(Main.java:63)
        at com.googlecode.dex2jar.v3.Main.main(Main.java:86)
Caused by: com.googlecode.dex2jar.DexException: while accept code in method:[Lcom/example/androidtest/Foo;.noUseMethod()V]
        at com.googlecode.dex2jar.reader.DexFileReader.acceptMethod(DexFileReader.java:684)
        ... 9 more
Caused by: java.lang.IllegalArgumentException: id out of bound
        at com.googlecode.dex2jar.reader.DexFileReader.getString(DexFileReader.java:527)
        at com.googlecode.dex2jar.reader.DexOpcodeAdapter.xlc(DexOpcodeAdapter.java:129)
        at com.googlecode.dex2jar.reader.DexCodeReader.acceptInsn(DexCodeReader.java:436)
        at com.googlecode.dex2jar.reader.DexCodeReader.accept(DexCodeReader.java:337)
        at com.googlecode.dex2jar.reader.DexFileReader.acceptMethod(DexFileReader.java:682)
        ... 9 more
Done
```

Figure 7.13 dex2jar failed to decompile the app with bad index.

except step 4. We first check the total number of strings `classes.dex` contains and then set the string index to the value that is bigger than the strings number. In our case, the `classes.dex` file has 8588 strings in total. Hence, let's modify the original opcode value `4A15` to `8C21` (hex format in little endian) in the file `classes.dex`. After rebuilding the modified app, we test it with dex2jar and baksmali. The results are shown in Figures 7.13 and 7.14. We can infer that both dex2jar and baksmali before version 1.4.2 (including) suffer from such bad index panic. However, the app after injecting bad indexes can still be installed and executed without any problems.

7.8.3 AXML Format Modification

`AndroidManifest.xml` will be converted to the binary format called AXML when the app released. All the strings in `AndroidManifest.xml` will be stored in the string chunk. There is a member in the string chunk's structure named `string_chunk_flag` that indicates the string encoding type. In most cases, this flag is set to 0x00, and each character of the string is encoded with fixed two bytes as shown in Figure 7.15. Android also accepts other string encoding types including UTF-8, where the string character is not encoded with two bytes as depicted in Figure 7.16. Tools

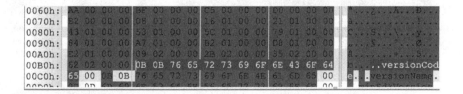

Figure 7.14 baksmali before version 1.4.2 failed to decompile the app with bad index.

Figure 7.15 String encoding with the string_chunk_flag set to 0x00.

Figure 7.16 String encoding with UTF-8.

such as AXMLPrinter and Apktool treat each string character as two bytes by default without taking the UTF-8 encoding type into consideration, and will crash in the face of apps whose strings in `AndroidManifest.xml` are encoded in UTF-8. We can then set the string encoding type of `AndroidManifest.xml` as UTF-8 and modify all the strings using the UTF-8 accordingly. The result of using Apktool to decompile such Android apps is shown in Figure 7.17.

7.8.4 Resource File Modification

Apktool is one of the most widely used tools for Android app repackaging. It compiles the resource according to the resource's suffix instead of checking the file's actual type. For example, if one file's suffix is "png," Apktool will compile the file based on PNG's scheme. Therefore, we could put a file

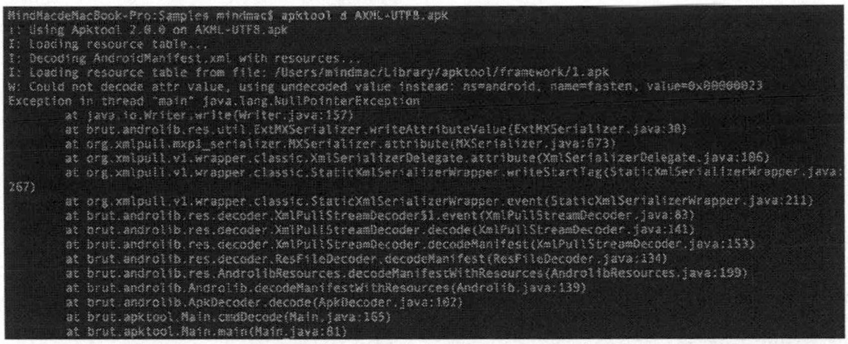

```
MindMacdeMacBook-Pro:Samples mindmac$ apktool d AXML-UTF8.apk
I: Using Apktool 2.0.0 on AXML-UTF8.apk
I: Loading resource table...
I: Decoding AndroidManifest.xml with resources...
I: Loading resource table from file: /Users/mindmac/Library/apktool/framework/1.apk
W: Could not decode attr value, using undecoded value instead: ns=android, name=fasten, value=0x00000023
Exception in thread "main" java.lang.NullPointerException
        at java.io.Writer.write(Writer.java:157)
        at brut.androlib.res.util.ExtMXSerializer.writeAttributeValue(ExtMXSerializer.java:38)
        at org.xmlpull.mxp1_serializer.MXSerializer.attribute(MXSerializer.java:673)
        at org.xmlpull.v1.wrapper.classic.XmlSerializerDelegate.attribute(XmlSerializerDelegate.java:106)
        at org.xmlpull.v1.wrapper.classic.StaticXmlSerializerWrapper.writeStartTag(StaticXmlSerializerWrapper.java:
267)
        at org.xmlpull.v1.wrapper.classic.StaticXmlSerializerWrapper.event(StaticXmlSerializerWrapper.java:211)
        at brut.androlib.res.decoder.XmlPullStreamDecoder$1.event(XmlPullStreamDecoder.java:83)
        at brut.androlib.res.decoder.XmlPullStreamDecoder.decode(XmlPullStreamDecoder.java:141)
        at brut.androlib.res.decoder.XmlPullStreamDecoder.decodeManifest(XmlPullStreamDecoder.java:153)
        at brut.androlib.res.decoder.ResFileDecoder.decodeManifest(ResFileDecoder.java:134)
        at brut.androlib.res.AndrolibResources.decodeManifestWithResources(AndrolibResources.java:199)
        at brut.androlib.Androlib.decodeManifestWithResources(Androlib.java:139)
        at brut.androlib.ApkDecoder.decode(ApkDecoder.java:102)
        at brut.apktool.Main.cmdDecode(Main.java:165)
        at brut.apktool.Main.main(Main.java:81)
```

Figure 7.17 Apktool fails to decompile the app whose `AndroidManifest.xml` is encoded in UTF-8.

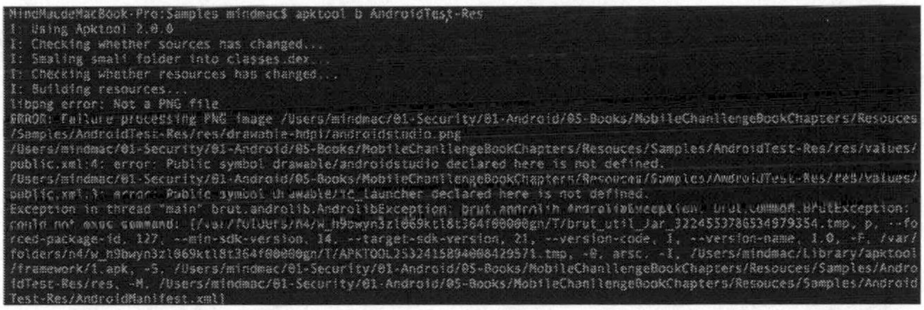

```
MindMacdeMacBook-Pro:Samples mindmac$ apktool b AndroidTest-Res
I: Using Apktool 2.0.0
I: Checking whether sources has changed...
I: Smaling smali folder into classes.dex...
I: Checking whether resources has changed...
I: Building resources...
libpng error: Not a PNG file
ERROR: Failure processing PNG image /Users/mindmac/01-Security/01-Android/05-Books/MobileChanllengeBookChapters/Resouces
/Samples/AndroidTest-Res/res/drawable-hdpi/androidstudio.png
/Users/mindmac/01-Security/01-Android/05-Books/MobileChanllengeBookChapters/Resouces/Samples/AndroidTest-Res/res/values/
public.xml:4: error: Public symbol drawable/androidstudio declared here is not defined.
/Users/mindmac/01-Security/01-Android/05-Books/MobileChanllengeBookChapters/Resouces/Samples/AndroidTest-Res/res/values/
public.xml:3: error: Public symbol drawable/ic_launcher declared here is not defined.
Exception in thread "main" brut.androlib.AndrolibException: brut.androlib.AndrolibException: brut.common.BrutException:
could not exec command: [/var/folders/n4/w_h9bwyn3zl069ktl8t364f00000gn/T/brut_util_Jar_3224553786534979354.tmp, p, --fo
rced-package-id, 127, --min-sdk-version, 14, --target-sdk-version, 21, --version-code, 1, --version-name, 1.0, -F, /var/
folders/n4/w_h9bwyn3zl069ktl8t364f00000gn/T/APKTOOL2532415894008429571.tmp, -0, arsc, -I, /Users/mindmac/Library/apktool
/framework/1.apk, -S, /Users/mindmac/01-Security/01-Android/05-Books/MobileChanllengeBookChapters/Resouces/Samples/Andro
idTest-Res/res, -M, /Users/mindmac/01-Security/01-Android/05-Books/MobileChanllengeBookChapters/Resouces/Samples/Android
Test-Res/AndroidManifest.xml]
```

Figure 7.18 Apktool fails to repackage the app containing a fake PNG file.

with a suffix "png," into our Android app project, where actually this file is not PNG (e.g., ICON). When we use Apktool to repackage this kind of Android apps, errors will occur like in the Figure 7.18. However, this problem can be easily fixed by renaming the corresponding file's suffix to its actual file type.

We have introduced a collection of antireverse engineering tools to protect our Android apps from being cracked. Please note that these tools may not work properly when the reverse engineering tools fix the corresponding bugs. Please refer to [21,28] for more antireverse methods.

7.9 Emulator Detection

Android emulators are always used for dynamic analysis during reverse engineering. Therefore, antiemulator techniques should also be taken into consideration to protect Android apps from being cracked. Android emulators can be distinguished from real devices in various categories. We will demonstrate some methods to detect the Android emulator environment at run time in this section.

7.9.1 android.telephony.TelephonyManager

Through `android.os.TelephonyManager` class, we can get a great deal of valuable information about the Android device, including telephone number, device ID, and voice mail number. Table 7.1 shows some related APIs belonging to this class. The information mentioned earlier

Table 7.1   APIs and Return Values on an Emulator

	API Name	Description	Return Value on Emulators
1	getLine1Number	Returns the phone number string for line 1	1555521 + emulator_port
2	getDeviceId	Returns the unique device ID	000000000000000
3	getSubscriberId	Returns the unique subscriber ID	310260000000000
4	getVoiceMailNumber	Returns the voice mail number	15552175049
5	getSimSerialNumber	Returns the serial number of the SIM	89014103211118510720

Table 7.2   System Properties on an Emulator

	Property Name	Return Value on Emulators
1	ro.product.brand	generic
2	ro.product.device	generic
3	ro.hardware	goldfish
4	ro.product.name	sdk
5	ro.build.host	android-test
6	ro.product.model	sdk
7	ro.product.name	sdk

extracted from the Android emulator differs from the one from a real Android device. For example, the value got by getLine1Number on an Android emulator starts with 1555521 and ends with the emulator's port number, hence, may be 15555215554 if the emulator's port number is 5554. The device id on an Android emulator is always 000000000000000. Table 7.1 lists all the return values on the Android emulator, whereas these values on real devices differ from each other. Therefore, these values can be used as fingerprints of the existence of the Android emulator.

7.9.2 System Properties

Android system properties contain device-specific information like the device brand, hardware, and model. Some system properties on the Android emulator are different from those on a real device. Table 7.2 shows the value of such system properties on the Android emulator. One can also utilize these system properties to detect if the Android app is running on an emulator. However, Android system does not export the API to get the system properties. We need to use Java reflection to work it out, and the target API is `android.os.SystemProperties.get`.

7.9.3 Characteristic Files

The Android emulator is based on quick emulator (QEMU), and thus there exist some QEMU-related files on the emulator. Table 7.3 demonstrates these files, the presence of which indicates

Table 7.3   QEMU-Related Files on the Android Emulator

	File Path
1	/dev/socket/qemud
2	/dev/qemu_pipe
3	/system/lib/libc_malloc_debug_qemu.so
4	/sys/qemu_trace
5	/system/bin/qemu-props
6	ro.product.model
7	ro.product.name

that the app is running on the emulator. There are also some studies focusing on Android emulator detection. Interested readers may refer to [20,22,27,30].

7.10 Antidebugging

Dynamically debugging is always used to help quickly understand the logic of the Android app during reverse engineering. There are kinds of debugging measures including Dalvik (smali) debugging, as demonstrated in [8,13], and native binary debugging using IDA Pro [26]. In this section, we will cover some methods to prevent Android apps from debugging.

7.10.1 Antidebugging at the Java Layer

In `AndroidManifest.xml`, the attribute `android:debuggable` indicates if the app can be debugged or not. If this attribute is set as true, the app cannot be connected with a debugger. While crackers can repackage the app and set the `android:debuggable` attribute as false to bypass this method. Also, they can customize the Android system with the system property `ro.debuggable` set to 1. Therefore, we can not only rely on the `android:debuggable` attribute and must go one step further. The attribute mentioned earlier can be checked at run time. Hence, we can dynamically check the attribute's value to check if it has been modified intentionally. The code snippets in Listing 7.9 show how to make such validation.

```
public boolean isDebugAttribModified(Context context){
    int flags = context.getApplicationInfo().flags;
    if(flags &= ApplicationInfo.FLAG_DEBUGGABLE != 0)
        return true;
      else
        return false;
}
```

Listing 7.9   Code example detecting debugger.

```
root@android:/ # cat /proc/551/status
Name:     ple.androidtest
State:    S (sleeping)
Tgid:     551
Pid:      551
PPid:     37
TracerPid:        0
Uid:      10049    10049    10049    10049
Gid:      10049    10049    10049    10049
FDSize:   256
Groups:   1015 1028
VmPeak:      171504 kB
VmSize:      171504 kB
```

Figure 7.19 `TracerPid` **value in the file** `/proc/[pid]/status`.

To detect the debugger connected to the app at run time, Android also provides an API named `isDebuggerConnected` in class `android.os.Debug`. The app can invoke this API at boot stage and directly exit once the debugger is detected.

7.10.2 Antidebugging at the Native Layer

The native code such as a shared library in an Android app can also be debugged with tools like IDA Pro. For the native debugging, the `ptrace` system call is mainly used for debugging. However, each process can only be attached using `ptrace` by only one tracer at a time. We can then utilize this feature against native debugging. What we need to do is to implement self-attach to our app's process. When a tracer does not attach the process, the `TracerPid` value contained in the file `/proc/[pid]/status` will be 0 as shown in Figure 7.19. A thread can then be created to monitor this value. If this value is not zero, we know that the process is attached.

7.11 Packer

Considering the situation that Android apps can be easily cracked, the researchers further developed packer products such as Bangcle [18], iJiami [16], Ali [17], and Qihoo 360 Mobile [19]. Most of the packers combine multiple antireverse methods to protect Android apps. For example, the original `classes.dex` data are encrypted and will only be decrypted at run time, reverse engineering

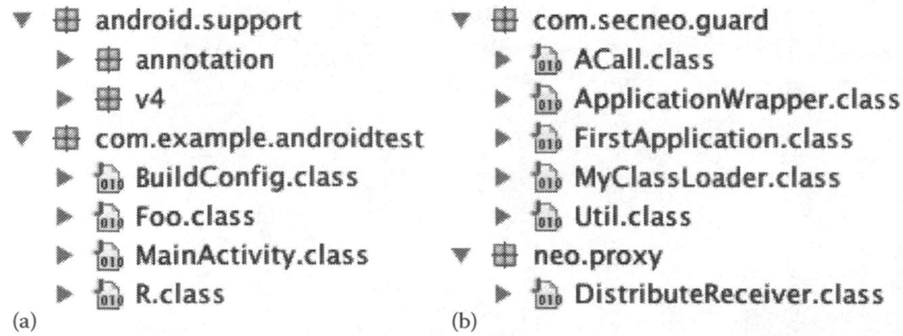

▼ ⊞ android.support
  ▶ ⊞ annotation
  ▶ ⊞ v4
▼ ⊞ com.example.androidtest
  ▶ BuildConfig.class
  ▶ Foo.class
  ▶ MainActivity.class
  ▶ R.class
  (a)

▼ ⊞ com.secneo.guard
  ▶ ACall.class
  ▶ ApplicationWrapper.class
  ▶ FirstApplication.class
  ▶ MyClassLoader.class
  ▶ Util.class
▼ ⊞ neo.proxy
  ▶ DistributeReceiver.class
  (b)

Figure 7.20 Decompiled results of the original and packed app: (a) decompiled results of original app and (b) decompiled results of packed app.

```
MindMacdeMacBook-Pro:Samples mindmac$ adb shell ps | grep "com.example.androidtest"
u0_a49    736   37    175448 34872 ffffffff 40033a40 S com.example.androidtest
u0_a49    754   736   5368   2148  ffffffff 400338e0 S com.example.androidtest
u0_a49    756   754   4488   980   ffffffff 4003292c S com.example.androidtest
```

Figure 7.21 Android app protected by Bangcle creates another two processes.

tools' bugs will be triggered to avoid reversing, and antidebugging methods are also taken into consideration. Take Bangcle for example—decompiled results of the original Android app and the processed one by Bangcle are shown in Figure 7.20. We can infer from the results that all the codes in the original app disappear. This is because Bangcle hides the original `classes.dex` and will reload it into the DVM at run time. When the protected app is launched on Android device, it will create another two processes as depicted in Figure 7.21. These three processes will attach to each other using ptrace to avoid debugging. Android apps protected by such packers are not easy to reverse, since most of the logic is processed at the native layer.

However, a recent study by Zhang et al. [32] shows that most of the packers mentioned earlier can be compromised. Consequently, the original `classes.dex` could be dumped for further analysis.

7.12 Summary

Due to its open source nature, Android system offers more flexibility to the developers at the cost of less trusted running environment to the apps. Moreover, the Android apps construction scheme leads to easy reverse engineering. The app repackaging is prevalent in the current Android market, resulting in a proliferation of app piracy and monetary loss for developers. We demonstrated several countermeasures to protect Android apps while there remains no silver bullet against Android apps reverse engineering. Reverse and antireverse are something like an arms race. Given the packing services provided in the market could be compromised as demonstrated in [32], we should make every effort to raise the bar for crackers. The methods mentioned earlier should be jointly used for app protection. For example, one can implement core functionalities using native code and further utilize some antidebugging methods. Please never forget to obfuscate the code before releasing. For convenience, one can also directly seek help from existing packer productions.

References

1. A. Apvrille. 2013. dextools. https://github.com/cryptax/dextools, accessed August, 2015.
2. A. Apvrille. Playing hide and seek with Dalvik executables. In *Hacktivity*, Budapest, Hungary, 2013.
3. A. Apvrille and R. Nigam. Obfuscation in Android malware, and how to fight back. In *Virus Bulletin*, pp. 1–10, 2014.
4. Arxan. State of security in the app economy: Mobile apps under attack. Vol. 2, 2013, Research Report. https://www.arxan.com/wp-content/uploads/assets1/pdf/State_of_Security_in_the_App_Economy_Report_Vol._2.pdf.
5. C. S. Collberg and C. Thomborson. Watermarking, tamper-proofing, and obfuscation-tools for software protection. *IEEE Transactions on Software Engineering*, 28(8):735–746, 2002.
6. M. Dalla Preda and R. Giacobazzi. Semantic-based code obfuscation by abstract interpretation. In *Automata, Languages and Programming*, Lisbon, Portugal, pp. 1325–1336. Springer, Berlin Heidelberg, 2005.

7. DexGuard. 2015. Dexguard. https://www.guardsquare.com/dexguard, accessed August, 2015.

8. V. Dorneanu. 2014. How to: Debug Android APKs with eclipse and DDMS. http://blog.dornea.nu/2014/08/21/howto-debug-android-apks-with-eclipse-and-ddms, accessed August, 2015.

9. J. Freke. 2015. smali/baksmali. https://github.com/JesusFreke/smali, accessed July, 2015.

10. Google. 2014. Android NDK. https://developer.android.com/ndk/index.html, accessed August, 2015.

11. Google. 2015. Proguard. http://developer.android.com/tools/help/proguard.html, accessed July, 2015.

12. Google. 2015. Signing your applications. http://developer.android.com/tools/publishing/app-signing.html, accessed July, 2015.

13. Hex-Rays. 2014. Debugging Dalvik programs with IDA. https://www.hex-rays.com/products/ida/support/tutorials/debugging_dalvik.pdf, accessed August, 2015.

14. iBotPeaches. 2015. A tool for reverse engineering Android APK files. https://ibotpeaches.github.io/Apktool/, accessed July, 2015.

15. IDC. Smartphone os market share, 2015, Q2. http://www.idc.com/prodserv/smartphone- os-market-share.jsp, accessed September, 2015.

16. iJiami Inc. 2015. http://www.ijiami.cn, accessed August, 2015.

17. Ali Ju Security. 2015. http://jaq.alibaba.com, accessed August, 2015.

18. Bangcle Inc. 2015. http://www.bangcle.com, accessed August, 2015.

19. Qihoo 360 Android Packer. 2015. http://jiagu.360.cn, accessed August, 2015.

20. F. Matenaar and P. Schulz. 2012. Detecting Android sandboxes. https://www.dexlabs.org/blog/btdetect, accessed August, 2015.

21. F. M. Patrick Schulz. 2013. Android reverse engineering & defenses. https://bluebox.com/wp-content/uploads/2013/05/AndroidREnDefenses201305.pdf, accessed August, 2015.

22. T. Petsas, G. Voyatzis, E. Athanasopoulos, M. Polychronakis, and S. Ioannidis. Rage against the virtual machine: Hindering dynamic analysis of Android malware. In *Proceedings of the Seventh European Workshop on System Security*, Amsterdam, Netherlands, p. 5. ACM, New York, NY, 2014.

23. ProGuard. 2015. Proguard. http://proguard.sourceforge.net, accessed July, 2015.

24. ProGuard. 2015. Proguard examples. http://proguard.sourceforge.net/manual/examples.html, accessed July, 2015.

25. V. Rastogi, Y. Chen, and X. Jiang. Droidchameleon: Evaluating Android anti-malware against transformation attacks. In *Proceedings of the Eighth ACM SIGSAC Symposium on Information, Computer and Communications Security (ASIA CCS '13)*, Hangzhou, China, pp. 329–334, ACM, New York, NY, 2013.

26. A. Righter. 2014. Remotely debugging Android binaries in IDA PRO. http://finn.svbtle.com/remotely-debugging-android-binaries-in-ida-pro, accessed July, 2015.

27. P. Schulz. Android emulator detection by observing low-level caching behavior. Online, December, 2013. http://bluebox.com/corporate-blog/android-emulator-detection, accessed July, 2015.

28. T. Strazzere. Dex education: Practicing safe dex. *BlackHat USA*, July, 2012.

29. R. Unuchek. 2013. The most sophisticated Android trojan. https://securelist.com/blog/research/35929/the-most-sophisticated-android-trojan/, accessed August, 2015.

30. T. Vidas and N. Christin. Evading Android runtime analysis via sandbox detection. In *Proceedings of the Ninth ACM Symposium on Information, Computer and Communications Security*, Kyoto, Japan, pp. 447–458. ACM, New York, NY, 2014.

31. Xavier. 2013. Nifty stuff that you can still do with Android. https://speakerdeck.com/hackitoergosum/hes2013-nifty-stuff-that-you-can-still-do-with-android-by-xavier-martin, accessed August, 2015.

32. Y. Zhang, X. Luo, and H. Yin. Dexhunter: Toward extracting hidden code from packed Android applications. In *Proceedings of the 20th European Symposium on Research in Computer Security*, Vienna, Austria, 2015.

33. M. Zheng, P. P. Lee, and J. C. Lui. ADAM: An automatic and extensible platform to stress test Android anti-virus systems. In *Detection of Intrusions and Malware, and Vulnerability Assessment*, Berlin, Germany, pp. 82–101. Springer, Berlin Heidelberg, 2013.
34. W. Zhou, Y. Zhou, X. Jiang, and P. Ning. Detecting repackaged smartphone applications in third-party Android marketplaces. In *Proceedings of the Second ACM Conference on Data and Application Security and Privacy*, San Antonio, TX, pp. 317–326. ACM, New York, NY, 2012.

MALWARE CLASSIFICATION AND DETECTION

Chapter 8

Efficient Classification of Android Malware in the Wild Using Robust Static Features

Hossein Fereidooni, Veelasha Moonsamy, Mauro Conti, and Lejla Batina

Contents

Abstract

The ubiquitous use of Android smartphones continues to threaten the security and privacy of users' personal information. Its fast adoption rate makes the smartphone an interesting target for malware authors to deploy new attacks and infect millions of devices. Moreover, the growing number and diversity of malicious applications render conventional defenses ineffective. Thus, there is a need not only to better understand the characteristics of malware families but also to generate features that are robust and efficient for classification over an extended period of time.

In this chapter, we propose a machine learning–based malware detection and classification methodology, with the use of static analysis as feature extraction method. Our tool, **uniPDroid** can be used to extract a plethora of informative features from our extensive dataset. We performed a malware family classification and obtained an average classification accuracy of 92%. We also present the empirical results for our cumulative classification that investigates how well features from old malware can contribute to the detection of new variants of both known and unknown malware.

8.1 Introduction

Since its first release in late 2008*, Android smartphones have continuously been replacing traditional mobile phones. The advent of such high-powered and affordable smart devices has redefined the way mobile phone users carry out their day-to-day activities. From checking emails to doing online banking, mundane tasks once conducted on a desktop only are now being executed *on the go*. According to Gartner[†], worldwide sale of Android smartphones in 2015 has reached more than 271 million devices, which accounted for 82.2% of the market share. Due to its popularity, the amount of malware targeting the Android platform has increased significantly in recent years. As such, malicious applications pose a significant threat to the smartphone platform security. In the first half of 2014, F-Secure[‡] reported that 295 new threat families or new variants of known families were collected. It is also worth mentioning that 294 out of these 295 families run on Android. Additionally, in the first quarter of 2015, Kaspersky's mobile security products detected 103,072 new malicious applications, a threefold increase from last quarter of 2014 [15].

On one hand, these statistics further prove that Android continues to be a favorite target for the majority of mobile threats as smartphones continue to replace traditional phones. On the other hand, the security of Android platform still requires thorough understanding, as demonstrated by the plethora of attacks in [11,13,14,25]. Thus, effective ways of enforcing security on such devices are still subject to investigation and there exists further room for improvement. To address the aforementioned security issue, we can leverage various techniques to analyze and detect Android malicious applications.

The techniques used to detect Android malware are similar to the ones used on other platforms. Detection techniques are essentially broken into (1) static analysis by analyzing a compiled file, (2) dynamic analysis by analyzing the runtime behavior, and (3) hybrid analysis by combining static and dynamic techniques [19]. *Static analysis* refers to extraction and analysis of information about an application from binary, source code, or other associated files. Static analysis can be performed before executing the application for the first time. However, this method is rendered ineffective by obfuscation techniques as it is not able to deal with a malware sample that changes its code without changing functionality, such as polymorphic malware.

On the other hand, *dynamic analysis* relies on execution of code in a virtual environment or sandbox to monitor the interaction of applications with the operating system. This approach comes with several drawbacks: (1) it is not clear how long the monitoring period should be in order to detect

---

* http://www.cnet.com/news/a-brief-history-of-android-phones/.
† http://www.gartner.com/newsroom/id/3115517.
‡ https://www.f-secure.com/documents/996508/1030743/Threat_Report_H1_2014.pdf.

important key events, (2) it is not always evident which conditions trigger the malicious behavior, and (3) dynamic analysis might be more resource consuming and computationally expensive than static analysis.

In the early days, malware detection and classification mechanisms employed only either static or dynamic analysis for feature extraction and malware prediction. However, as malicious programs continued to evolve in complexity and to deploy sophisticated attacks, there was a need for more robust frameworks. Thus, applying a hybrid method, which is a combination of static and dynamic analysis as shown in [24], when building the feature vector space is considered as one way of dealing with this problem. It should be noted that selecting a hybrid method when dealing with smartphone malware is not a popular method as this technique requires high computational resources and could impact negatively on the desired seamless interaction between the user and the device.

In a nutshell, static analysis is beneficial on memory-limited Android-powered devices because the malware is not executed and only analyzed. Additionally, static analysis makes use of reverse engineering tools to extract information from an application. For these reasons, we will concentrate on the lightweight approach and, thus, advocate for static analysis through the use of machine learning (ML).

ML techniques to detect mobile malware have been extensively investigated, leveraging a few characteristics of the mobile applications (e.g., call graphs [16], permissions [6], or both API calls and permissions [2,22]), and the results obtained were promising. Classification approaches have also been proposed to model and approximate the behaviors of Android applications and discern malicious apps from benign ones. The detection accuracy of a classification method depends on the quality of the features (e.g., how specific the features are [10]). Grace et al. [18] proposed a classification method with pure static features (data and control-flow analysis) that gives a false negative (FN) rate of 9%. Zhou et al. [38] extracted hybrid features, which are a combination of static and dynamic features, obtaining a better FN rate of 4.2%.

Although it is critical to distinguish malicious applications from clean ones, it is also important to efficiently classify malware into their correct families. Malware authors often redistribute repackaged versions of existing malware, and therefore, by correctly classifying the original malware, it becomes easier for antivirus engines to detect repackaged versions. Moreover, the features used to classify malware should also be robust and relevant over a long period of time as out-of-date features would allow malware samples to evade detection and classification mechanisms. To address the aforementioned issues, we focus solely on malicious applications to *first* investigate how to efficiently and accurately classify malware samples into their correct families and, *second*, generate robust feature sets that will stand the test of time and still be relevant over a period of years; this is tested through the experimental work referred to as cumulative classification.

In this chapter, we propose a malware classification method to (1) leverage an extensive coverage of applications' behavioral characteristics than the state of the art; (2) integrate decision-making through multiple classifiers; and (3) utilize the robustness of extracted features to detect and classify newly discovered malware. Specifically, we utilize a large number of features, extracted statically, from our extensive dataset comprising 15,884 samples. We extract intents, actual permissions used by an application, critical API calls, Linux system commands, and some other features that could possibly indicate the presence of malicious behaviors in an application. In order to build our classifier, we utilize the *eXtreme Gradient boosting* (XGboost*.) classifier, which is an ensemble method where weaker learners are combined to make a stronger learner. XGboost contains a modified version of the gradient boosting algorithm and can automatically do parallel computation

---

* https://github.com/dmlc/xgboost.

with open multi-processing (OpenMP), and it is much faster than the existing gradient boosting algorithm. Our aim is to maximize the accuracy scores of our classifier in terms of F1-score, recall, and precision.

In particular, our main contributions can be summarized as follows:

- We presented an Android malware detection and classification method that uses several informative features with good discriminative power to categorize malicious apps under their respective family names. We designed and built a tool named uniPDroid, written in Python programming language to extract the features such as intents, permissions used by an app, critical API calls, Linux system commands, and some other features that might indicate capability of performing malicious activities by an app.
- We performed an extensive static analysis on a large-scale well-labeled dataset of 15,884 Android applications. The dataset includes malware developed within a seven-year period, from year 2009 to 2015, and collected from different well-known and reliable repositories.
- We used several ML classification algorithms to discover the most highly performing one in terms of accuracy and speed. We leveraged boosting techniques to obtain as much detection and classification performance as possible for Android malware detection in the wild. Our experimental evaluations show that our proposed detection method is very effective and efficient. It obtained a true positive rate in detecting malware applications as high as **92%**.

This chapter is organized as follows: In Section 8.2, we present the related work in the area of malware detection and classification. Section 8.3 provides an extensive description of the proposed classification framework, including the dataset collection and preprocessing, feature extraction and selection, and evaluation metrics used. In the next section, we then present the experimental work for malware family-based classification followed by the work on cumulative classification in Section 8.5. Finally, we conclude in Section 8.6.

8.2 Literature Review

ML techniques have been extensively used for detection of malware on mobile devices [6,23,35]. Figure 8.1 presents an overview of the general framework of a standard ML technique classification. In the remainder of this section, we present some of the existing work in the area of Android malware classification.

The authors in [27] applied clustering techniques in malware detection of Android applications. They extracted the features of the applications from the application's extensible markup language (XML) file, which contains permissions requested by apps then applied unsupervised ML techniques to detect malware applications automatically. Similarly, Arp et al. [5] presented Drebin, an on-device malware detection tool utilizing ML-based methods on features such as requested hardware components, permissions, names of application components, intents, and API calls. Gascon et al. [16] presented a method that disassembles applications and extracted their function call graphs using the Androguard framework. They also proposed a learning-based method for the detection of malicious Android applications. Their method employed an explicit feature map inspired by the neighborhood hash graph kernel to represent applications based on their function call graphs.

Allix et al. [3] have used several ML classifiers to build a set of features in the form of control flow graphs (CFG) of applications to classify benign from malicious applications. The authors focused exclusively on the history aspect of datasets used in their experiment rather than malware detection

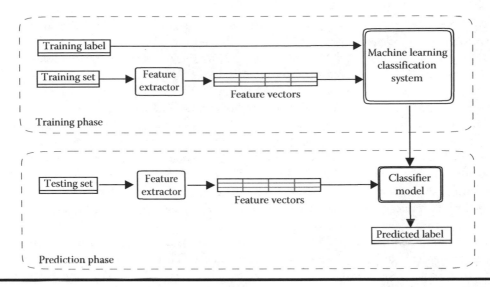

Figure 8.1   General classification methodology.

performance. Elish et al. [10] proposed a classification approach to detect malware by extracting a data dependence graph representing interprocedural flows of data. The authors extracted a data-flow feature on how user inputs can trigger sensitive API invocations.

Elish et al. [9] suggested a solution to detect Android malware collusions by constructing Inter-Component Communication (ICC). The authors constructed ICC maps to capture pairwise communicating ICC channels of 2,644 Android applications. Wolfe et al. [34] extracted the frequencies of all possible *n*-byte sequences in the Android application's bytecode as features and trained several classification algorithms to classify benign applications from malicious ones. The authors used a dataset comprising of 3,869 Android applications. Sahs et al. [26] presented an ML-based framework for Android malware detection using support vector machines (SVM) algorithms. The authors exploited a single-class SVM model derived from benign samples. They used the Android permissions in the manifest files and CFGs of applications from the dataset. Crowdroid [7] collects behavioral-related data directly from users via crowdsourcing and evaluates the data with a clustering algorithm.

Shabtai et al. [32] proposed a new method for categorizing Android applications through ML techniques. To represent each application, their method extracts different feature sets including the frequency of occurrence of the printable strings, the different permissions of the application itself, and the permissions of the application extracted from the Android Market. Abela et al. [1] presented AMDA, an automated malware detection system for the Android platform. The authors extracted features such as system calls from benign and malware applications to provide baseline behavior datasets to feed machine learners. Test applications are then passed through the behavior-based module for identification of presence of malicious payloads. Similarly, RobotDroid [36] is a framework that detects smartphone malware based on SVM active learning algorithm. The authors in [29] designed an anomaly detection system that extracts the strings contained in application files in order to detect malware. Their proposed method is based on features that were extracted from string analysis of the application.

Martinelli et al. [12] proposed CAMAS, a framework for the analysis and classification of malicious Android applications, through pattern recognition on execution graphs. They extracted a subset of frequent subgraphs of system calls that are executed by most of the malware. The resulting vector of the subgraphs is given to a classifier that returns its decision in terms of whether or not a malware has been detected. DroidAnalytics [37] is a malware analytic system for malware collection, signature generation, and association based on similarity scores by analyzing the low-level system at the application, class, or method level.

Sato et al. [30] proposed another detection method for Android malware. In particular, they used only manifest files to detect malware. The proposed method extracts six types of information from manifest files such as permission, intent (action, priority, and category), process name, and number of redefined permission and then uses them to detect Android malware. DroidMat presented by Wu et al. [35] exploits permissions, intents, ICCs, and API calls to distinguish malicious apps from benign ones. The detection performance was evaluated on a dataset of 1,500 benign and 238 malicious applications and compared with the Androguard risk ranking tool, with respect to detection metrics such as accuracy rate.

Peiravian and Zhu [22] presented an ML approach including SVM, decision trees (DT), and bagging predictor to detect malicious Android applications. They trained and tested a classifier by using extracted permissions and API calls as features to identify whether an application is potentially malicious or not. Koundel et al. [20] designed a naive Bayes classifier to classify applications using various attributes of an application, such as the permissions used by an application, battery usage, and rating acquired by the application on Android market. MAMA [28] presents manifest analysis for malware detection in Android. It extracts several features from the Android manifest of the applications to build ML classifiers such as K-nearest neighbors, DT, SVM, and Bayesian networks.

The literature presented in this section provides an overview of the existing work in the field of Android malware versus cleanware detection and ML-based classification methodologies. In our work, we focused solely on Android malware, proposing a novel ML-based methodology that can efficiently and, with high accuracy, assign malware samples to their correct family names. We argue that it is not only important to detect malicious applications, but also to label them correctly as malware authors often repackage existing malware. Hence, redetecting these repackaged samples becomes easier if the correct family names are used. Additionally, we analyzed the robustness of our extracted features used by the proposed methodology by performing a cumulative malware classification. We verified how features are efficiently extracted from old malware samples in terms of detecting and classifying newly discovered malware.

8.3 Proposed Classification Framework

This section provides extensive details on how the experimental dataset was collected and pre-processed, feature extraction and selection, and a description of the classification models and evaluation metrics used for the empirical results. In Section 8.3.1, we describe the composition of our experimental dataset, followed by an explanation of the different types of features extracted in Section 8.3.2. In Section 8.3.3, we elaborate on the methodology used for selecting the most representative features used by our classification model—Section 8.3.4. Finally, in the last subsection, we provide more details on the evaluation metrics used for our empirical results.

8.3.1 Dataset Collection and Preprocessing

In this subsection, we provide further detail on the composition of our experimental dataset. In order to conduct an extensive analysis, we collected a set of large well-labeled Android malicious

Table 8.1   Dataset Composition

Repository	Number of Samples
Genome [39]	1,260
Drebin [5]	5,560
M0Droid [21]	193
Virustotal [33]	8,871
Total	15,884

applications. The dataset used in our evaluation is composed of 15,884 malicious applications collected from the following existing work in the literature: [5,21,33,39]. The samples were released over a period of seven years, starting from 2009 until 2015. Table 8.1 shows the details of the dataset composition.

To perform malware classification using supervised ML classification algorithm (e.g., XGBoost classifier), we are required to provide a well-labeled dataset. To find the class label associated with each malware sample in our dataset, we wrote several scripts in Bash and Python programming languages. We submitted each malware sample to Virustotal [33] and made a query to get the malware family name, as shown in Listings 8.1 and 8.2. Virustotal then returned an analysis report for the given file in the form of JSON object as depicted in Listing 8.3. We then parsed the JSON object and performed text processing to extract the related family names. The names were then used as class labels since there is no agreed-upon malware naming convention among antivirus (AV) companies.

In order to decide on the family name for each class label, we took into account the family names of top eight AV engines.* Leveraging these top eight AVs and based on majority voting role, we extracted the selected malware family names. The AVs that we exploited are among the top AV engines used on the Android platform and are namely: MicroWorld-eScan, BitDefender, Kaspersky, Avira, AVG, Emsisoft, AVware, and F-Secure. The reason for considering only these eight AVs is because (1) they are among the top AV engines dedicated to the Android Platform, (2) we observed

```
1  #imports
2  import simplejson
3  import urllib
4  import urllib2
5
6  url = "https://www.virustotal.com/vtapi/v2/file/report"
7  parameters = {"resource":APK-hasH-name,"apikey":apikey}
8  data = urllib.urlencode(parameters)
9  req = urllib2.Request(url, data)
10 response = urllib2.urlopen(req)
11 json-object = response.read()
12 print json-object
```

Listing 8.1   Example of a Python script for submitting malware samples to Virustotal.

---

* http://www.av-comparatives.org/wp-content/uploads/2014/03/security_survey2014_en.pdf.

```
1   {"scans": {
2   "Kaspersky":{"detected":true,"version":"15.10","result":
       "Trojan-Spy.AndroidOS.Adrd.a",..},
3   "BitDefender":{"detected":true,"version":"7.2","result":
       "Android.Trojan.Adrd.A",..},
4   "Emsisoft":{"detected":true,"version":"3.5.0.642","result":
       "Android.Trojan.Adrd.A",.. },
5   "F-Secure":{"detected":true,"version":"11.0.19100.45",
       "result":"Trojan:Android/Adrd.A",..},
6   "Avira":{"detected":true,"version":"8.3.2.4","result":
       "ANDROID/Spy.Adrd.D.Gen",..},
7   .
8   .
9   "AVG":{"detected":true,"version":"16.0.0.4489","result":
       "Android/Adr",..},
10  "resource": "4de0d8997949265a4b5647bb9f9d42926bd88191",
       "total": 54, "positives": 38,
11  "md5": "77b0105632e309b48e66f7cdb4678e02",...}
```

Listing 8.2   Example of a JSON file produced by Virustotal.

```
1   from androguard.core.bytecodes.dvm import *
2   from androguard.core.bytecodes.apk import *
3   from androguard.core.analysis.analysis import *
4
5   a = APK("app.apk")
6   d = dvm.DalvikVMFormat( a.get_dex() )
7   z = d.get_strings()
8   for i in range(len(z)):
9     if z[i].startswith('android.intent.action.'):
10       intents = z[i]
11       intentList.append(intents)
```

Listing 8.3   Example of a Python script for extracting intents from Dalvik bytecode.

that these AV outperform others in most cases, particularly when detecting malware, and (3) we did not further complicate the text processing phase by increasing the number of AV engines.

8.3.2 *Feature Extraction*

Android applications are written in Java, compiled to Java bytecode, and then converted into platform-specific Dalvik bytecode. This bytecode can be efficiently disassembled and provides us with useful information about features used in an application. We mainly extracted the features from bytecode and converted these features into binary feature vectors, which are made up of 560 features. Each feature vector is comprised of the features described in the following:

■ *Intents*: The intent is an abstract description of an operation to be performed and allowing information about events to be shared among different components and applications. We extract all intents in Android app as a feature set because malware often listen to specific intents. Listing 8.3 shows the snippet used to extract intents from an application.

■ *Used permissions*: A significant part of Android's built-in security is its permissions system. Permissions allow an application to access potentially dangerous API calls. Many applications need several permissions to function properly, and a user must accept them at install time. The used permission provides a more in-depth view on the behavioral characteristics of an application. We extract and include them to the feature set (e.g., `INTERNET`, `ACCESS_FINE_LOCATION`, `INSTALL_PACKAGES`). Listing 8.4 describes the permissions extraction process.

■ *System commands*: Malware use system commands to run root exploit code or download and install additional executable files. Since system command can provide us with valuable information to detect malicious behavior, we extract and include them in the feature set. The authors in [31] listed the most commonly used system commands in malicious applications (e.g., `chmod`, `su`, `mount`, `sh`, `killall`, `reboot`, `mkdir`, `ln`, `ps`). These commands are executed after the malware gains root privilege on the device. Listing 8.5 shows how the system commands are extracted from Dalvik bytecode.

■ *Suspicious API calls*: We extracted the API calls that are frequently seen in malware samples and can result in malicious behavior. In order to obtain a deeper understanding of the functionality of an application, we collected these API calls and included them in the feature set (e.g., `openFileOutput`, `sendTextMessage`, `getPackageManager`, `getDeviceId`, `Runtime.exec`, `Cipher.getInstance`). The authors in [31] mentioned the most commonly used API calls in malicious applications. Listing 8.6 shows the snippet of code used to extract suspicious API calls.

```
1   ...
2   # the APK
3   a = APK("app.apk")
4   # the classes.dex
5   d = dvm.DalvikVMFormat( a.get_dex() )
6   # the analyzed classes.dex
7   dx = analysis.uVMAnalysis( d )
8
9   Permission_dexFile = dx.get_permissions( [] )
10  for i in Permission_dexFile:
11      permList.append(i)
```

Listing 8.4   Example of a Python script for extracting permissions from Dalvik bytecode.

```
1   ...
2   a = APK("app.apk")
3   d = dvm.DalvikVMFormat( a.get_dex() )
4   z = d.get_strings()
5   # to back trace unix commands
6   suspicious_cmds=["su","mount","reboot","mkdir"
7                    ,...]
8   for i in range(len(z)):
9     for j in range(len(suspicious_cmds)):
10        if suspicious_cmds[j]==z[i]:
11          cmdList.append(suspicious_cmds[j])
```

Listing 8.5   Example of a Python script for extracting system commands from Dalvik bytecode.

```
1  a = APK("app.apk")
2  d = dvm.DalvikVMFormat( a.get_dex() )
3  z = d.get_strings()
4  suspicious_APIs=["getSimSerialNumber",
5      "getSubscriberId","getDeviceId",...]
6  for i in range(len(z)):
7    for j in range(len(suspicious_APIs)):
8      if suspicious_APIs[j]==z[i]:
9        APIsList.append(suspicious_APIs[j])
```

Listing 8.6   Example of a Python script for extracting suspicious API calls from Dalvik bytecode.

◼ *Malicious activities*: We considered different malicious behaviors seen in malware applications. We investigate whether an Android application is capable of performing such malicious activities through Dalvik bytecode analysis. We consider different kinds of information that malicious applications are able to harvest from smartphones. In Listing 8.7, we describe how to search for features that are prone to perform malicious activities. The following are some of these features discussed briefly:

— Reading the international mobile station equipment identity (IMEI)
— Loading native, dynamic, and reflection code
— Accessing files on an SD card
— Reading location information through GPS/WiFi
— Intercepting data network activities
— Making phone calls and disabling incoming SMS notifications
— Retrieving information of the application installed
— Recording audio and capturing video
— Opening a TCP/UDP socket
— Performing encryption and message digest algorithms

```
1  ...
2  # Searching for Doing Cipher
3  a = APK("app.apk")
4  d = dvm.DalvikVMFormat( a.get_dex() )
5  dx = analysis.uVMAnalysis( d )
6  getIN=dx.tainted_packages.search_methods
7  ("Ljavax/crypto/Cipher","getInstance",".")
8  ScrKey=dx.tainted_packages.search_methods
9  ("Ljavax/crypto/spec/SecretKeySpec","<init>",".")
10 Cipherini=dx.tainted_packages.search_methods
11 ("Ljavax/crypto/Cipher","<init>", ".")
12 CipherDO=dx.tainted_packages.search_methods
13 ("Ljavax/crypto/Cipher","doFinal", ".")
14 if ((getIN)and(Cipherini)and(CipherDO))or(ScrKey):
15   potential_misBhve.append('Does Cipher')
```

Listing 8.7 Example of a Python script for extracting potential misbehavior from Dalvik bytecode.

8.3.3 Feature Selection

We should consider that a large number of features, some of which are redundant or irrelevant, may present several problems such as misleading the learning algorithm, overfitting, and increasing model complexity. Feature selection is a process that automatically selects features in a dataset that contribute most to the prediction results. The benefits of performing feature selection before modelling the data are to reduce overfitting, to improve accuracy, and to reduce training time. We used a technique leveraging ensemble of randomized DT, that is, extra trees classifier for determining the feature importances [17]. We exploited extra trees classifier to compute the relative importance of each attribute to help better the feature selection process. We used a metatransformer, SelectFromModel [17], for selecting features based on importance weights as shown in Listing 8.8. This feature transformer can be used along with any estimator that has a `feature_importances_` attribute after fitting. If the corresponding features' importance values are below the user-defined threshold parameter (e.g., Mean), the features are considered as unimportant and, consequently, are discarded. Figure 8.2 shows the most important features (total of 101 binary features) that we used to train and evaluate our classification algorithms.

8.3.4 Classification Models

XGBoost [8] is the abbreviation for eXtreme gradient boosting. It is a gradient boosting tree method. *Gradient* refers to the use of gradient descent, which can be used as a way to find a local minimum of a function, and *boosting* is a technique, which consists of the fact that a set of weak learners is stronger than a single strong learner. XGboost algorithm uses a differentiable loss function to calculate the adjustments needed to be made to a consecutive successor learner in an iterative learning sequence. The algorithm can automatically do parallel computations with OpenMP (an API for writing multithreaded applications), and it is much faster than existing gradient boosting algorithm. Listing 8.9 provides an excerpt of the source code for XGBoost.

The different parts of the proposed classification methodology, explained in previous subsections, can be summarized in Figure 8.3. We extended the Androguard tool [4] and built uniPDroid, a static analysis tool written in Python programming language. Our proposed method uses this tool to extract several informative features representing characteristics of the application and leverages several Python ML libraries to build the best performing classifier, XGBoost, in order to perform classification task. In particular, the system consists of two modules: (1) feature extraction module and (2) machine learning classification module. The feature extraction module includes three components. The `uniPDroid.py` is the main component within this module extracting informative features from an application, while Androguard and `Androlyze.py`

```
1  #To build a forest
2  clf = ExtraTreesClassifier(n_estimators=600)
3  clf = clf.fit(X_train, y_train)
4  #To compute the feature importances
5  importances=clf.feature_importances_
6  # To reduce 560 features to 101
7  model = SelectFromModel(clf, prefit=True)
8  X_train_new = model.transform(X_train)
```

Listing 8.8 Example of a Python script used for feature selection process.

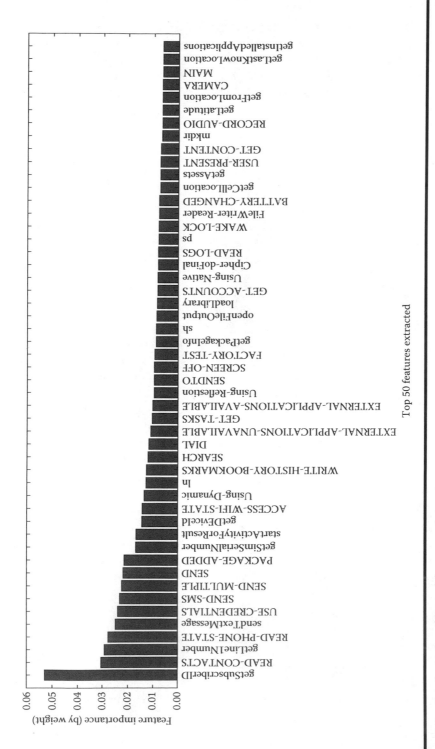

Figure 8.2 Key features extracted from our dataset.

```
1   import numpy as np
2   import xgboost as xgb
3   from sklearn.metrics import  classification_report
4
5   def train():
6     data_train = np.genfromtxt(open("train.csv","r"), delimiter=",")
7     y_train = data_train[:,0]
8     X_train = data_train[:,1:]
9     xg_train = xgb.DMatrix(X_train, label=y_train)
10    data_test = np.genfromtxt(open("test.csv","r"), delimiter=",")
11    y_test = data_test[:,0]
12    X_test = data_test[:,1:]
13    xg_test = xgb.DMatrix(X_test, label=y_test)
14    # setup parameters for xgboost
15    param = {}
16    param['objective'] = 'multi:softmax'
17    param['eta'] = 0.1
18    param['max_depth'] = 6
19    param['silent'] = 1
20    param['nthread'] = 4
21    param['num_class'] = '8 # Number of classes starting from 0
22    watchlist = [ (xg_train,'train'), (xg_test, 'test') ]
23    num_round =  260
24    bst = xgb.train(param, xg_train, num_round, watchlist);
25    # get prediction
26    y_pred = bst.predict( xg_test );
27    print classification_report(y_test, y_pred)
28
29  if __name__ == '__main__':
30    train()
```

Listing 8.9   Example of code for the machine learning classifier, eXtreme gradient boosting.

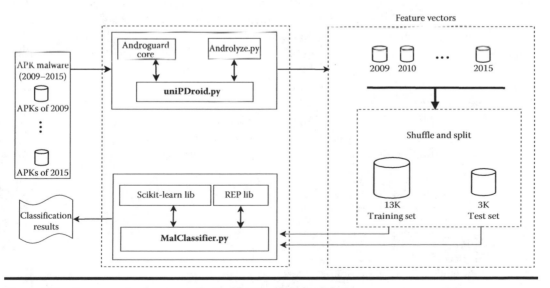

Figure 8.3   Framework of proposed classification methodology.

are auxiliary components providing support for performing feature extraction task. The ML classification module leverages several ML packages to perform classification. The main component within this module is the `MalClassifier.py`. The `Scikit-learn` and `REP` packages provide different classification algorithms and some helper functions for performance evaluation.

8.3.5 Evaluation Metrics

Table 8.2 introduces the metrics that we considered in order to assess the performance of the ML classification algorithms in class imbalance problem, that is, the total number of a class of positive data is far less than the total number of negative data. The highest precision means that an algorithm returns substantially more relevant results than irrelevant ones, while the highest recall means an algorithm returns the most of the relevant results. The F1-score combines precision and recall: it is the harmonic mean of precision and recall. We elaborate further on our empirical results in Section 8.6.

8.4 Malware Family-Based Classification

In this experiment, we carried out family-by-family malware classification. To this end, we grouped 15, 884 Android malware apps in our repository into 204 different malware families. To perform an efficient and effective classification task and have sufficient samples to feed our proposed ML classification algorithm, we discard malware families that include less than 10 samples and consequently, ended up with 78 malware families. We shuffled and split the whole data points into training and testing sets, 80% and 20%, respectively. We leveraged the XGBoost classification algorithm to perform a classification task over the 78 different malware families. Tables 8.3 through 8.5 show the malware families used in our experiments as well as the infection risks associated with each malware family.

Before conducting classification task, in order to achieve a high accuracy in performance, we fine-tuned the hyperparameters of our classification algorithm (e.g., learning rate, the number of DT, and their maximum depth) through grid search procedure combined with five fold cross-validation over the training set. Having the best parameters selected, we trained our classifier on the training set (using 13,000 samples) and tested its performance against 3,000 unseen samples in classifier point of views (i.e., on the testing set). Table 8.4 shows the classification results (F1-score) for each malware family, and Table 8.6 illustrates the overall accuracy measures in terms of precision, recall, and F1-score over the 78 malware families.

Table 8.4 shows the results of classification per malware family, number of samples, the year that those samples have been developed, and the percentage of malware families represented in our dataset. According to the table, the malware families such as SMSReg, FakeInstall, SMSPay,

Table 8.2   Performance Metrics

Metric	Description	Formula
Precision	Measure of exactness or quality	$\dfrac{T_P}{T_P + F_P}$
Recall	Measure of completeness or quantity	$\dfrac{T_P}{T_P + F_N}$
F1-score	Harmonic mean of precision and recall	$2 \times \dfrac{Precision \times Recall}{Precision + Recall}$

Table 8.3   Infection Risks Associated with Each Malware Family

Family Name	Infection Risks
AdFlex	An advertisement library that may compromise your personal information
ADRD	Steals private information
Adwo	An advertisement library that may compromise your personal information
Agilebinary	A spyware accessing the file system and retrieving app data
AirPush	A very aggressive ad network and compromises your personal information
Andup	Steals personal information
AppQuanta	An advertisement library that may compromise your personal information
Asroot	Uses asroot root exploit
AutoSMS	Attempts to steal sensitive data by seizing incoming SMS messages and forwards them to a remote site
BaseBridge	Sends premium-rate SMS to predetermined numbers
Boxer	Sends SMS to premium-rated numbers
Cobbler	A monitoring tool and wipes the SD card's contents and everything stored in the device
DDLight	Collects information about the device and sends back to a remote server
Dianjin	An advertisement library that may compromise your personal information
Dianle	Interrupts the normal operations and gains access to private information
Dougalek	Steals personal information and uploads these data to a remote server
Downloader	Gains root access and downloads additional malicious apps
DroidSheep	Captures and hijacks unencrypted web sessions
Dropper	Interrupts the normal operations and gains access to private information
Ewalls	Steals information from the mobile device
Exploid	Exploits vulnerabilities to gain root privileges on devices
FakeApp	Downloads configuration files to display advertisements and collects information from the compromised device
FakeBank	Opens a back door and steals information from the compromised device
FakeDoc	Installs additional applications
FakeInstall	Pretends to be an installer for a legitimate app, sends premium-rate SMS
FakeTimer	Sends personal information to a remote server and opens pornographic websites

(Continued)

Table 8.3 (*Continued*) Infection Risks Associated with Each Malware Family

Family Name	Infection Risks
Feejar	Sends SMS to premium-rated numbers
Geinimi	Opens a back door and transmits private information
Gepew	Attempts to replace installed apps with trojanized versions
GingerBreak	A root exploit for Android 2.2 and 2.3
GingerMaster	Utilizes a root exploit and provides root-level access
GoldDream	Steals information from Android devices
GoneSixty	Steals private information
Hamob	An advertisement library that may compromise your personal information
HiddenAds	Does not have an icon and runs in a stealth mode and displays various advertising messages
Igexin	An advertisement library that may compromise your personal information
InfoStealer	Secretly collects and uploads sensitive information
JSmsHider	Opens a back door and sends information to a specific URL
Kmin	Attempts to send data to a remote server
Kuguo	An advertisement library that may compromise your personal information
KungFu	Forwards confidential information to a remote server
LeadBolt	An advertisement library that may compromise your personal information
Lovetrap	Sends SMS to premium-rated numbers and steals information
Mecor	Monitors and compromises your personal information
Metasploit	Exploits vulnerabilities to gain root privileges on devices
Minimob	Compromises personal information and distributes via spam email
Mobclick	Aggressively pushes unwanted ads and steals personal information
MobileTX	Steals information from the compromised device and may sends SMS to a premium-rate number
Mseg	Steals private data and secretly sends SMS to premium-rated numbers
MTK	Interrupts the normal operations and gains access to the private information
Mulad	Generates income by injecting ads into legitimate free apps
NickiSpy	Gathers information from infected user's smartphone and uploads the data to a specific URL

(Continued)

Table 8.3 (*Continued*)   Infection Risks Associated with Each Malware Family

Family Name	Infection Risks
NoiconAds	Compromises personal information
Pentr	A spyware and hack tool that enables penetration testing
RuFraud	Sends SMS to premium-rated numbers
SecApk	An advertisement library that compromises your personal information
SLocker	Encrypts images, documents, and videos in the SD card to later ask for a ransom to decrypt the files
SMSKey	Interrupts the normal operations and gains access to the private information
SmsPay	Mimics a legitimate app and requires an activation fee through SMS
SMSReg	Registers the infected user to nonfree services
SMSSend	Reaps profit by silently sending SMS to premium-rate numbers
SmsSpy	Attempts to steal sensitive data by seizing incoming SMS and forwards them to a remote site
SMSZombie	Exploits a vulnerability in the mobile payment system used by China Mobile
SndApps	Compromises your personal information
SpyHasb	Monitors phone calls, SMS, and GPS locations
SpyPhone	Steals personal data
Steek	A fraudulent app that advertises an online income solution and steals privacy-related information and sends SMS
Tekwon	Interrupts the normal operations and gains access to the private information
Utchi	An advertisement library that may compromise your personal information
Vdloader	Steals personal information
Viser	Opens a back door by use of the system loopholes to introduce some adware, browser extensions, spyware, or ransomware
Wallap	Promises access to a wide collection of wallpapers and uses ad libraries to generate revenue
Waps	An advertisement library that may compromise your personal information
Wapz	An advertisement library that may compromise your personal information
Youmi	An advertisement library that may compromise your personal information
YZHCSMS	Sends SMS to a premium-rate number
Zdtad	An advertisement library that may compromise your personal information
Zsone	Sends SMS to premium rated numbers

Table 8.4   The Number of Malware Samples, Year Developed, and Classification Results of 78 Malware Families from Our Experimental Dataset

Family Name	Samples	Year Developed	Percentage of Apps	Classification F1-Score (%)
AdFlex	68	2013	0.40	86
ADRD	59	2010	0.37	100
Adwo	388	2011	2.4	83
Agilebinary	10	2010	0.06	100
AirPush	787	2010	4.9	93
Andup	18	2013	0.11	100
AppQuanta	39	2013	0.24	100
Asroot	12	2009	0.07	100
AutoSMS	46	2013	0.28	75
BaseBridge	608	2010	3.8	97
Boxer	21	2010	0.13	77
Cobbler	15	2011	0.09	100
DDLight	124	2011	0.78	100
Dianjin	91	2012	0.57	91
Dianle	54	2012	0.33	77
Dougalek	22	2012	0.13	93
Downloader	75	2012	0.47	83
DroidSheep	11	2011	0.06	100
Dropper	123	2014	0.77	95
Ewalls	43	2009	0.27	100
Exploit	41	2010	0.25	100
FakeApp	104	2011	0.65	80
FakeBank	84	2014	0.52	96
FakeDoc	130	2011	0.81	100
FakeInstall	1729	2011	10.8	98
FakeTimer	21	2012	0.13	100
Feejar	12	2014	0.07	50

(Continued)

Table 8.4 (*Continued*)   The Number of Malware Samples, Year Developed, and Classification Results of 78 Malware Families from Our Experimental Dataset

Family Name	Samples	Year Developed	Percentage of Apps	Classification F1-Score (%)
Geinimi	152	2010	0.95	100
Gepew	13	2014	0.08	100
GingerBreak	14	2011	0.08	67
GingerMaster	489	2011	3	90
GoldDream	126	2011	0.8	77
GoneSixty	15	2011	0.09	100
Hamob	35	2012	0.22	80
HiddenAds	44	2014	0.28	91
Igexin	42	2011	0.26	91
InfoStealer	209	2010	1.3	91
JSmsHider	11	2009	0.07	100
Kmin	187	2010	1.1	99
Kuguo	84	2012	0.52	50
KungFu	1051	2011	6.6	98
LeadBolt	178	2011	1.1	77
Lovetrap	11	2010	0.07	100
Mecor	10	2015	0.06	100
Metasploit	23	2014	0.14	100
Minimob	14	2013	0.09	40
Mobclick	101	2010	0.63	71
MobileTX	69	2011	0.43	100
Mseg	20	2011	0.12	67
MTK	97	2013	0.61	100
Mulad	1008	2012	6.3	99
NickiSpy	11	2010	0.07	100
NoiconAds	882	2014	5.5	99
Pentr	13	2011	0.08	67

(Continued)

Table 8.4 (Continued) The Number of Malware Samples, Year Developed, and Classification Results of 78 Malware Families from Our Experimental Dataset

Family Name	Samples	Year Developed	Percentage of Apps	Classification F1-Score (%)
RuFraud	21	2011	0.13	93
SecApk	59	2012	0.37	50
SLocker	22	2014	0.13	100
SMSKey	34	2011	0.21	100
SmsPay	1331	2010	8.4	88
SMSReg	1916	2010	12.3	88
SMSSend	487	2010	3	84
SMSSpy	207	2010	1.3	89
SMSZombie	18	2012	0.11	100
SndApps	23	2011	0.14	100
SpyHasb	13	2010	0.08	100
SpyPhone	23	2010	0.14	91
Steek	28	2011	0.17	91
Tekwon	16	2013	0.10	86
Utchi	26	2012	0.16	100
Vdloader	17	2012	0.10	77
Viser	36	2012	0.22	100
Wallap	88	2012	0.55	92
Waps	570	2011	3.5	78
Wapz	231	2012	1.5	75
Youmi	588	2010	3.7	82
YZHCSMS	59	2010	0.37	100
Zdtad	396	2015	2.5	99
Zsone	31	2011	0.19	86

Kungfu, and Mulad have the biggest share of malware samples in the entire dataset, 12.3%, 10.8%, 8.4%, 6.6%, and 6.3%, respectively. The worst classification results, 40%, belongs to the Minimob family with 14 samples. It is obvious that by increasing the number of samples in the training set our proposed ML classification algorithm will be expected to perform the training procedure better. It can be noted in Table 8.4; as the size of the training set for each malware family increases (e.g., number of samples in each family), the accuracy (F1-score) gets better. In other words, with a few

Table 8.5   Classification Report (%) for Test Set (Unseen Samples)

	Precision	*Recall*	*F1-Score*	*Support*
Avg/Total	92	92	92	3000

Table 8.6   Cross-Validation Result for Training Set

	Train$_{cv}$ *Mean Error Rate*	Test$_{cv}$ *Mean Error Rate*
CV 10-fold	0.033359 ($\pm$ 0.000760)	0.091460 ($\pm$ 0.007298)

amount of samples, it is not reasonable to expect to achieve good prediction accuracies from the classification algorithm.

Additionally, the average accuracy in terms of precision, recall, and F1-score for all 78 malware families are reported in Table 8.5. We conducted a 10-fold cross-validation experiment to compute mean error rates for both training and testing sets. Table 8.6 shows the results obtained from this experiment. For the 10-fold cross-validation, the data are randomly partitioned into 10 equal-sized subsamples. Of the 10 subsamples, a single subsample is retained as the validation data for testing the model (*Test$_{cv}$*), and the remaining 9 subsamples are used as training data (*Train$_{cv}$*). The process is then repeated 10 times, with each of the 10 subsamples used exactly once as the validation data. The 10 results from the folds can then be averaged to produce a single estimation.

Comparing the F1-score, as shown in Table 8.5, which has been obtained from evaluating our proposed classifier against unseen samples with *Test$_{cv}$*, mean error rate and the prediction from cross-validation (which is equal to 92% accuracy), we can draw this conclusion that our ML classification algorithm is never overfitted and is able to predict unseen samples with high accuracy.

8.5 Cumulative Classification

In this experiment, we accumulated Android malware apps and carried out cumulative classification where the classification results are continuously updated as new malware samples are discovered. The number of malware used in our experiment is 15,884 samples. Figure 8.4 depicts the number of malware collected by month within the period of 2009 and 2015, and Figure 8.5 shows the cumulative graph of the malware apps collected each month for that same period. In our cumulative classification, we used 56 different malware groups.

To generate the first malware group, MG_1, we take the malware apps from June 2009 to September 2010, which comprises 124 samples in order to have an initial set of samples enough to perform classification. The second data group, MG_2, contains the malware from June 2009 up to October 2010; this is achieved by adding malware belonging to upcoming month to previous months to generate the next malware group). For MG_3, we take the malware from June 2009 up to November 2010. The process is repeated until all the malware in the dataset are incorporated into the malware groups. Finally, we ended up with having 56 groups, MG_1 to MG_{56} altogether, as shown in Figure 8.5.

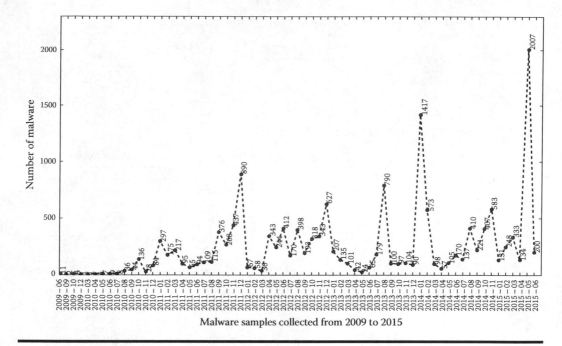

Figure 8.4 Malware number per month.

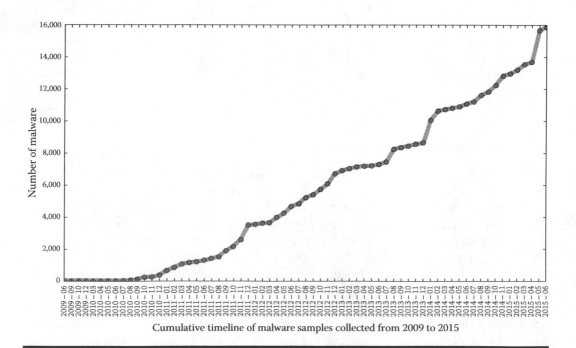

Figure 8.5 Malware number per month.

We trained the classification algorithm, XGBoost, on each malware group, MG_i, and tested its performance against malware belonging to upcoming months. We should take this point into account that malware belonging to the next month is unseen for the classification algorithm. We computed accuracy measures in terms of precision, recall, and F1-score (Figure 8.6). The aim was to investigate how features of old malware samples can be of help to classify new variant of both known and unknown malware families.

We performed cumulative classification to investigate how well the old malware can help us to detect new malware. In other words, how old malware can contribute to detecting new variants of both known and unknown malware families. As for the accuracy measures obtained from cumulative classification, Figure 8.6, at some points (e.g., February 2015) the accuracy measures drops. The reason for such a decrease in classifier performance is that we have trained the ML algorithm in certain time, on datapoints belonging to past up to that time, and we evaluate its performance against future datapoints. In the testing dataset, there exist some samples that are considered as zero-day malware in the wild (i.e., recently developed malware). The ML classification algorithm has not been trained on such samples and has no idea about these malware samples that have completely different patterns in terms of features. Consequently, the classifier cannot predict the correct label of these samples based on its past experience. As it can be seen, in the next round of cumulative classification by adding the old samples and enriching the training set, we let the classifier learn more about past data, and as a result, the classifier might perform better during the classification stage.

8.6 Conclusion

In this chapter, we proposed an ML-based malware detection and classification methodology together with the application of static analysis on an extensive dataset of Android applications. To this end, we designed a tool, uniPDroid, to extract as many informative features as possible from our dataset. We considered mainly features from the Dalvik bytecode. The features extracted were converted into feature vectors, each containing 560 binary features. We then applied feature selection on the aforementioned extracted features, which led to the selection of 101 informative binary features suitable to feed our proposed classification methodology.

Moreover, we performed an extensive grid search analysis along with a 10-fold cross-validation to tune the hyperparameters of the classification algorithm to maximize the prediction accuracy. We performed family-by-family classification and obtained an average accuracy score of 92% in classification of unseen malware. In addition to this, we conducted a cumulative classification in order to investigate how well old malware can contribute to the detection of new variants of both known and unknown (zero-day) malware. We achieved an reasonable accuracy rate, hence proving the robustness of the features extracted.

As future work, we will extend our Android application analysis and combine static analysis with features extracted from dynamic analysis to compensate the limitation associated with static analysis and get the best of both static and dynamic analyses. We will extract more features related to the behavior of Android applications such as CPU and memory consumption, network traffic activities, Interprocess Communications, and system calls made by applications so as to interact with Android OS. In addition to these, we will make use of classification algorithms such as SVM along with different kernels to deal with structured data (e.g., string, set, graph).

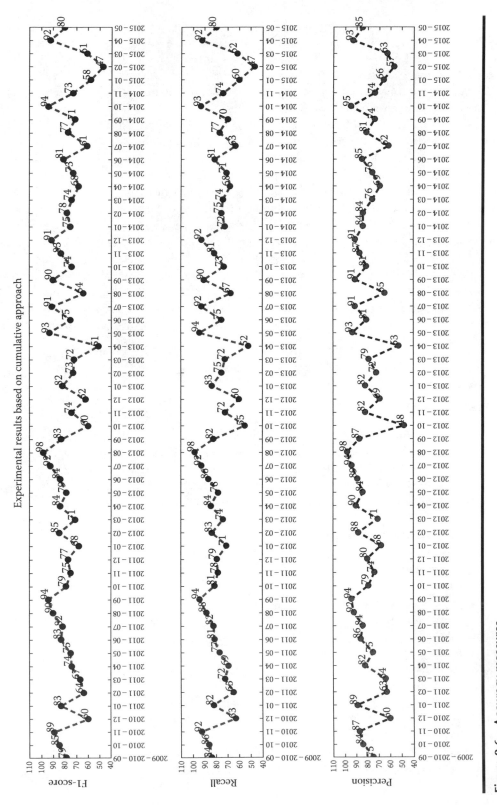

Figure 8.6 Accuracy measures.

Acknowledgments

Lejla Batina and Veelasha Moonsamy are supported by the Technology Foundation STW (project 13499 - TYPHOON & ASPASIA) from the Dutch government.

Mauro Conti is supported by a Marie Curie Fellowship funded by the European Commission (agreement PCIG11-GA-2012-321980). This work is also partially supported by the EU TagItSmart! Project (agreement H2020-ICT30-2015-688061), the EU-India REACH Project (agreement ICI+/2014/342-896), the Italian MIUR-PRIN TENACE Project (agreement 20103P34XC), and by the projects, "Tackling Mobile Malware with Innovative Machine Learning Techniques," "Physical-Layer Security for Wireless Communication," and "Content Centric Networking: Security and Privacy Issues" funded by the University of Padua.

References

1. K. J. Abela, D. K. Angeles, J. R. D. Alas, R. J. Tolentino, and M. A. Gomez. An automated malware detection system for Android using behavior-based analysis—AMDA. *International Journal of Cyber-Security and Digital Forensics (IJCSDF)*, 2(2):1–11, 2013.

2. M. Alazab, V. Moonsamy, L. Batten, P. Lantz, and R. Tian. Analysis of malicious and benign Android applications. In *Proceedings of the 32nd International Conference on Distributed Computing Systems Workshops (ICDCSW 2012)*, pp. 608–616, Macau, China, June 2012.

3. K. Allix, T. F. Bissyandé, J. Klein, and Y. Le Traon. Are your training datasets yet relevant? *Engineering Secure Software and Systems*, 8978:51–67, 2015.

4. Androguard. https://github.com/androguard/androguard/, accessed July 30, 2016.

5. D. Arp, M. Spreitzenbarth, M. Hübner, H. Gascon, and K. Rieck. Drebin: Efficient and explainable detection of Android malware in your pocket. In *Proceedings of the 2014 Network and Distributed System Security Symposium (NDSS 2014)*, pp. 1–15, San Diego, CA, February 2014.

6. Z. Aung and W. Zaw. Permission-based Android malware detection. *International Journal of Scientific and Technology Research*, 2(3):228–234, 2013.

7. I. Burguera, U. Zurutuza, and S. Nadjm-Tehrani. Crowdroid: Behavior-based malware detection system for Android. In *Proceedings of the First ACM Workshop on Security and Privacy in Mobile Devices (SPSM 2011)*, pp. 15–26, Chicago, IL, October 2011.

8. Tianqi Chen. https://github.com/dmlc/xgboost, accessed July 30, 2016.

9. K. Elish, D. Yao, and B. Ryder. On the need of precise inter-app ICC classification for detecting Android malware collusions. In *Proceedings of Mobile Security Technologies (MoST 2015)*, pp. 1–5, San Diego, CA, May 2015.

10. K. O. Elish, X. Shu, D. Yao, B. Ryder, and X. Jiang. Profiling user-trigger dependence for Android malware detection. *Computers & Security*, 49:255–273, March 2015.

11. E. Kalige and D. Burkey. A case study of Eurograbber: How 36 million euros was stolen via malware, Dec., 2012. https://www.checkpoint.com/download/downloads/products/whitepapers/Eurograbber_White_Paper.pdf, accessed July 30, 2016.

12. A. Saracino, F. Martinelli, and D. Sgandurra. Classifying Android malware through subgraph mining. In *Proceedings of Sixth International Workshop on Autonomous and Spontaneous Security (SETOP 2013)*, pp. 1–15, Egham, U.K., September 2013.

13. P. Faruki, A. Bharmal, V. Laxmi, V. Ganmoor, M.S., Gaur, M. Conti, and M. Rajarajan. Android security: A survey of issues, Malware penetration, and defenses. *IEEE Communications Surveys & Tutorials*, 17(2):998–1022, 2015.

14. E. Fernandes, B. Crispo, and M. Conti. FM 99.9, Radio virus: Exploiting FM radio broadcasts for malware deployment. *IEEE Transactions on Information Forensics and Security*, 8(6):1027–1037, 2013.

15. M. Garnaeva, V. Chebyshev, D. Makrushin, and A. Ivanov. IT threat evolution in Q1 2015. https://securelist.com/analysis/quarterly-malware-reports/69872/it-threat-evolution-in-q1-2015/, accessed July 30, 2016.

16. H. Gascon, F. Yamaguchi, D. Arp, and K. Rieck. Structural detection of Android malware using embedded call graphs. In *Proceedings of the 2013 ACM Workshop on Artificial Intelligence and Security* (*AISec 2013*), Berlin, Germany, pp. 45–54, 2013.

17. GitHub. Scikit-learn. https://github.com/scikit-learn/, accessed July 30, 2016.

18. M. Grace, Y. Zhou, Q. Zhang, S. Zou, and X. Jiang. Riskranker: scalable and accurate zero-day Android malware detection. In *Proceedings of the 10th International Conference on Mobile Systems, Applications, and Services* (MobiSys 2012), pp. 281–294, Lake District, U.K., June 2012.

19. N. Idika and A. P. Mathur. *A Survey of Malware Detection Techniques*. Purdue University, West Lafayette, IN, 2007.

20. D. Koundel, S. Ithape, V. Khobaragade, and R. Jain. Malware classification using naives Bayes classifier for Android OS. *The International Journal of Engineering and Science* (*IJES*), 3:59–63, 2014.

21. M0droid. http://m0droid.netai.net/modroid/, accessed July 30, 2016.

22. N. Peiravian and X. Zhu. Machine learning for Android malware detection using permission and API calls. In *Proceedings of the 2013 IEEE 25th International Conference on Tools with Artificial Intelligence* (*ICTAI 2013*), Herndon, VA, pp. 300–305, November 2013.

23. A. Reina, A. Fattori, and L. Cavallaro. A system call-centric analysis and stimulation technique to automatically reconstruct Android malware behaviors. In *Proceedings of the Sixth European Workshop on Systems Security* (*EuroSec 2013*), pp. 1–6, Prague, Czech Republic, April 2013.

24. B. P. S. Rocha, M. Conti, S. Etalle, and B. Crispo. Hybrid static-runtime information flow and declassification enforcement. *IEEE Transactions on Information Forensics & Security*, 8(8):1294–1305, 2013.

25. G. Russello, M. Conti, B. Crispo, and E. Fernandes. MOSES: Supporting operation modes on smartphones. In *Proceedings of the 17th ACM Symposium on Access Control Models and Technologies* (*SACMAT 2012*), pp. 3–12, Newark, NJ, June 2012.

26. J. Sahs and L. Khan. A machine learning approach to Android malware detection. In *Proceedings of the 2012 European Intelligence and Security Informatics Conference* (*EISIC 2012*), pp. 141–147, Odense, Denmark, August 2012.

27. A. A. Samra, O. A. Ghanem, and K. Yim. Analysis of clustering technique in Android malware detection. In *Proceedings of the Seventh International Conference on Innovative Mobile and Internet Services on Ubiquitous Computing* (*IMIS 2013*), pp. 729–733, Taichung, Taiwan, July 2013.

28. B. Sanz, I. Santos, C. Laorden, X. Ugarte-Pedrero, J. Nieves, P.G. Bringas, and G. Álvarez. Mama: Manifest analysis for malware detection in Android. *Cybernetics and Systems, Intelligent Network Security and Survivability*, 44:469–488, 2013.

29. B. Sanz, I. Santos, X. Ugarte-Pedrero, C. Laorden, J. Nieves, and P. G. Bringas. Anomaly detection using string analysis for Android malware detection. In *Proceedings of the Sixth International Conference on Computational Intelligence in Security for Information Systems* (*CICIS 2013*), pp. 1–10, Salamanca, Spain, September 2013.

30. R. Sato, D. Chiba, and S. Goto. Detecting Android malware by analyzing manifest files. *Proceedings of the Asia-Pacific Advanced Network*, 36:23–31, 2013.

31. S-H. Seo, A. Gupta, A. M. Sallam, E. Bertino, and K. Yim. Detecting mobile malware threats to homeland security through static analysis. *Journal of Network and Computer Applications*, 38:43–53, 2014.

32. A. Shabtai, U. Kanonov, Y. Elovici, C. Glezer, and Y. Weiss. Andromaly: A behavioral malware detection framework for Android devices. *Journal of Intelligent Information Systems*, 38(1):161–190, 2012.

33. VirusTotal. http://www.virustotal.com, accessed July 30, 2016.

34. B. Wolfe, K. Elish, and D. Yao. High precision screening for Android malware with dimensionality reduction. In *Proceedings of the 13th International Conference on Machine Learning and Applications (ICMLA 2014)*, pp. 21–28, Detroit, MI, December 2014.

35. D. J. Wu, C. H. Mao, T. E. Wei, H. M. Lee, and K. Wu. DroidMat: Android malware detection through manifest and API calls tracing. In *Proceedings of the Seventh Asia Joint Conference on Information Security (Asia JCIS 2012)*, pp. 62–69, Tokyo, Japan, August 2012.

36. M. Zhao, T. Zhang, F. Ge, and Z. Yuan. RobotDroid: A lightweight malware detection framework on smartphones. *Journal of Networks*, 7(4):1–8, 2012.

37. M. Zheng, M. Sun, and J. C. S. Lui. DroidAnalytics: A signature based analytic system to collect, extract, analyze and associate Android malware. In *Proceedings of the 12th IEEE International Conference on Trust, Security and Privacy in Computing and Communications (TrustCom 2013)*, pp. 163–171, Melbourne, Victoria, Australia, July 2013.

38. Y. Zhou, Z. Wang, W. Zhou, and X. Jiang. Hey, You, Get off of my market: Detecting malicious apps in official and alternative Android markets. In *Proceedings of the 19th Network and Distributed System Security Symposium (NDSS)*, pp. 1 13, San Diego, CA, February 2012.

39. Y. Zhou and X. Jiang. Dissecting Android malware: Characterization and evolution. *2012 IEEE Symposium on Security and Privacy (SP)*, San Francisco, CA, pp. 95–109, May 20–23, 2012.

Chapter 9

An Empirical Analysis of Android Banking Malware

Andi Fitriah A. Kadir, Natalia Stakhanova, and Ali A. Ghorbani

Contents

Abstract

Information technology's rapid evolution was always closely followed by the sophistication of malware. With ubiquitous shift to the mobile platforms, the rise of mobile malware and, in particular, banking malware came as no surprise. In general, any financial operation on the mobile platform potentially exposes a user to a variety of threats including data leakage, theft, and financial loss. Driven by financial profits, banking malware leverages user's cluelessness, openness of mobile platforms, and lack of security measures. In this work, we aim to give insight into mobile banking malware and explore unique characteristics of its communication patterns. Given the popularity of Android platform, in this work, we focus on Android banking malware detected since the first appearance of Android platform in 2008. Through static and dynamic analysis combined with visualization, we analyze patterns of benign and malicious URLs employed by malware, their common characteristics, encoding trends, and the relationships with other types of malware. Through our study, we reveal methods (e.g., hidden encryption techniques) currently adopted by attackers to avoid detection. As a part of this study, we compile and offer to the research community a dataset containing 973 samples representing 10 Android banking malware families.

9.1 Introduction

In recent years, mobile banking attacks have seen significant attention [5,18,19,25]. With ubiquitous shift to financial gain, mobile banking malware emerged as the fastest-growing threat of all attacks targeting Android platform, as shown in Figure 9.1. In 2014, over 588,000 Android users worldwide became victims of malware exploiting phones for financial gain. This constituted an increase of 71% compared to 2013 [40].

Today, the mobile platforms face a range of security challenges. One of the primary concerns is the amount of sensitive information stored on mobile phone and, traditionally, are not available on stationary computers (e.g., location information, user's activities, financial information). This situation coupled with the widespread adoption of mobile platforms for online banking creates a new lucrative target for the underground world. The other challenge relates to an increasing number of opportunities for new context-aware mobile malware to access and exfiltrate information typically not monitored by traditional detection systems. Indeed, the resource-constraint environment of

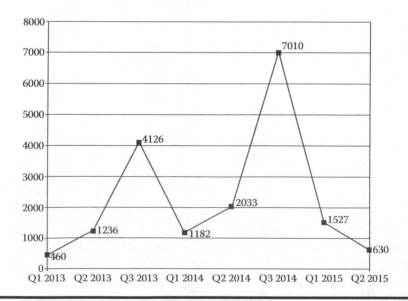

Figure 9.1 Android banking malware detected from 2013 to 2015. (Based on the analysis of data provided by [11–17]).

smartphones that are unable to afford computationally intensive operations presents significant challenges to the development of solutions for their detection.

To address an increasing problem, the research community came forward with a number of solutions [22,24,26,35,53]. Ranging from fundamentally new strategies to incremental improvements of traditional already-existing security controls, these studies search for an ultimate solution. We believe that this solution is not possible without a comprehensive understanding of the challenges.

In this work, we aim to address this gap and explore unique characteristics of mobile financial malware and its communication patterns in a comprehensive manner. Specifically, we offer an insight into the communication of Android mobile banking malware families detected since the first appearance of the Android platform. Through static and dynamic analysis combined with visualization, we reveal the common characteristics employed by the banking malware, their encoding trends, the URL patterns and similarity, and the relationships with other types of malware. Through our study, we also reveal methods (e.g., hidden encryption techniques) currently adopted by attackers to avoid detection.

The contribution of our work is threefold. First, we conduct a thorough investigation of the Android banking malware families, its characteristics and communication behavior. Second, we extract and visualize all URLs: benign or malicious, encrypted or unencrypted, and encoded or unencoded. This analysis allows us to demonstrate the significant relationships between banking families and helps us to illustrate the malware behavioral trends and characteristics. Finally, we release the accumulated dataset containing 973 banking malware samples to the research community.

The rest of the chapter is organized as follows: Section 9.2 describes the overview of Android banking malware and the collected dataset. Section 9.4 discusses the characteristics of each banking malware family. Section 9.5 analyzes the impact of Android banking malware. Similarly, Sections 9.6 and 9.7 present the discovered trends of banking malware. Section 9.8 presents the similarity of banking malware with other types of malware. Finally, Section 9.9 concludes the chapter with some remarks about the implication of the work.

9.2 Android Banking Malware

9.2.1 Android Background

An Android application is written in Java language and compiled into a *.dex* file that can be run by the Dalvik virtual machine on an Android platform. The apps are packaged in an .apk file containing the executable *.dex* file, manifest.xml file that describes the content of the package including the permissions information, optional native code (in the form of executable or libraries) that usually is called from the *.dex* file, file with a digital certificate authenticating an author, and resources that the app uses (e.g., image, sound files) [33]. Each .apk file is annotated with additional information, the so-called metadata, such as the app creation date and time, version, and size. The *.dex* file is a binary container for the code and the associated data. It includes a header containing metadata about the executable followed by identifier lists that contain references to strings, types, prototypes, fields, methods, and classes employed by the executable. The final part of the *.dex* file is the data section that contains the code and the data (i.e., URLs). The structure of *.dex* file is shown in Figure 9.2.

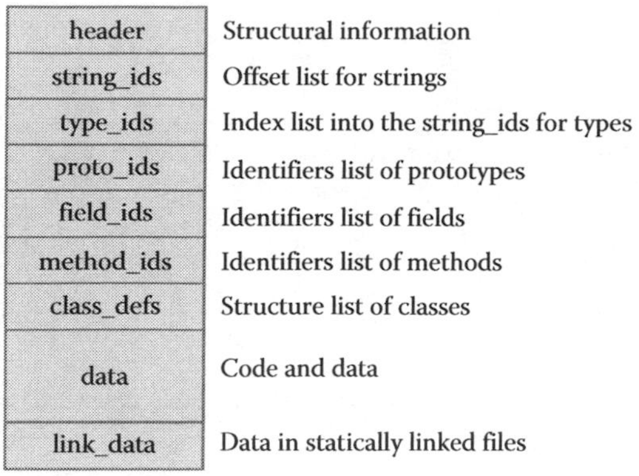

header	Structural information
string_ids	Offset list for strings
type_ids	Index list into the string_ids for types
proto_ids	Identifiers list of prototypes
field_ids	Identifiers list of fields
method_ids	Identifiers list of methods
class_defs	Structure list of classes
data	Code and data
link_data	Data in statically linked files

Figure 9.2   The layout of a *.dex* file.

9.2.2 Dataset

To give a comprehensive evaluation of Android banking malware, we gathered a large collection of Android banking malware samples representing 10 banking malware families. Our accumulated dataset combines samples from the Android Genome Malware project [53], malware security blogs [43], VirusTotal collection [1], as well as samples provided by anti-malware vendors and other researchers. Overall, our dataset includes 973 unique samples* spanning a period of 2010 (the first appearance of Android banking malware) to 2015. To ensure correct labeling of samples and accuracy of analysis, all obtained malware apps were inspected by VirusTotal [1]. We believe

Table 9.1   Overview of the Collected Data

Malware Family	Total Samples	Discovered Year	The Year of the Earliest Sample (the .dex File Year)
Bankbot	136	2015	2008
Binv	2	2014	2014
Sandroid	61	2014	2008
Wroba	152	2014	2008
FakeBank	151	2014	2008
SMSspy	131	2013	2014
ZertSecurity	4	2013	2012
Citmo	3	2012	2012
Spitmo	191	2011	2008
Zitmo	142	2010	2008
Total	973		

---

* The uniqueness was judged by different hash values.

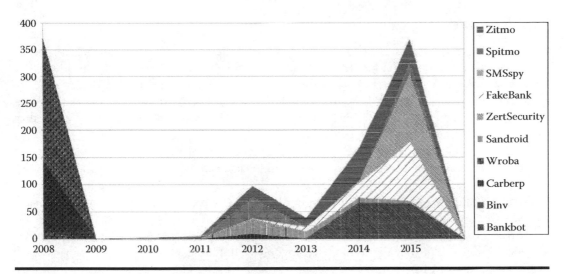

Figure 9.3   The yearly breakdown of the collected Android banking malware families.

our dataset reflects the state of the art of Android banking malware. Table 9.1 shows the list of collected malware along with discovery time of the earliest sample. Figure 9.3 illustrates the cumulative growth of malware samples. A close analysis of available samples revealed a few samples with time stamps that contradicted an official discovery time of a family. For example, the first banking malware, zeus-in-the-mobile (*Zitmo*), was first announced in 2010; there are few samples that have creation date in 2008 (Figure 9.3), which potentially indicates that although widely spread these families were released soon after appearance of the Android platform.

For comprehensive analysis, we complemented this malware set with 103 legitimate banking applications collected from Google Play market [21]. These official banking apps are associated with 17 different countries (Australia, Canada, China, Denmark, Germany, India, Indonesia, Malaysia, New Zealand, Nigeria, Russia, Saudi Arabia, Singapore, Sweden, Thailand, the United States, Vietnam). These countries were reported as the countries with the highest infection rates in the world [8].

Figure 9.4 depicts the cumulative growth of benign samples in our dataset. The graph shows that the number of Android applications is increasing yearly, which indicates that most of the banks have started adopting mobile banking.

We will release the accumulated dataset to the research community at ISCX [54].

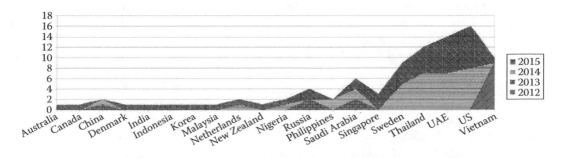

Figure 9.4   The yearly breakdown of the collected Android legitimate banking apps.

9.3 Related Work

In 2004, Guo et al. [34] predicted that mobile malware would be used for attacks against telecom networks, call centers, spam, identity theft, and wiretapping. It is ironic now since this prediction became a reality in a very short period of time. Essentially, mobile malware was almost nonexistent before the official release of the Android platform in 2008, and those few studies that were conducted focused on other platforms popular at that time. In fact, the first computer worm that infected mobile phones was targeting Symbian.

With the rapid advancement of Android devices, researchers focused their attention on Android malware. A broad overview of mobile malware characteristics were offered by Alzahrani and Ghorbani [25] and Zhou and Jiang [53]. The work by Zhou et al. was one of the early studies in domain that aimed to give researchers an understanding of mobile malware through systematic characterization of the Android malware from various aspects.

At that time, one of the main concerns was a timely detection of Android malware and one of the first attempts to provide that was offered by Bose et al. [28]. The work presented a behavioral detection framework based on logical ordering of app actions. This study was quickly followed by a series of more advanced detection approaches focused on developing detection and mitigation techniques in various areas, for example, mobile botnet detection [30,36,44,50,51], detection of privacy violations (TaintDroid [31], MockDroid [27], VetDroid [52]), and security policy violations [45,48].

Driven by profits, generic mobile malware has quickly evolved on becoming more and more focused on extracting profits. Although this already became a real concern for the industry [32, 37,40,42], there exists only a few studies on the academic side. The first one is presented by Jung et al. [39], which tested some of the major Android-based banking apps to verify whether a money transfer could be made to an unintended recipient through a repackaging attack. The experimental results showed that this repackaging attack is possible without having to illegally obtain any of the sender's personal information, such as the sender's public key certificate, the password to their bank account, or their security card. The recent work, which was published in 2015 [46], analyzed the behavior of an Android banking malware family called *BadAccents*. The authors described in detail the techniques that have been used by this malware family, that is, static and dynamic code-analysis techniques for Android applications.

Apart from that, there exist several studies on the computer-based financial malware, which are related to our work. Riccardi et al. [47] presented work-in-progress research aimed at creating a system for mitigating financial botnets. The introduced architecture promoted information sharing among law enforcement authorities, ISPs, and financial institutions. A recent work by Tajalizadehkhoob et al. [49] explored the incentives and strategies of attackers by analyzing the instructions sent to machines infected with Zeus malware between 2009 and 2013. The authors highlighted that on average code similarity is well over 90% across all Zeus versions. This suggests heavy code reuse, selling, or perhaps stealing among hackers. Another study looked at the life cycle of Zeus botnet, its attack behavior, topology, and technology based on two versions 1.2.7.19 and 2.0.8.9 [38].

None of these studies offer a comprehensive understanding of Android financial malware necessary for building effective defenses against financial mobile malware. In our study, we fill this gap and offer an empirical study of Android banking malware followed by a similarity analysis between the Android banking malware and other types of malware.

9.4 Overview of Android Banking Malware

A financial malware, that is, banking malware is a term usually used to describe the emerging trend of using specialized malware that has been designed to scan a computer system, mobile devices, or entire network, to gain information associated with financial transactions. Traditionally, financial malware is employed by hackers to commit banking fraud. Financial malware has managed to bypass secure information technologies developed specifically to protect the monetary assets of financial institutions and their customers. However, the modern definition of financial malware today ranges from keyloggers to spyware to ransomware to botnets. For example, the most advanced banking malware such as *Zeus* is not only capable of stealing the financial information but also acting as a botnet.

We have provided a characterization of the Android banking malware families, focusing on their timeline, that is, year of discovery and origin, propagation methods, attack strategies, and their special characteristics:

Evolution: As listed in Table 9.2, the *year* refers to the first appearance of the malware (year of discovery) and the *market origin* means the original location of the malware when it was first

Table 9.2   Android Banking Malware Characteristics

Malware Family	Year	Market Origin	Target Country	Information Stealing	TAN Theft	Malicious Download	Through SMS	Botnet Attack	Via Fake Application
				Propagation and Attack Types					
BankBot	2015	Third-party		✓	✓		✓		✓
Binv	2014	Google Play	Brazil	✓	✓	✓		✓	
Sandroid	2014	Third-party	MiddleEast				✓	✓	✓
Wroba	2014	Google Play	Korea				✓	✓	✓
FakeBank	2013	Google Play	Iran	✓	✓	✓			✓
SMSspy	2013	Third-party	Spain				✓		✓
ZertSecurity	2013	Third-party	German	✓					
Citmo	2012	Google Play	Russia	✓	✓				
Spitmo	2011	Third-party		✓	✓	✓			✓
Zitmo	2010	Third-party	Europe			✓	✓	✓	

discovered, either found in Google official market, that is, Google Play, or a third-party market such as Anzhi and Yandex.

The evolution of banking malware can be traced through the following four stages:

1. Genesis (2008–2012)—The introduction period of Android banking malware. The revolution started primarily with the release of the traditional desktop banking malware in the downsized mobile versions (Zitmo, Spitmo, Citmo).
2. Middle age (2013)—These malware families feature simplistic modifications, recompiled the source code with improved infection and distribution strategies, for example, ZertSecurity, SMSspy, and FakeBank.
3. Industrial era (2014)—During this year, we saw malware emerging with innovative techniques based on new infection strategies and payloads, for example, Wroba, Sandroid, and Binv.
4. Modern times and beyond (2015–onward)—So far, this is the most advanced stage that covers the recently discovered malware such as Bankbot. We now see infection and distribution advanced techniques with payloads covering a wide range of activities including classic areas of file system destruction, dropping other malcode, and stealing data.

Propagation methods: While there are a number of various possibilities for infection (e.g., SMS, MMS, Bluetooth, Wi-Fi, and infrared), mobile banking malware present in our set employs only a social engineering technique as its propagation method. Social engineering is a method that requires the target user's participation to succeed. For instance, the attacker tricks a compromised user to reveal banking account details that are then transferred to the attackers. We discovered the following social engineering medium in our dataset:

1. Through SMS—An SMS is sent to mobile devices with a fake URL that asks users to download a security certificate that is in fact the malicious payload. It also intercepts messages that are sent by banks to customers.
2. Via fake application—A fake application that masquerades as the legitimate one but silently performs malicious actions. Commonly, these trojanized apps with malicious content are distributed through third-party markets.

Attack strategies: We discovered the following attack types of Android banking malware in our dataset:

1. Botnet infection—A mobile bot is a type of malware that runs automatically once installed on a mobile device to gain complete access to the device and its contents as well as providing control to the botnet creator. It starts communicating with and receiving instructions from one or more command and control (C&C) servers. Mobile botnets take advantage of unpatched exploits to provide hackers with root permissions over the compromised mobile device, enabling hackers to send e-mail or text messages, make phone calls, access contacts and photos, and more.
2. Personal information stealing—The malware is harvesting various information on the infected phones, including SMS messages, phone numbers, and user accounts.
3. Transaction authentication number (TAN) theft—A TAN is used by online banking services as a form of single-use one-time passwords to authorize financial transactions. TANs provide

additional security because they act as a form of two-factor authentication. TAN theft is a known attack targeting mobile banking services.

4. Malicious download—A malicious application applies the traditional drive-by download attacks to mobile space. This attack tricks users to download a particular smartphone application, which is claimed to better protect online banking activities. But the downloaded application is actually a malware, which can collect and send mTANs or SMS to a remote server.

Target country: It is interesting to note that Android banking malware tends to be focused on specific geographical areas. For instance, 80% of the malware families are targeting a specific country such as Brazil, Korea, Iran, Spain, German, and Russia. Table 9.2 summarizes the characteristics of the Android banking malware.

9.4.1 Brief Summary of Malware Families

BankBot: This malware family is a Trojan-banker type that steals the user's confidential banking information. This Trojan can imitate legitimate online banking applications to get access to clients' confidential data; by sending a special SMS message to a phone number (associated with the users bank account). Another variant of this family can scan the system for the presence of banking applications, download its fake version as an update package, and uninstall original programs. Moreover, the malware can intercept incoming SMS messages, steal phone book information, gather data regarding the infected mobile device, and even uninstall some antiviruses [6].

Binv: This malware is a classical banking-Trojan that targets the Brazilian users of Android devices by masquerading its malicious applications as legitimate banking applications. The Brazilian mobile banking users are particularly exposed to the attack due to the lack of authentication system (i.e., two-factor authentication) for banking applications. The malware is very simple; it was designed to steal login credentials in a classic phishing scheme. This malware was created using a free development tool known as App Inventor, which does not require any particular skill to create a mobile application. The malware was originally published on the Google Play store with the name Governo Federal (Federal Government) [19].

Citmo: The cybercriminals behind Carberp began its operations in 2009 but did not actually migrate to the mobile platform until 2012, when a Carberp-in-the-mobile, or Citmo, is found on Google Play masqueraded as mobile applications. Citmo was primarily targeting Russian banks and financial institutions (Sberbank and Alfa bank). However, in December 2012, Citmo was also targeting several banks in the United States and Canada. Citmo has joined the ranks of Zeus-in-the-mobile (Zitmo) and SpyEye-in-the-mobile (Spitmo) in targeting SMS messages sent by banks to authenticate online banking transactions. The Citmo Android Trojan works in almost the same way as ZitMo. It can hide particular SMS messages, or resend them to the attackers command server or to other phone numbers. Even though the cybercriminals associated with Carberp have been arrested by Russian and Ukranian authorities, the Carberp malware is not over as Carberp's source code was leaked in 2013. Similar to Zeus, anyone can modify Carberp. Thus, the new variants can appear in Europe, the United States, or Latin America. The malware was also published on the Google Play store [10].

FakeBank: This malware name is self-explanatory, a fake banking application that targets Iranian users. FakeBank is a Trojan horse for Android devices that opens a back door and steals information from the compromised device. Once installed on a victim's phone, the application monitors

SMS activity for incoming verification messages from the Iranian bank it seeks to imitate. This malware copies the information into another message and sends it out via SMS to the attackers. Additionally, it is able to infect a connected Windows PC and tricks the user to exchange legit banking apps against malicious ones. It was also published on the Google Play store [7].

Sandroid: This malware disguises itself as a bank dynamic token generator that targets middle east banks, including Riyad Bank, SABB (formerly the Saudi British Bank), AlAhliOnline (National Commercial Bank), Al Rajhi Bank, and Arab National Bank. It is running another background service that monitors users incoming SMS. Every incoming SMS is silently sent to a certain number through SMS. It also actively silently sends SMS commands from a remote website. The remote malicious user may then initiate unauthorized transaction without user's consent and may incur financial loss [2].

SMSspy: This malware is a classical banking Trojan that targets the Spanish users of Android devices by masquerading its malicious applications as legitimate banking applications. Once installed on a victim's phone, the application monitors SMS activity for incoming verification messages and is able to intercept and modify banking authentication codes (mTAN messages) [7].

Spitmo: This malware is also known as SpyEye-in-the-mobile and was discovered in 2011 after the SpyEye source code was leaked. Spitmo is a banking Trojan that steals information from the infected smartphone. The Trojan also monitors and intercepts SMS messages from banks (mTAN messages) and uploads them to a remote server. In 2009, SpyEye surpassed Zeus and evolved into a pricey, user-friendly software program. It was sold, updated, and copyrighted like a legit business application. However, in January of 2014, the man who created SpyEye pleaded guilty, and he was sentenced by a U.S. district judge [7].

Wroba: The Korean malware Wroba spreads via alternative application stores. Once it infects a device, Wroba behaves very aggressively. It searches for mobile banking applications, removes them, and uploads counterfeit versions. This malware hides its main malicious activity within a package that is encrypted and hidden within itself. Wroba malware is used to capture login credentials for bank accounts and banking information and other data to monetize the attack. The inner malicious package is present in the original package as an asset file and is decrypted using DES before it can be loaded and the malicious functions called [3].

ZertSecurity: This malware family is a Trojan-banker type that steals confidential banking authentication codes (mTAN messages). Once installed on a victim's phone, the application tricks a compromised user to insert his banking account details. The Trojan also monitors and intercepts SMS messages from banks (mTAN messages) and uploads them to a remote server [7].

Zitmo: This malware is also called Zeus-in-the-mobile and was discovered at the end of September 2010. Zitmo targeted various mobile platforms such as Symbian, Windows Mobile, Blackberry, and Android. Its functionality consists of the ability to upload all incoming SMS messages (with mTAN also) to a remote web server, the ability to forward SMS messages from a particular number, as well as the ability to change the C&C. The malicious application passes itself off as a security tool from the *Trusteer* company. If a user installs the malicious application, then the *Trusteer Rapport* icon will appear in the main menu [4].

9.5 Analyzing the Impact of Android Banking Malware

Transitioning from static to dynamic analysis: The sophistication of Android banking malware is increasing at a rapid pace. Although similar techniques are continuously used, the method in which these techniques are applied is constantly changing. Based on our analysis, we found a

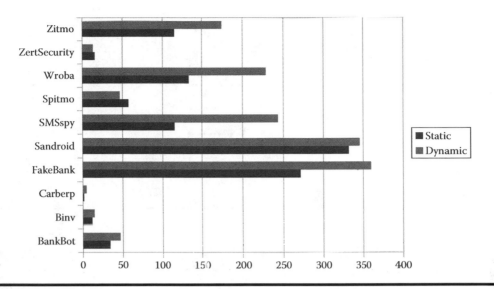

Figure 9.5 Static versus dynamic: URL extraction.

transitioning trend (from a static analysis to a dynamic ones) in analyzing the banking malware samples. The amount of data that can be extracted using static analysis such as the C&C URLs, encryption key, and password has been decreasing over the years. As such very little to almost no information can be identified through static analysis of samples dating 2011 and beyond. While with malware evolution this is somewhat expected, it emphasizes the importance of dynamic analysis. Figure 9.5 shows the comparison number of the URLs that we have extracted using two types of analysis.

From traditional desktop to mobile version: Out of 10 malware families in our dataset, 3 of them have evolved from the traditional desktop versions. Zitmo, Spitmo, and Citmo were discovered in 2010, 2011, and 2012, respectively. All of these three families have utilized the social engineering technique, advising a victim to download the application from an online store. Once installed, the app steals one-time SMS authentication passwords. This period of 2010–2012 is the period of active reuse of existing old malware. In other words, mobile banking malware of this period was proof of concept malware.

9.6 Analyzing the Encryption and Obfuscation Technique

The use of obfuscation in malware is often necessitated by a wide variety of reverse engineering tools that help in malware analysis. This obfuscation ranges from various syntactic code mutations such as junk code insertion and variable renaming to code encryption.

As such, we checked the presence of traces of three commonly used obfuscation products (ProGuard, DexGuard, APK Protect) in our dataset. We followed the same procedure detailed in [9]. For instance, strings such as *a/a/a;->a* in the *smali* code is a strong indication that the sample has been obfuscated using ProGuard, while the repetitive use of non-ASCII characters in *smali* code is an identification of samples obfuscated with DexGuard. The list of the obfuscation tools employed by apps in our dataset is given in Table 9.3. Among families that were analyzed,

Table 9.3   Overview of Obfuscation Methods Used in Android Banking Malware

	Total APK	Dexguard (Total APK)	Proguard (Total APK)	APK Protect (Total APK)
Bankbot	136	136	12	1
Binv	2	2	0	0
Citmo	3	3	0	0
FakeBank	151	151	10	8
Sandroid	61	61	7	54
SMSspy	131	130	4	3
Spitmo	191	191	0	0
Wroba	152	150	28	3
Zert	4	4	0	0
Zitmo	142	139	6	10
Total	973	967	67	76

six malware families (Bankbot, FakeBank, SMSspy, Sandroid, Wroba, Zitmo) adopted more than one obfuscation method in hiding their malicious activities (see Figure 9.6).

Throughout the analysis, we also revealed five types of encryption/encoding algorithms that have been used in banking malware: advanced encryption standard (AES), data encryption standard (DES), RSA, Base64, and keyed-hash message authentication code (HMAC) (see Table 9.4). As the results show, the majority of banking malware is using RSA and Base64. Out of 10 malware families, 7 of them are using more than one encryption technique. For instance, FakeBank uses all five encryption methods in their malicious applications. Figure 9.7 visualizes the sharing of the encryption algorithms for each malware family.

9.7 Analyzing the URLs

Traditionally, URLs are contained in the file's metadata (e.g., the application links for updates) and are embedded in malware code either as plain text or obfuscated strings. URLs from metadata are easily extracted with a simple, regular expression-based search. We refer to these extracted metadata URLs as "built-in URLs".

To obtain URLs, we employ two types of analysis: static and dynamic. The combination of static and dynamic analysis helps us analyze the behavior of banking malware. For instance, a sophisticated banking malware would try to hide its malicious intent and, therefore, would avoid putting C&C URLs as metadata, where they can easily be extracted by static analysis. This hidden malicious intent can be only discovered through the dynamic analysis.

In the static analysis, we leveraged regular expressions and a set of keywords to indicate URL presence (e.g., *http, password, key*). We looked at the similarity of keyword pattern between families. For instance, searching the *const-string* keyword can give us more results of URLs than the plain *http* keyword. By examining the existing pattern for each banking families, we managed to extract more

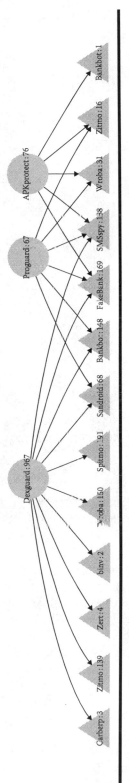

Figure 9.6 Examples of sharing obfuscation methods.

Table 9.4 Overview of the Employed Encryption/Encoding

Malware Family	Encryption/Encoding Type				
	AES	DES	RSA	Base64	HMAC
BankBot	0	1	2	13	0
Binv	0	0	1	2	0
Citmo	0	0	1	0	0
FakeBank	6	7	9	22	1
Sandroid	0	3	28	45	1
SMSspy	56	72	116	118	0
Spitmo	0	0	0	5	0
Wroba	0	1	3	49	0
ZertSecurity	4	0	0	0	0
Zitmo	27	6	13	14	0
Total	93	90	173	268	2

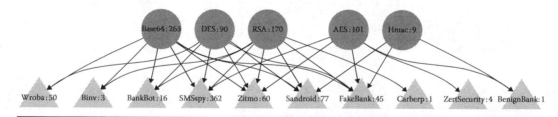

Figure 9.7 Examples of sharing encryption algorithms.

URLs even the encoded ones. This is obtained by using the binary code searching. Search of binary code, however, requires disassembly of .apk file and analysis of .dex file bytecode. Using baksmali disassembler, we retrieve bytecode from each sample and string portion of its data section.

For the dynamic analysis, we employed the ANUBIS platform [20]. ANUBIS provides both static and dynamic analysis reports that cover the following aspects of .apk files: activities, services, broadcast receivers, required permissions, used permissions, features, URLs, file operations, network operations, cryptography operations, started services, and native libraries loaded. In this research, we are only interested in analyzing the URLs and the cryptography operations. We have extracted all the URLs from the collected .xml generated by ANUBIS. Following the described procedure, we extracted thousands of URLs from our collected dataset (see Table 9.5). The total unique URLs of all banking families for both static and dynamic analysis is 2577. Removing 550 duplicative URLs, we ended up with 2027 unique URLs for analysis.

Figure 9.8 depicted an overview of the URL category that we found in our URL analysis. We identify seven main categories of the URLs: advertisement, files, top-level domain (TLD), popular

Table 9.5 Overview of the Extracted URLs

Malware Family	Total Samples	Static Analysis (URL)		Dynamic Analysis (URL)		Duplicative URL
		Total	Unique	Total	Unique	
Bankbot	136	141	34	1,013	47	9
Binv	2	36	12	36	15	10
Citmo	3	3	2	9	5	0
FakeBank	151	1,346	272	2,177	361	169
Sandroid	61	1,866	332	1,794	346	126
SMSspy	131	9,608	116	11,247	244	47
Spitmo	191	103	58	265	47	14
Wroba	152	1,426	135	835	229	54
ZertSecurity	4	16	16	64	14	16
Zitmo	142	396	116	596	175	105
Total	973	14,941	1094	18,036	1483	550

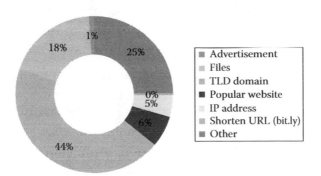

Figure 9.8 Overview of URL category.

website, IP address, shorten URL, and other categories. The subcategories of these URLs are listed in Table 9.6. There are 11 TLD domains found in the URL (.biz, .cn, .co, .com, .gov, .info, .net, .org, .ru, .us, .kr) and about 14 different types of files identified (APK, cgi, css, html, image, json, jsp, mp3, pdf, php, txt, video, xml, zip). We have checked all these 2027 URLs with VirusTotal [1] and found that 539 of them are detected as malicious by VirusTotal. The result of the malicious URLs will be shown in the next section. These malicious URLs, especially the blog's URLs, indicate the various use of C&C channel in the banking malware attacks.

Out of 10 banking families, only 3 families (Spitmo, SMSspy, and Sandroid) are using the advertisement URL in their malicious applications as listed in Table 9.7. This is conducted by filtering the extracted URLs with the *ads* keyword. As shown in Table 9.7, Sandroid family contains many ads URLs if compared to other families. We suspect that the Sandroid samples that we have

Table 9.6   Overview of URLs

No.	Category	Total	Total Malicious
1	Popular website	245	80
	Amazon	6	0
	Baidu	54	49
	Blogspot	21	21
	Dropbox	6	6
	Facebook	9	0
	Github	9	2
	Google	63	2
	Linkedin	2	0
	Paypal	1	0
	Twitter	23	0
	Wikipedia	13	0
	Youtube	2	0
2	TLD domain	1685	428
	.biz	4	2
	.cn	117	30
	.co	94	28
	.com	1166	285
	.gov	5	0
	.info	1	0
	.net	148	68
	.org	127	5
	.ru	17	9
	.us	4	0
	.kr	2	1
3	Files	701	459
	APK	74	19
	Cgi	17	3
	Css	8	4

(Continued)

Table 9.6 (*Continued*)   Overview of URLs

No.	Category	Total	Total Malicious
	Html	65	32
	Image	170	15
	Json	20	17
	Jsp	18	6
	Mp3	5	0
	Pdf	3	0
	Php	249	93
	Txt	6	5
	Video	22	7
	Xml	8	1
	zip	3	3
4	Advertisement	39	20
5	IP address	173	111
6	Shorten URL (bit.ly)	6	4
7	Other types	945	119

in our dataset are mostly repackaged versions. Repackaging is one of the most common techniques used by adversaries to inject malicious payloads such as adware.

Moreover, we have identified that the banking malware exploits the domain name system (DNS) by adopting the domain generation algorithm (DGA) and the URL obfuscation techniques. Le et al. identified several URL obfuscation types [41]:

- Type I—Obfuscation of host with an IP address
- Type II—Obfuscation of host with another domain
- Type III—Obfuscation with large hostname
- Type IV—Unknown or misspelled domain

Our set features all four types of obfuscation. Table 9.8 shows the URL obfuscation techniques commonly used by the attackers.

Specifically, out of the 2027 unique URLs that we have extracted, 173 URLs are obfuscated with IP addresses; 111 of them are labeled as malicious according to VirusTotal.

One interesting aspect of the analyzed banking malware families is the sharing of resources among malware families. This is clearly visible in Figure 9.9 that displays all the extracted URLs collected during both static and dynamic analysis. Through static analysis, we found only two families sharing URLs (one-to-one relationship) (between Sandroid and FakeBank, Wroba and Zitmo). On the other hand, we identified a many-to-many relationship in dynamic analysis.

Table 9.7   Overview of the Extracted Advertisement (ad) URLs

Family	URL
Spitmo	http://schemas.android.com/apk/lib/com.google.ads
SMSspy	http://ads.wapx.cn/action/pop_ad/ad?ap
Sandroid	http://ads.mdotm.com/ads/receiver.php?referrer=
Sandroid	http://ads.openfeint.com
Sandroid	http://backoffice.adviator.com/ads/madss_api.php
Sandroid	http://ads.mopub.com/m/ad
Sandroid	http://ads.mopub.com/m/imp
Sandroid	http://ads.mp.mydas.mobi/appConfigServlet?apid=
Sandroid	http://androidsdk.ads.mp.mydas.mobi/getAd.php5?sdkapid=
Sandroid	http://beta.airpush.com/images/adsthumbnail/48.png
Sandroid	http://ads.mopub.com/
Sandroid	http://ads.wapx.cn
Sandroid	http://ad.leadboltapps.net/clk
Sandroid	http://ads.dt.mydas.mobi/getAd.php5?asid=
Sandroid	http://incorporateapps.com/spoty/spotyliteAds.php
Sandroid	http://relay.mobile.toboads.com

Table 9.8   Commonly Used URL Obfuscation Techniques

Type	Descriptive Examples
I	http://126.68.92.179/xinhan.apk
II	http://twitter.com/share?url=https://www.cgd.pt/Corp
III	http://d36hc9ptsltjmz.cloudfront.net/static/templates_app/8/style.css
IV	http://ykjbvkl.iego.net/appHome/&lt/string&gt

9.8 Android Banking Malware Similarity

To provide a comprehensive evaluation of Android banking malware, the malicious banking applications were compared with the legitimate apps. Table 9.9 summarizes extracted information from the legitimate banking applications: the encryption technique, obfuscation tool, and the URLs employed by the legitimate applications. We identified that the benign applications adopt AES, Base64, and RSA. These applications also use Proguard and Dexguard as their obfuscation tools. As expected, the benign applications utilize the popular social networks such as Facebook and Twitter

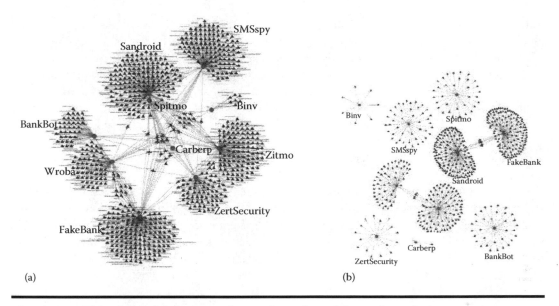

Figure 9.9   URL relationship between the (a) dynamic and (b) static analysis.

Table 9.9   Overview of the Extracted Information of the Legitimate Banking Applications

No.	Category	Total
1	Encryption technique	94
	AES	8
	DES	0
	Base64	0
	RSA	0
	HMAC	7
2	Obfuscation tool	71
	Proguard	66
	Dexguard	5
	APKprotect	0
3	URLs	218
	Facebook	34
	Twitter	50
	Google	102
	Ads	32

and advertisement technology in their applications. We also checked the URL similarity between the benign and the malicious banking applications and found 25 similar URLs. But the similar URLs are all the normal URLs like google.com and apache.org.

We analyzed the URL similarity between the banking dataset with the Android botnet dataset [23]. This botnet dataset contains 14 botnet families: Anserverbot, Bmaster, DroidDream, Geinimi, MisoSMS, NickySpy, NotCompatible, Pjapps, Pletor, Rootsmart, Sandroid, Tigerbot, Wroba, and Zitmo. We excluded three repetitive malware families that we have with our dataset (Sandroid, Wroba, Zitmo). Overall, we found 173 same URLs between the datasets. These URLs are mainly the advertisement, social networking sites, and famous websites such as *google.com* and *schemas.android.com*.

We also looked at the maliciousness of the extracted URLs. In this case, we compared all of the extracted URLs using service provided by VirusTotal. Since the VirusTotal service incorporates a large collection of antivirus scanners, which use different detection strategies. Out of the 2577 unique URLs we extracted, 539 of them were detected as malicious. Table 9.6 summarizes the list of malicious URLs based on the category. Interestingly, some famous file-sharing services such as Dropbox and Github are labeled malicious when provided as complete URL.

9.9 Conclusion

In this work, we have conducted a comprehensive analysis of the Android banking malware; we have explored the unique patterns of benign and malicious URLs employed by banking malware, their common characteristics and encryption trends. We have also demonstrated the major difference between Android malicious banking apps and the benign ones and showed their relationship with other types of malware. Through this study, we discovered that the Android banking malware are actually sharing some URLs with other types of malware including the URLs of the advertisement technology and some popular websites. We confirmed that Android banking malware are evolving; thus, the samples from 2013 until the recent ones are more dynamic. Most of the malware encrypts various types of data including the URLs, the method names, the file path, and the content of the payloads to evade detection. We were also able to identify the variety of encryption techniques used by attackers. This was achieved by extracting the strings and visualizing each .apk files from its malware family as well as the mappings between the URLs.

9.9.1 Implication of Our Studies

Based on our analysis and results, we have identified three factors that should be taken into account when developing techniques for Android banking malware detection:

1. Behavioral similarity of Android banking malware. Our results revealed a significant relationship among banking malware in terms of their resources, techniques, and configurations. We noticed that most .apk files (regardless of family) adopt the same practice: (1) recycle the same resources of URLs, (2) adopt similar techniques in generating the URLs (DGA, URL obfuscation techniques), and (3) apply the same configuration of encryption (encryption key and algorithm). These behavioral relationships should be taken into account for developing Android banking malware detection techniques.

2. Evolution of Android banking malware. Our dataset offers a comprehensive set of Android banking malware covering samples from the first appearance in 2010 to the recent one in 2015. Our analysis revealed that Android banking malware has become extensively sophisticated over time. Based on the 10 banking malware families that we have analyzed, we discover that the first group of botnets from 2010 to 2012 such as Zitmo, Spitmo, and Citmo have utilized multiple techniques (encrypting URLs, exploiting DNS, utilizing public blog) to evade detection. We suspect that these years were a period of transition from the traditional desktop banking malware in the downsized mobile versions where adversaries are just implementing and testing the possible techniques to prove their concept. Later in 2013 and 2014, we saw malware emerging with innovative and advanced techniques based on new infection strategies and payloads, which are capable of carrying payloads, including file system destruction, dropping other malcode, and stealing data.

3. URL similarity. Our result shows the following: (1) There is approximately 1% similarity between the benign and the malicious banking URLs. But the similar URLs are all the normal URLs like google.com and apache.org. (2) Similarity between the Android banking malware and the Android botnet URLs is about 7%. these URLs are mainly the advertisement, social networking sites, and famous websites such as *google.com* and *schemas.android.com*. (3) According to VirusTotal, 20% of the extracted URLs are malicious URLs. Interestingly, some well-known file sharing services such as Dropbox and Github are labeled malicious when provided as complete URLs.

Acknowledgments

This research has been partially supported by the New Brunswick Innovation Foundation under the research grant RAI 2015-090 and the Atlantic Innovation Foundation (AIF) under the project #201212. The authors also gratefully acknowledge the funding from the International Islamic University Malaysia (IIUM).

References

1. Virus Total Teamv. 2016. VirusTotal-Free online virus, malware and URL scanner. Retrieved from https://www.virustotal.com/en/, accessed August 1, 2015.
2. K. Clerix. 2014. Android botnet targets Middle East banks. Retrieved from http://krebsonsecurity.com/2014/04/android-botnet-targets-middle-east-banks/, accessed August 13, 2015.
3. R. Nigam. 2015. A timeline of mobile botnets. Retrieved from https://www.virusbtn.com/virusbulletin/archive/2015/03/vb201503-mobile-botnets, accessed August 13, 2015.
4. D. Maslennikov. 2011. Zeus-in-the-mobile for Android. Retrieved from https://securelist.com/blog/virus-watch/29258/zeus-in-the-mobile-for-android-10/, accessed August 13, 2015.
5. Norsecorp. 2015. Android users targeted by sberbank mobile banking trojan. Retrieved from http://darkmatters.norsecorp.com/2015/06/15/android-users-targeted-by-sberbank-mobile-banking-trojan/, accessed July 11, 2015.
6. Dr.Web. 2015. Android.bankbot in virus library. Retrieved from http://vms.drweb.com/search/?q=Android.BankBot&lng=en/, accessed July 11, 2015.
7. Spreitzenbarth. 2015. Current Android malware. Retrieved from http://forensics.spreitzenbarth.de/android-malware/, accessed July 11, 2015.

8. NQ Mobile Inc. 2015. NQ mobile: Mobile malware trends. Retrieved from http://ir.nq.com/phoenix.zhtml?c=243152&p=irol-newsArticle&ID=1963833, accessed July 11, 2015.

9. Axelle Apvrille. 2014. Obfuscation in Android malware, and how to fight back. Retrieved from https://www.virusbtn.com/virusbulletin/archive/2014/07/vb201407-Android-obfuscation#citation.11, accessed July 11, 2015.

10. F. Y. Rashid. 2012. Carberp trojan goes mobile: 'Citmo' found targeting Russian banks. Retrieved from http://www.securityweek.com/carberp-trojan-goes-mobile-citmo-found-targeting-russian-banks, accessed July 13, 2015.

11. D. Maslennikov. 2013. IT threat evolution in Q1 2013. Retrieved from https://securelist.com/analysis/quarterly-malware-reports/36660/it-threat-evolution-q1-2013/, accessed July 13, 2015.

12. V. Chebyshev, D. Emm, M. Garnaeva, and R. Unuchek. 2014. IT threat evolution in Q1 2014. Retrieved from https://securelist.com/analysis/quarterly-malware-reports/59417/it-threat-evolution-q1-2014/, accessed July 13, 2015.

13. M. Garnaeva, V. Chebyshev, D. Makrushin, and A. Ivanov. 2015. IT threat evolution in Q1 2015. Retrieved from https://securelist.com/analysis/quarterly-malware-reports/69872/it-threat-evolution-in-q1-2015/, accessed July 13, 2015.

14. IT threat evolution in Q2 2013. Retrieved from https://securelist.com/analysis/quarterly-malware-reports/37163/it-threat-evolution-q2-2013/, accessed July 13, 2015.

15. D. Emm, R. Unuchek, V. Chebyshev, M. Garnaeva, and D. Makrushin. 2014. IT threat evolution in Q2 2014. Retrieved from https://securelist.com/analysis/quarterly-malware-reports/65340/it-threat-evolution-q2-2014/, accessed July 13, 2015.

16. V. Chebyshev, C. Funk, and M. Garnaeva. 2013. IT threat evolution in Q3 2013. Retrieved from https://securelist.com/analysis/quarterly-malware-reports/57885/it-threat-evolution-q3-2013/, accessed July 13, 2015.

17. D. Emm, M. Garnaeva, V. Chebyshev, R. Unuchek, D. Makrushin, and A. Ivanov. 2014. IT threat evolution in Q3 2014. Retrieved from https://securelist.com/files/2014/11/KLQ32014REPORT.pdf, accessed July 13, 2015.

18. L. Caetano. 2014. Mcafee: From Russia with malware: Svpeng mobile banking trojan. Retrieved from https://blogs.mcafee.com/consumer/watch-out-for-svpeng-trojan, accessed June 12, 2015.

19. P. Paganini. 2015. Security affair 2015: 11 percent of mobile banking apps includes harmful code. Retrieved from http://securityaffairs.co/wordpress/33212/malware/mobile-banking-apps-suspect.html, accessed June 12, 2015.

20. iseclab. 2015. Anubis: Web based malware analysis for unknown binaries. Retrieved from https://anubis.iseclab.org/, accessed June 7, 2015.

21. Google. 2015. Android apps on Google Play. Retrieved from https://play.google.com/store/apps, accessed June 7, 2015.

22. Abdelrahman, O.H., Gelenbe, E., Görbil, G., and Oklander, B. Mobile network anomaly detection and mitigation: The nemesys approach. In *Information Sciences and Systems 2013*, Baltimore, MD, pp. 429–438. Springer, New York, NY, 2013.

23. A. Kadir, A.F., Stakhanova, N., and Ghorbani, A.A. Android botnets: What URLs are telling us. In *Ninth International Conference on Network and System Security (NSS'15)*, New York. Springer, New York, 2015.

24. Alzahrani, A.J. and Ghorbani, A.A. SMS mobile botnet detection using a multi-agent system: Research in progress. In *Proceedings of the First International Workshop on Agents and CyberSecurity*. (ACySE'14), Paris, France, pp. 2:1–2:8. ACM, New York, 2014. http://doi.acm.org/10.1145/2602945.2602950.

25. Alzahrani, A.J., Stakhanova, N., Gonzalez, H., and Ghorbani, A.A. Characterizing evaluation practices of intrusion detection methods for smartphones. *Journal of Cyber Security and Mobility*, 3(2): 89–132, 2014.

26. Arabo, A. and Pranggono, B. Mobile malware and smart device security: Trends, challenges and solutions. In *Proceedings of the 2013 19th International Conference on Control Systems and Computer Science (CSCS'13)*, Bucharest, Romania, pp. 526–531. IEEE Computer Society, Washington, DC, 2013. http://dx.doi.org/10.1109/CSCS.2013.27.

27. Beresford, A.R., Rice, A., Skehin, N., and Sohan, R. Mockdroid: Trading privacy for application functionality on smartphones. In *Proceedings of the 12th Workshop on Mobile Computing Systems and Applications*, Phoenix, AZ, pp. 49–54. ACM, New York, NY, 2011.

28. Bose, A., Hu, X. Shin, K.G., and Park, T. Behavioral detection of malware on mobile handsets. In *Proceedings of the Sixth International Conference on Mobile Systems, Applications, and Services*, Breckenridge, CO, pp. 225–238. ACM, New York, NY, 2008.

29. Chebyshev, V. and Unuchek, R. Mobile malware evolution: 2013, July 2014. Retrieved from http://www.securelist.com/en/analysis/204792326/Mobile_Malware_Evolution_2013, accessed June 7, 2015.

30. Choi, B., Choi, S.K., and Cho, K. Detection of mobile botnet using VPN. In *2013 Seventh International Conference on Innovative Mobile and Internet Services in Ubiquitous Computing (IMIS)*, Taichung, Taiwan, pp. 142–148. IEEE, New York, NY, 2013.

31. Enck, W., Gilbert, P., Han, S., Tendulkar, V., Chun, B.G., Cox, L.P., Jung, J., McDaniel, P., and Sheth, A.N. Taintdroid: An information-flow tracking system for realtime privacy monitoring on smartphones. *ACM Transactions on Computer Systems (TOCS)*, 32(2): 5, 2014.

32. Erturk, E. Two trends in mobile security: Financial motives and transitioning from static to dynamic analysis, 2005. CoRR abs/1504.06893. Retrieved from http://arxiv.org/abs/1504.06893, accessed June 12, 2015.

33. Gonzalez, H., Kadir, A.A., Stakhanova, N., Alzahrani, A.J., and Ghorbani, A.A. Exploring reverse engineering symptoms in Android apps. In *Proceedings of the Eighth European Workshop on System Security (EuroSec'15)*, Bordeaux, France, pp. 7:1–7:7. ACM, New York, 2015. http://doi.acm.org/10.1145/2751323.2751330.

34. Guo, C., Wang, H.J., and Zhu, W. Smart-phone attacks and defenses. In *HotNets III*, San Diego, CA, 2004.

35. Hasan, R., Saxena, N., Haleviz, T., Zawoad, S., and Rinehart, D. Sensing-enabled channels for hard-to-detect command and control of mobile devices. In *Proceedings of the Eighth ACM SIGSAC Symposium on Information, Computer and Communications Security. (ASIA CCS'13)*, Hangzhou, China, pp. 469–480. ACM, New York, 2013.

36. Hua, J. and Sakurai, K. A SMS-based mobile botnet using flooding algorithm. In *Information Security Theory and Practice: Security and Privacy of Mobile Devices in Wireless Communication*, Crete, Greece, pp. 264–279. Springer, New York, NY, 2011.

37. IBM Software. 2014. Financial malware explained. http://www-01.ibm.com/common/ssi/cgi-bin/ssialias?htmlfid=WGW03086USEN, accessed June 12, 2015.

38. Ibrahim, L.M. and Thanon, K.H. Analysis and detection of the zeus botnet crimeware. *International Journal of Computer Science and Information Security*, 13(9): 121, 2015.

39. Jung, J.H., Kim, J.Y., Lee, H.C., and Yi, J.H. Repackaging attack on Android banking applications and its countermeasures. *Wireless Personal Communications*, 73(4): 1421–1437, 2013.

40. Virus News. 2014. Interpol & Kaspersky lab: 60% of Android attacks use financial malware. http://www.kaspersky.com/about/news/virus/2014/sixty-per-cent-of-Android-attacks-usefinancial-malware, assessed June 7, 2015.

41. Le, A., Markopoulou, A., and Faloutsos, M. Phishdef: URL names say it all. In *INFOCOM, 2011 Proceedings IEEE*, Shanghai, China, pp. 191–195. IEEE, NJ, 2011.

42. M.-E. M. Leveille. 2014. TorrentLocker - Ransomware in a country near you. http://www.welivesecurity.com/2014/12/16/torrentlocker-ransomware-in-a-country-near-you/, accessed June 7, 2015.

43. Mila. 2015. Contagio mobile: Mobile malware mini dump. Retrieved from http://contagiominidump. blogspot.ca/, accessed July 11, 2015.
44. Mulliner, C. and Seifert, J.P. Rise of the ibots: Owning a telco network. In *2010 Fifth International Conference on Malicious and Unwanted Software* (*MALWARE*), Nancy, France, pp. 71–80. IEEE, NJ, 2010.
45. Nauman, M. and Khan, S. Design and implementation of a fine-grained resource usage model for the Android platform. *International Arabian Journal of Information Technology*, 8(4): 440–448, 2011.
46. Rasthofer, S., Asrar, I., Huber, S., and Bodden, E. How current Android malware seeks to evade auto- mated code analysis. In *Information Security Theory and Practice*, Crete, Greece, pp. 187–202. Springer, New York, NY, 2015.
47. Riccardi, M., Oro, D., Luna, J., Cremonini, M., and Vilanova, M. A framework for financial botnet analysis. In *eCrime Researchers Summit* (*eCrime*), Dallas, TX, pp. 1–7. IEEE, NJ, 2010.
48. Schreckling, D., Posegga, J., and Hausknecht, D. Constroid: Data-centric access control for Android. In *Proceedings of the 27th Annual ACM Symposium on Applied Computing*, Riva (Trento), Italy, pp. 1478– 1485. ACM, New York, NY, 2012.
49. Tajalizadehkhoob, S., Asghari, H., Gañán, C., and Van Eeten, M. Why them? Extracting intelligence about target selection from Zeus financial malware. In *Proceedings of the 13th Annual Workshop on the Economics of Information Security* (*WEIS 2014*), State College PA, WEIS, June 23–24, 2014.
50. Vural, I. and Venter, H. Mobile botnet detection using network forensics. In *Future Internet-FIS 2010*, Berlin, Germany, pp. 57–67. Springer, New York, NY, 2010.
51. Zeng, Y., Shin, K.G., and Hu, X. Design of sms commanded-and-controlled and P2P-structured mobile botnets. In *Proceedings of the Fifth ACM Conference on Security and Privacy in Wireless and Mobile Networks*, Tucson, AZ, pp. 137–148. ACM, New York, NY, 2012.
52. Zhang, Y., Yang, M., Xu, B., Yang, Z., Gu, G., Ning, P., Wang, X.S., and Zang, B. Vetting undesirable behaviors in Android apps with permission use analysis. In *Proceedings of the 2013 ACM SIGSAC Con- ference on Computer & Communications Security*, Berlin, Germany, pp. 611–622. ACM, New York, NY, 2013.
53. Zhou, Y. and Jiang, X. Dissecting Android malware: Characterization and evolution. In *2012 IEEE Symposium on Security and Privacy* (*SP*), San Francisco, CA, pp. 95–109, May 2012.
54. ISCX. 2016. ISCX dataset. http://www.unb.ca/research/iscx/dataset/index.html, accessed January 5, 2016.

MOBILE NETWORK SECURITY

Chapter 10

Physical Layer Security in the Last Mile Technology of Mobile Networks

Özge Cepheli, Volker Lücken, Güneş Karabulut Kurt,
Guido Dartmann, and Gerd Ascheid

Contents

Abstract

Mobile networks have become very widely used in recent years, with expanding possibilities for usage ranging from personal life to business needs. This increased usage has brought more security problems and the importance of maintaining security has been also raised. Mobile networks consist of two main parts, including wired backhaul and wireless last mile. Wired backhaul is the part between base station and the core network. It is a highly reliable network with high data rates. Security in this part is very important. In its cable-based physical layer—the part where physical signals are carried—data are hard to acquire, as physical protection of cables and devices is possible. The last mile is the last part where the user is served. This link has to be wireless for the mobility of users. Wireless medium has an open nature; hence, wireless links are more vulnerable to physical layer attacks compared to their wired counterparts. In this chapter, a general understanding will be given on why wireless technologies are often chosen as a last mile technology and why maintaining security is a challenge. Moreover, the current and future solutions to protect the wireless last mile from physical layer attacks will be explained.

10.1 Introduction

Mobility has become a very important part of today's communication networks. Over the years, users have changed their preferred platform to access data by using mobile technologies instead of conventional desktop devices. From user statistics, it is easy to see that the share of mobile and tablet users are in an increasing trend for website access [1]. As an example, Figure 10.1 shows the percentage of users that accessed websites that use StatCounter [1] by using desktop, mobile, and tablet devices. Here, we can easily see that users prefer to use mobile platforms lately, and the trend shows that the increase of usage is likely to continue in the near future. As wireless technologies are evolving, users demand higher data rates, increased reliability, and security for their mobile connections. However, maintaining those requirements are not easy as mobile networks are especially vulnerable to failures, intrusion, and eavesdropping. Also, the possibility of movement of users, subnets, and even base stations increases the challenge. The wireless medium, which is used to carry the signals to the mobile user, is usually less reliable, rapidly changing, and open to eavesdropping attacks. In addition to all these challenges, the mobile network should survive intentional and unintentional (natural) threats.

The selection of the wireless communication channel for future network deployments is almost certain despite its insecure nature, when compared to the wired counterparts. The wireless channel, however, introduces new threat types that are not addressed with classical security solutions that

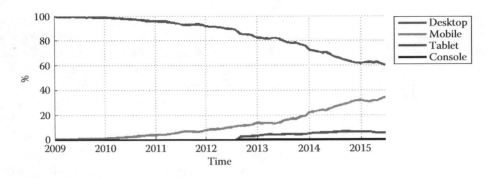

Figure 10.1 Worldwide platform usage statistics acquired from StatCounter Global Stats between December 2008 and June 2015 [1]. The share of mobile and tablet users are in an increasing trend.

mainly target the security of the wired backhaul links. To address the security threats due to the wireless channel, physical layer security measures that are custom designed according to the channel status are utilized.

Considering wireless communication networks, there are many natural challenges such as channel impairments and limited transmission bandwidth. Moreover, there can be adversaries that try to perform various attacks to capture or prevent the communication between legitimate parties. Physical layer security solutions address detection and mitigation of these attacks along with considering the wireless channel attributes.

The basic system that can be defined for physical layer security consists of three nodes, one legitimate transmitter, one legitimate receiver, and one eavesdropper, which are often referred to as Alice, Bob, and Eve, respectively. By making use of this basic system, most of the physical layer techniques can be analyzed. Note that this basic model may represent the wireless last mile of a bigger communication network, for example, a mobile network where Alice is the base station and Bob is the mobile user for the downlink. Throughout the chapter, the same system will be considered for addressing the physical layer security issues of mobile networks. We will analyze the last mile technology of the mobile networks as a separate wireless network apart from the rest of the whole network that is often wired.

In this chapter, security in the wireless last mile technologies will be described in detail. Section 10.2 introduces the last mile concept explaining the importance of the technology and the security challenges when wireless technologies are used as the last mile. Section 10.3 explains all the fundamentals of wireless channels and attacks to give a complete picture. Section 10.4 explains the security measures used today, and Section 10.5 details the security measures in the literature that are the candidates for being a part of the future installations. Conclusions are given in Section 10.6.

10.2 Last Mile Technologies

In order to provide a diverse set of services, an end-to-end communication link consists of many components addressing different communication functions. The scope and quality of these services are derived by the requirements and demands of the *end user*; the user that needs and will use services. The major entities that provide services to enable an end user to reach another are *network service providers* (NSPs) and *Internet service providers* (ISPs). NSPs construct global networks and lease bandwidth to regional NSPs, which offer the bandwidth to local ISPs. Local ISPs provide and manage services to end users. Hence, the overall network is maintained as separate blocks and various technologies can be used in each block. According to this structure, Internet backbone, ISP network, and end-user network can be designed and operated almost independently.

There are many different media that are used when a communication system is considered as a whole between a sender and a receiver. However, it is possible to divide the communication process into smaller modules for easier analysis. In commercial mobile networks, the *backhaul* of any network uses a highly reliable bandwidth-rich physical medium for the transmission of the communication signals. This implies making use of a wired infrastructure, such as fiber optic cables.

The standard that is used to establish the final connectivity link between the service providers and the end user is referred to as *last mile technology*. Unlike the backhaul, the last mile technology also depends on the requirements of the end users, where coverage and cost may become more important than reliability and data transmission rate. Wireless technologies are prominent candidates as the last mile technology. They provide a good compromise, enabling user mobility and ease of use, which are very important on new-generation mobile technologies. The usage of mobile devices and mobile communications is rapidly increasing among technology users [2,3]. The fact that the wireless technologies cannot achieve the reliability and data transmission rate of a wired counterpart becomes a secondary concern.

10.2.1 Choosing Wireless as a Last Mile Technology

Wireless networks have become a major part of communication networks today and also appear to be a strong component of the future communication networks. Besides fourth-generation (4G) and fifth-generation (5G) mobile systems, the family of IEEE 802.11 standards are being employed by a large fraction of broadband users to connect their computers to Internet for the last mile. This means that wireless technologies are very likely to be utilized as the last mile technology in the future as well. The main reasons for the selection of wireless technologies are their mobility and ease of use. In fact, according to Mary Meeker's KPCB 2015 Internet Trends report [4], the percentage of the time spent on mobile digital media in the United States is already higher at 51% when compared to desktop computers by 42%. The use of wireless technologies comes at a price of bandwidth scarcity and security issues. The disadvantage associated with lower data transmission rate is easily overlooked as newer wireless technologies have higher data rates. However, the security issues are critical in wireless technologies due to the open nature of wireless links. Besides the security threats from upper layers of Open Systems Interconnection (OSI) model, using wireless medium brings particular vulnerabilities to the physical layer. Note that besides the apparent use of wireless medium in mobile networks, even for broadband Internet connections the wireless technologies are being considered [5,6]. Hence, when we mention wireless physical layer security, most of the commercial networks like mobile networks and future broadband networks are addressed.

10.2.2 Importance of Last Mile for Security

As already mentioned earlier, the main security challenge of wireless technologies as the last mile is caused by the open nature of the wireless channel. In conventional cable networks, the physical layer medium consists of cables and the signal transmission over a cable between two nodes is considered to be safe. This is an accurate assumption as the security of a cable could be easily achieved by physically making the cable unreachable, for example, by using locked server rooms or underground cables. An eavesdropping attack, which is performed by illegitimate users to capture data by listening to the link, is quite unlikely to be successful when the cables are unreachable. However, in wireless communications, instead of using cables to transmit the signals, antennas are used to transmit and receive the signals. The signals experience alterations in power and phase after leaving the transmitter antenna before reaching the receiver antenna. However, when wireless channels are used to transmit signals, the signal can reach anywhere within the transmission range, that is, where an acceptable signal to interference plus noise ratio (SINR) level is available. Service providers have to keep the SINR above an acceptable level for their users, which is a constraint for successful reception of the signals on the receiver side. However, the SINR level for illegitimate users may also be enough for them for successfully capturing the transmitted data, which leads to a major security issue considering eavesdropping attacks. For signals transmitted from an antenna, there is a spatial pattern that signals are spread, and the signal strength decreases by distance, which will be explained in detail in the next section. *Antenna range* is the spatial definition of signal strength's being above some level. This defines the locations that the signal can be received successfully. Every user within the transmission range, including the illegitimate receivers such as eavesdroppers, can capture the transmitted signals.

A typical mobile network with a wired backhaul is shown in Figure 10.2. In the figure, the wireless transmitter antenna is referred to Alice and is connected to the core network by a wired link. Alice is the legitimate transmitter, transmitting signals to the legitimate receiver Bob, whereas the signals are also received by illegitimate users, referred to as Eves. The received power of signals from transmitter antenna versus distance is also illustrated in the figure, which is caused by the inherent properties of wireless channels, like path loss and shadowing. These properties will be thoroughly explained in the next section. In the given setting, it is clear that security of the overall network is affected by the security of the last mile. Although the core network security threats can be addressed by the conventional security solutions, the security of the wireless last mile remains a challenge, which should be addressed with physical layer security measures that are custom designed according to the wireless channel. Throughout this chapter, the current approaches will be introduced along with classical solutions and main concepts to analyze this challenge.

10.3 Physical Layer Security Fundamentals

10.3.1 Properties of Wireless Channels

In a basic wireless communication scenario, electrical signals are converted into electromagnetic waves by using an antenna and these electromagnetic waves are broadcast from the transmitter. These waves are then captured by a receiving antenna and signals are obtained at the receiver side. The impact of the physical phenomena that signals experience while traveling from transmitter to receiver is taken into account in wireless channel models. There are different channel models that are frequently being used in the literature such as Rayleigh, Rician, or Nakagami-m fading channel models [7], as well as Stanford University Interim (SUI) [8] or 3GPP WIM2 [9] models. Channel effects as path loss, shadowing, fading (small scale and large scale) and Doppler shift are usually included in these models. Interference, which is the additional electromagnetic signal received along with the intended signal, is another major cause for signal quality degradation and is usually modeled independently.

Looking from a system level, in order to study the eavesdropping case, Wyner [10] introduced the wiretap channel, which is a special channel scheme where the eavesdropper's channel

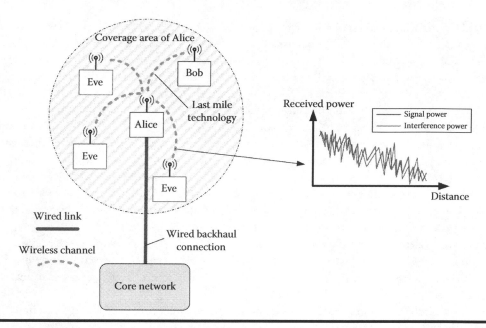

Figure 10.2 Wireless communication as a last mile technology. The wireless channel characteristics increase the security threats due to the open nature of the channel. In the figure, there is a legitimate wireless transmitter, called Alice, who is connected to the core network by a wired connection and use the wireless medium for the last mile connection to communicate with the legitimate receiver, Bob. As wireless medium has an open nature, illegitimate receivers are also able to capture the signals from Alice. The power of the received signal on Eve is dependent on the distance and fluctuates due to channel characteristics such as multipath fading and shadowing. Interference from other resources are also received in the same way. Note that interference has a big effect on the signal quality. In this case, why not use it as a security countermeasure? We can whisper nonsensical signals to Eve's ear to improve secrecy—and these signals are called *artificial interference*.

is a degraded version of the channel of the legitimate user. In the study [11], the authors considered a general independent channel condition by eliminating the degraded eavesdropper channel assumption and studied the transmission. Note that the studies were not considering wireless channels directly but, however, became fundamental studies in physical layer security in wireless networks as channel definitions not only comply with wireless channel models but also perfectly cover eavesdropping in wireless channels. Since multiantenna technologies are now frequently used, many studies have been conducted with various antenna configurations, such as single-input-multiple-output (SIMO) [12], multiple-input-single-output (MISO) [13–15], and multiple-input-multiple-output (MIMO) [16–19] channels.

1. *Thermal noise*: Both wired and wireless communication systems are subject to random fluctuations on the signal received levels, which are caused by many natural sources. Referred to as *thermal noise* or *additive white Gaussian noise*, this phenomenon is generally modeled with a Gaussian (normal) distribution.
2. *Path loss and finding*: *Path loss* refers to the weakening of the signals as they propagate through space. This weakening is caused by the distance between transmitter and receiver. In addition

to the path loss, the magnitude and phase of the received signals over the wireless channel may rapidly change in time, frequency, and space [20]. Referred to as *fading*, this effect is usually modeled as a random process. There are two main classes of fading, namely, the *large-scale fading* and the *small-scale fading*. Large-scale fading is caused by path loss and shadowing by large objects (such as buildings and hills). The large-scale fading is typically frequency independent and frequently modeled by log-normal shadowing. Small-scale fading is generally caused by the interference generated by the existence of multiple transmission paths between the transmitter and receiver, and it is frequency dependent. Most frequently used small-scale fading models are Rayleigh, Rician, Weibull, Nakagami fading models.

It is also very useful and widely used to classify fading according to the rapidness of the changing compared to the signal. When the channel conditions change faster than the symbol duration, it is called as *fast fading*, and similarly a channel is called as *slow fading* when the channel conditions change slower than the symbol period. Fast fading can occur due to the relative motion between the transmitter and receiver objects, which is known as Doppler spreading.

3. *Multiuser interference*: When modeling the wireless channel effects, it is essential to consider *multiuser interference*, which is the signals of users or technologies that are captured by receiver. Basically, the receiver antenna captures not only the signals of the intended transmitter, but also a superposition of signals of all the transmitters using the same frequency band that have the receiver in their antenna range. These signals, also subject to fading and path loss, act as an impairment. They are often very strong, severely limiting the SINR values, possibly causing communication disruptions. Additionally, other than natural ambient interference sources, interference can be generated deliberately by an adversary. In this case, this approach is referred to as a jamming attack.

The impact of wireless channel is very important when security is considered as the performance of physical layer attacks is usually based on accurate channel estimation for both attacker and legitimate user channels.

10.3.2 Attacks

In wireless communication networks, the adversaries attack one or more of the four main system security requirements, which are secrecy, authentication, data integrity, or robustness. *Secrecy* defines the discreteness of the data between the origin and destination. *Authentication* is the act of confirming that the destination of the data has the access rights. *Data integrity* refers to the completeness and originality of the data during its life cycle. Lastly, *robustness* defines the ability of a communication system to remain operational under degrading effects. Major attack types are listed in Figure 10.3 along with their target requirements. Rather than being a complete list of all attacks, this table can act as a basic reference for attacks. It is a very hard task to cover all attacks as new attack types are being discovered every day. Following the latest discovered attacks is a crucial requirement for having up-to-date information on current security vulnerabilities of any system.

Physical layer–targeted attacks can be classified into two groups: passive or active attacks. In *passive attacks*, the adversary does not provide input to the system, making these types of attacks very difficult to detect. On the other hand, during *active attacks*, the adversary uses a transmitter, actively interfering to the network. These attacks are usually easier to detect; however, this does not imply that they are easy to prevent. Physical layer–targeted attacks can also be grouped by their targeted security requirements, namely, the secrecy, authentication, data integrity, and robustness.

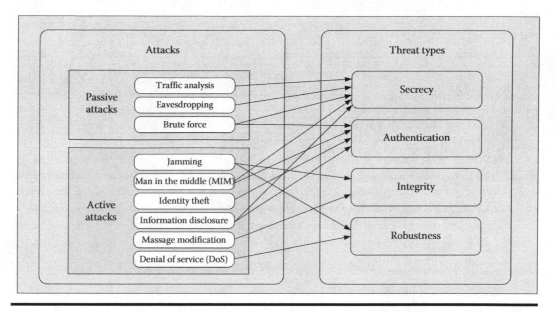

Figure 10.3 Physical layer attack types and targeted system security requirements. Attacks are classified according to the activity level of the adversary and the threat types.

10.3.2.1 Secrecy Attacks

Major attack types against data secrecy are eavesdropping and traffic analysis attacks. *Eavesdropping* is the act of secretly listening to the private conversation of others without their consent, which projects to gathering of wireless communication data by nonlegitimate users. Eavesdropping attacks are typically very easy to perform and very challenging to detect due to their passive nature. Such attacks can be executed by properly tuned reception and decryption of the encrypted data, in case data are encrypted. The eavesdropping attack can be implemented in either real-time or non-real-time fashion. The non-real-time eavesdropping is more critical as the adversary can use brute-force–based approaches with high computational complexity to capture the data that may take a longer time to compute.

Also a passive attack, *traffic analysis* is a similar version of eavesdropping attack, where the nonlegitimate user cannot intercept the communication data but gathers the traffic information, like transmitter/receiver identities or data rates. Usually, traffic analysis attack is performed where the secret key used for encryption cannot be obtained.

The main classical countermeasure to eavesdropping attacks is encryption. There is no particular classical security measure for traffic analysis. The recently proposed beamforming and artificial noise–based physical layer security approaches can aid combat of passive secrecy attacks, including traffic analysis attacks [21–25].

10.3.2.2 Authentication Attacks

Authentication is the process of confirming legitimacy of a transmitter. Most frequently, encountered authentication attack types are bruteforce, eavesdropping, man in the middle (MIM), and identity theft attacks. These attacks are also detectable in the physical layer.

In *MIM attacks,* the adversary makes independent connections with the target nodes (performing two-way communication) and transmits messages among them, impersonating each destination making them believe that they are communicating with directly to each other. MIM attacks also include eavesdropping attacks. In fact, during an MIM attack, the entire conversation is controlled by the attacker. Beyond the secrecy violation, it is clear that MIM attacks can be very dangerous to systems as the attacker gets authenticated and it is able to enter the system or change the communication data in a harmful manner. In an *authentication cloning attack*, an unauthorized user pretends to be a legitimate user by deceiving the authentication system. An authentication cloning attack can be implemented in many ways, including capturing the authentication sequences that are based on physical layer attributes. For example, an attacker can imitate its location or channel information as the legitimate user and get authenticated to access resources.

Identity (ID) theft attacks are usually performed by capturing and detecting network traffic data and identifying node with network privileges. Most wireless systems allow some kind of ID filtering to allow only authorized devices with specific IDs to gain access and utilize the network. ID information can also be gathered by executing brute-force attack, which means trying all the possible ID key options.

10.3.2.3 Data Integrity Attacks

Data integrity attacks compromise the trustworthiness of transmitted data over the communication life cycle. Frequently observed data integrity attacks are message modification and jamming attacks. Attackers can transmit fake control, management, or data frames over wireless channel to mislead the receiver. *Message modification* is the general class of attack types that are based on additions or deletions to actual data by adversaries. *Jamming* attacks are based on transmitting signals to disrupt communication link by limiting the SINR. Jamming attacks can result in partial disruptions as well. Authentication-based attacks can also lead to data integrity problems as altering data is possible after the authentication of the attacker.

In order to detect data integrity attacks, integrity checks such as key-based techniques or predetermined packet headers can be performed. Note that although such attacks may not always be detected, integrity checks are still an efficient way to deal with physical layer–related errors during transmission. Such errors can also be combatted using error control coding or automatic repeat request techniques.

10.3.2.4 Robustness Attacks

Robustness in wireless networks mainly implies the strength of the communication system against channel impairments. The major robustness attacks are denial of service (DoS) attacks. A *DoS attack* targets exhaustion of network resources to disrupt communication among legitimate users. Jamming is the most frequently observed DoS attack type at physical layer. DoS attacks may also be executed by a number of distributed adversaries to reduce their detection probability and are named as *distributed DoS* (*DDoS*) attacks and considered as one of the most challenging security issues in current communication systems. DDoS are attacks considered as one of the most challenging security issues in current communication systems.

Countermeasures of DoS or DDoS attacks are not clear as these attacks can be executed in various ways. Anomaly detection systems are used to determine if an attack is being held for any of the network resources. In order to prevent a detected DoS attack, the resource usage of attackers is blocked or a backup resource is used. If a DoS attack is detected, the network controller node usually prevents the adversaries by blocking their resource usage. Another approach to enhance the robustness of a system is to diversify network resources by using backup resources. These backup resources can be used if one resource is under attack. For example, if a jammer is detected, the wireless network may switch to a different carrier frequency to avoid the quality degrading effects.

Usually, systems are designed to have backup communication lines in different networks to avoid connection losses. Robustness can also be achieved in the device side; wireless systems can be designed to have backup devices that can be switched to if the master device is under a physical attack.

10.3.3 Performance Metrics of Physical Layer Security

The fundamental issues of secure channel capacity have drawn much attention in the information theory community in recent years. Most of these works focus on *secrecy capacity*, which refers to the maximum rate of secret information sent from a wireless node to its destination in the presence of eavesdroppers. This metric is very useful as it stands as an information theoretical measure of the security of a channel. Using such a metric, it is possible to compare channels and to measure how much security is gained using a security countermeasure. It is shown by Wyner [10] that the perfect secrecy capacity is the difference of the capacities for the two users in discrete memoryless channels. This result has been generalized to Gaussian channels by Leung et al. [26] then by [27], considering the full channel state information (CSI) case. The secrecy capacity under full-CSI assumption is adopted as an upper bound for the secrecy capacity when only the CSI of the legitimate receiver is known at the transmitter. The authors in [27] also proposed a low-complexity on/off power allocation strategy that achieves near optimal performance with only the main channel CSI. This scheme was shown to be asymptotically optimal as the average signal-to-noise ratio (SNR) goes to infinity. The authors in [28] extended the previous studies considering imperfect CSI case. Based on an information-theoretic formulation of the problem, the role of fading is also characterized in which two legitimate partners communicate over a quasistatic fading channel and an eavesdropper observes their transmissions through a second independent quasistatic fading channel. In [28], the authors defined the secrecy capacity in terms of outage probability and provide a complete characterization of the maximum transmission rate at which the eavesdropper is unable to decode any information. The results of this study are generalized into a multiple eavesdropper case in [29] and the secrecy capacity of the system is analyzed in terms of outage probability and outage capacity. In [12], the authors defined the ergodic secrecy capacity and find the optimal power allocation at the transmitter that achieves the secrecy capacity for the full-CSI and no-CSI cases along with the analytical expression for the lower bound of ergodic secrecy capacity is presented. They also provided the analytical expression for secure outage probability to study the secure outage performance of their proposed model. In the following text, we classify the main metrics for physical layer security performance:

10.3.3.1 Information Theoretical Metrics

Information theoretical metrics are widely used in the literature for the mathematical assessment of security. The upper bound of perfectly secret transmission rate from the legitimate transmitter node to a legitimate destination node is defined by the secrecy capacity [10]. The *probability of outage in secrecy capacity* (*OSC*) is another important concept in the information theoretical analysis of physical layer security approaches. OSC is defined as the probability that the instantaneous secrecy capacity is less than a target secrecy rate.

10.3.3.2 Quality of Service–Related Metrics

Clearly, the SINR value of the channel is directly affecting both the secrecy capacity and the OSC. However, SINR also dictates the performance of a communication link, hence is considered a QoS-related performance metric [21]. A sufficiently high (low) SINR value enables robust transmission with a maximum (minimum) desired error level. Hence, when considering a transmission from Alice, lower bounding the SINR of Bob and upper bounding the SINR of Eve can lead to reliable and secure communication. The SINR can be changed by using the physical layer security techniques, such as beamforming approaches, as will be described in Section 10.5. Note that the SINR is reduced to SNR in the absence of multiuser interference.

The received signals are demodulated and decoded to bits at both Bob and Eve. At the output of the decoder, it is possible to calculate (or estimate) the *bit error rate (BER)*, which is one of the primary performance measures for robustness of digital communication systems. BER is always related to SINR; however, modulation and coding techniques have different BER performance on the same SINR level. Usually, a minimum BER requirement is defined for a successful communication, depending on the desired application. If BER of a system is below a minimum required level, a communication link cannot be properly established. As a result, it can be seen that forcing a unsatisfactory BER on unauthorized nodes can actually improve the network security. Hence, BER can also be used to define the QoS and the physical layer security level of a system. When considering BER as a performance metric, the impact of the selected modulation and coding schemes also needs to be taken into account. Also, note that as a function of BER, the frame or packet error rates can be considered as performance metrics of physical layer security.

10.4 Current Physical Layer Security Solutions

Security is hard to maintain on operational networks with large volumes of data, analysis, and detection of security threats become very challenging when delay and data rates are important. There are several classical tools to protect networks against attackers. Ideally, all tools work collectively, minimizing maintenance requirements while improving the security level [30]. The most common way to protect a network is to make use of tools such as antivirus and antispyware software, firewalls, intrusion detection systems (IDSs), intrusion prevention systems (IPSs), and virtual private networks (VPNs) for upper layers along with encryption and spread spectrum–based eavesdropping mitigation techniques in the lower layers.

10.4.1 Network Security Solutions

Generally, network security solutions target the network layer of the OSI reference model. At its most basic, an IDS provides passive protection by observation of data network from a monitoring port (such as a tapped communications port); comparing the observed data traffic patterns with the predefined rules and patterns. This may not always be an effective solution for today's fast-evolving networks due to rapidly changing attack traffic patterns.

Jammer detectors are a good example of physical layer IDS, where special devices are used to detect the existence and the source of a jamming attack [31]. There are many commercial off-the-shelf jammer detectors available today.

IPSs simply add more security throughout the network by not solely monitoring and alerting but also by stopping the traffic flow that may be malicious or harmful in a proactive approach.

Also note that the probability of false positive decisions when using the IPS approach for data networks may erroneously cause communication disruption.

In addition to IDS and IPS tools, recently, web application firewalls (WAFs) are starting to be deployed for defending against attacks. Currently, WAFs are very effective at certain attack types where IPSs may not successfully prevent (such as HTTP request-based attacks). WAFs lack providing security in lower layers while their focus is only on the application layer.

10.4.2 Encryption Solutions

In encryption-based security solutions, data are encrypted using a *secret key*, enabling only authorized recipients to decrypt. The algorithm that carries out the encryption is called the *cipher* and the ciphers depend on some auxiliary information, called a secret key. Secret key is usually a randomized codeword, shared between the transmitter and the recipient and used to encode the original data called plaintext into ciphertext. Decryption can be successful if the transmitter and the receiver have the same key pair, and any person who seizes the key can solve all encrypted communication.

Encryption is possible in all layers of the OSI model, and the approach is usually named according to the target layer. Application layer (layer 7), network layer (layer 3), and data link layer (layer 2) encryption techniques are the most common methods that are used in today's networks.

Encryption is the key concept for *virtual private networks (VPNs)*, which is the extension of a private network across a public network. VPN is used to establish a secure connection between two sites, where the sites connect through a public network. The traffic is secured by using encryption and tunneling protocols so that the connection stays private from other users in the public network.

In lower layers, security is mostly maintained by encryption solutions in the data link layer. Encrypted data should be decrypted before its content can be revealed by an attacker. In order to decrypt the data, one should have the secret key in hand, which is only possessed by the legitimate receiver [32]. However, the key can be acquired by a *brute-force attack* or *exhaustive key search*, which is simply trying every possible key combination until obtaining the correct key. Note that every encryption key can be discovered within a significant time range that is dependent on the processing power and the length of the key. Hence, if data should be secure for a long time, a long key should be used for encryption, which causes more power consumption for the encryption and decryption processes, and affects the battery life of mobile devices.

The selection of encryption technique is a sensitive subject. Usually, the best option is to use a strong well-known encryption standard, which is already tested for many years against a diverse set of attack types. Novel encryption algorithms come with the risk of having security holes that may enable attackers to easily retrieve the secret key. Moreover, longer keys should be selected for increased security.

10.4.3 Spread Spectrum–Based Eavesdropping Mitigation Techniques

For many years, secure wireless communication systems are being developed, many with a military context as background. One of the earliest proposals of spread spectrum techniques was the frequency hopping patent from Antheil and Lamarr in 1941 [33], which describes a simple frequency hopping system for the wireless control of torpedoes. Based on a predefined synchronized sequence of carrier frequencies, which are stored on paper punch cards, the system was intended to provide secrecy and jamming robustness. Later, more sophisticated spread-spectrum techniques like *code division multiple access* (CDMA) have been developed for a use in both military and also personal wireless communications. For civil wireless communications, the security aspect was mostly a minor factor in the design of spread spectrum waveforms, and practically, security was not fully realized. Still, the waveforms used have the capability to significantly increase the security (secrecy and jamming-robustness) performance of a communication link in principle. Especially with the more recent focus on secure and private personal communications in mind, these waveforms can prove suitable.

The definition of spread spectrum techniques is that the transmitted signal uses a bandwidth that is independent of and significantly larger than the information bit rate, and further, demodulation can be performed by correlation of the receive signal with the spreading signal [34,35]. In the field of spread spectrum techniques, different types of waveforms and realizations exist. The most important ones are the *frequency hopping spread spectrum* (FHSS) and *direct sequence spread spectrum* (DSSS) types. FHSS, which already was employed in the initially presented system from 1941, is realized by using a time-dependent carrier frequency of the transmit signal. Each frequency position is held for a specific time, the so-called dwell time [36]. DSSS, in contrary, uses a modulation of the information signal with a spread sequence, which usually has a higher data rate than the information signal itself, and therefore leads to a frequency spreading.

The widely used frequency hopping offers a protection against casual eavesdropping and jamming attacks, but does not offer a significant secrecy increase in practice, as especially in a situation with only a single hopping signal in the channel. It can be easily tracked and the hopping sequence can be recovered afterward. In the final communication system, the hopping sequence is preshared or exchanged between the legitimate users in prior. Even with the practical disadvantages mentioned, a real system with a hopping rate can especially hinder jamming attacks, where the hopping steps cannot be predicted in a live scenario. Further, the impact of narrow-band interference can be reduced.

DSSS techniques are very relevant in current mobile communication standards. In the Universal Mobile Telecommunications System (UMTS) 3G standard, Wideband CDMA (W-CDMA) is used, which is a very high bandwidth version of CDMA. From the secrecy point of view, however, there are some reservations because of implementation issues of the spreading sequence in the standard [37–39]. Also, CDMA is used in many security-critical military applications as it both strongly reduces the risk of interception by eavesdropping and the risk of jamming. CDMA is realized by multiplying the input data with a *pseudorandom noise* sequence. Multiple access is then implemented by using orthogonal sequences (spreading codes) for each of the users. This offers an inherent separation, even with all users occupying the same frequency band and transmitting at the same time. Using specific long and more complex code sequences, high secrecy levels can be reached [37–39].

One main disadvantage is that for the current plannings for the 5G CDMA techniques only play a minor role. This is why some alternative techniques, for example, for the prospective 5G waveform *filter bank multicarrier* (FBMC), are under development and described later in this chapter.

10.5 Future Physical Layer Security Solutions

As physical layer security is an emerging topic, there are many new techniques and approaches that have been proposed that are not yet implemented in the current security solutions. We group these approaches into three major titles as signaling-based solutions [21–23,40–44], filter-based solutions [45,46], and physical layer key generation techniques [47–52]. The considerations on implementation of these new techniques in future mobile networks will also be discussed at the end of the section.

10.5.1 Signaling-Based Solutions

10.5.1.1 Beamforming

Beamforming is a multiantenna technique that enables the transmitter to spatially focus signals by adaptively adjusting the amplitude and/or phase of each array element. With transmit beamforming, one can utilize the spatial shaping to degrade the reception performance of eavesdroppers [40]. Beamforming recently became a viable countermeasure for wireless physical layer security as reception of signals by eavesdroppers can be physically limited [21–23].

Early implementations of beamforming are based on *switched beam* techniques that relied on the selection of predetermined beam patterns according to channel characteristics. A more dynamic way to spatially form transmission patterns is to use *adaptive beamforming* systems, where the beamforming coefficients are adaptively calculated in order to reach a specific beam pattern. As a result of beamforming, transmitters can physically focus signals only on the legitimate receiver, hence increasing the secrecy capacity and SINR at Bob, as shown in Figure 10.4.

In order to successfully deploy beamforming-based security solutions, the CSI of the legitimate receiver should be available at the transmitter. If ideal CSI of the eavesdroppers are also present at the transmitter, then it becomes possible to calculate the optimum beamforming coefficients to maximize secrecy capacity [23]. However, channel information may not be perfectly known, the information can be partial as in [14,53], delayed as in [13], or imperfect as in [15,16,54]. Even if Alice does not possess any information about Eve's channel, beamforming can be used to focus the signals on the legitimate receiver, which increases the overall security [21,55,56].

10.5.1.2 Artificial Noise

Artificial noise (AN) or *artificial interference (AI)* is another countermeasure that is commonly used as a physical layer security solution in a complementary fashion to beamforming. AN approaches mainly imply transmission of deliberately generated noise signals, in addition to information bearing signals. AN can be generated as legitimate transmitter based as in [41,42] or legitimate receiver based as in [57]. A new approach is to use friendly jammers instead of legitimate users, as in [18] and [58]. Using AN can be also interpreted as a jamming attack toward Eve. Clearly, effective usage

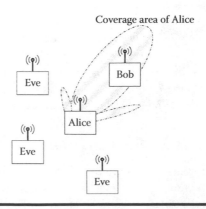

Figure 10.4 Adaptive beamforming can be done by Alice with a multiantenna transmitter by calculating the necessary phase components of each antenna component to create the antenna pattern focused on Bob.

Table 10.1 Assumptions about Eavesdropper Locations

CSI perfectly known	[10,26,27,40,41,57]
CSI partially known	[14,53]
Delayed CSI	[13]
Imperfect CSI	[15,16,54]
Direction known	[17]
CSI unknown	[55,56]

of AN requires beamforming; otherwise, AN would affect all the users including the legitimate receiver, Bob.

By using beamforming, it is possible to optimize both information signal and AN to maximize security [23–25]. Hence, in order to deploy AN approaches optimally, CSI of Eve should be available at Alice. In this case, Alice can deploy adaptive AN that only targets the eavesdropper, as shown in Figure 10.4. However, if Eve's CSI is not available, AN can still be used in an isotropic manner, which implies sending AN everywhere but the legitimate receiver.

Availability of location or channel information of eavesdroppers is an important point of the system model. The practical challenge of gathering information about eavesdroppers results in several approaches to be held in related studies. Major approaches can be grouped (with examples) as in Table 10.1.

Locations of eavesdroppers can be also modeled as a random process as in [59], where the locations of eavesdroppers are modeled by a Poisson point process, and as in [60], where the authors study the optimum location of an eavesdropper from a secrecy capacity perspective in multiterminal networks with power control. Also in [61], the authors consider end-to-end secure communication in a large wireless network, where the locations of eavesdroppers are uncertain.

AN and beamforming techniques are frequently resorted to in theoretical works but their practical considerations are quite recent. This is caused by their high processing requirements and intolerance to CSI estimation errors. However, with increasing processing capabilities of end-user devices, we expect to see them in the near future.

10.5.1.3 Interference Shaping–Based Solutions

Interference shaping can be done by making use of beamforming in a multiuser environment in order to calculate the optimal transmission schema for each user to increase overall performance [43]. Considering the security perspective, it is possible to spatially shape the existing interference to avoid or reduce AN usage that requires additional transmit power. The natural ambient interference of multicell networks can be tailored not only to maximize the SINR of legitimate users as in [44], but also to enhance secrecy against eavesdropper attacks by simultaneously degrading the SINR of the eavesdropper. To achieve such a system, a max–min beamforming problem is proposed in [23] that maximizes both the SINR of legitimate users and the inverse SINR of eavesdropper, with a sum power constraint. It is shown that the interference-shaping based techniques accomplish to enhance secrecy in a more energy-efficient way than the conventional AN-based physical layer security systems, often with the cost of higher processing resource requirements.

10.5.2 Filter-Based Solutions

More recently, new techniques for waveform-based physical layer security have been investigated, which are still in an initial state, but already provide promising results for the implementation in future wireless communication systems. Two techniques are presented here, which are based on the time-frequency-localization and matching of transmission filters. The first technique presented is an online eavesdropping mitigation technique that is based on the optimization of transmit and receive filters on the sides of the legitimate communication partners in a single-carrier transmission. The second technique is called *filter hopping* (which has no relationship to the similar-sounding *frequency hopping*) and employs the *filter bank multicarrier* (FBMC) waveform, which is a candidate waveform for 5G mobile communications. Based on specific energy-dispersing transmit filters, which are overlapping and varied over the time-frequency-lattice (TFL), the technique allows a strong secrecy capacity increase in a wireless communication system.

10.5.2.1 Single-Carrier Online Eavesdropping Mitigation Techniques

In [45], an optimization-based matched filter design with improved secrecy is presented. Figure 10.5 presents the concept of this approach. The legitimate transmitter and receiver use the same algorithm for the optimization of the matched filter pair. Both the receiver and the transmitter use the same algorithm; therefore, the optimization on both sides results in the same set of filter coefficients. The filters are designed such that the signal-to-interference ratio (SIR) is maximized at the

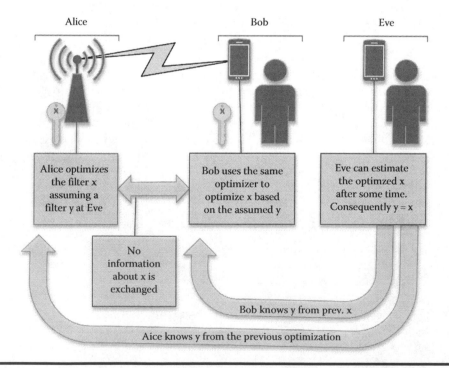

Figure 10.5 An Idea optimization-aided matched filter design. (From Dartmann, G. et al., Filter optimization aided interference management with improved secrecy, in *IEEE Vehicular Technology Conference (VTC Fall)*, September 2014, pp. 1–6.)

legitimate receiver. Furthermore, each filter should have a sufficient stop-band attenuation. Secrecy is achieved by a matched filter optimization against an eavesdropper by a reduction of their SIRs. The concept is designed for so-called online attacks where an eavesdropper tries to get access to an established communication link. After some time, the eavesdropper could potentially estimate the matched filter such that the secret information can be decoded at this illegitimate receiver. Therefore, the legitimate receiver must redesign their matched filter pair after a fixed time period to avoid a reception of the secret information by an eavesdropper in the vicinity. The new design matched filter is, therefore, designed against the filter estimated by the eavesdropper, which is the filter of the previous optimization cycle.

An advantage of this concept compared to beamforming-based approaches is that the optimization does not need any CSI of the illegitimate link. A weakness of this concept is the potential risk that an eavesdropper can estimate the algorithm itself. This problem can be solved by random-constraints based on a preshared key set similar to encryption keys. Instead of preshared keys, the legitimate users can also use the channel as a random source in case the eavesdropper is sufficiently far away. With these keys, randomization constraints, random filter change period, and different initial solutions can be used to design randomized matched filters.

A different approach is a so-called offline optimization. In this case, a large set of random-ized matched filters can be shared by the legitimate links. Alice and Bob randomly exchange their matched filter pair. The selected filter pair must be orthogonal to the previously selected filter pair to ensure a low SIR at the eavesdropper. An open question is: How many of these orthogonal filter can be designed with the given set of secrecy and random constraints? The work [46] is an extension of [45] and based on a offline optimization of large filter set.

10.5.2.2 Multicarrier Techniques: FBMC Filter Hopping

The second technique presented is the filter hopping method [46] for systems using the FBMC waveform. FBMC is a multicarrier waveform like cyclic prefix OFDM (CP-OFDM), but it features a per-subcarrier symbol shaping using a prototype filter, which can be efficiently realized by a polyphase filter bank. A major difference is the lattice structure, which is different to the one of OFDM/QAM. For the FBMC case, the lattice structure is called *offset QAM* (OQAM). This means that instead of complex-valued QAM symbols, twice the rate of real-valued PAM symbols are trans-mitted. Furthermore, each subcarrier is filtered with the prototype filter, whose frequency response is plotted in Figure 10.6. Finally, the CP is also omitted as it is not necessary due to the localization properties of the symbols in the TFL. For comparison, the TFLs for both the CP-OFDM and the FBMC waveform are shown in Figure 10.7.

Due to the pulse shaping with the prototype filter, which is a major degree of freedom of the FBMC waveform, the symbols are spread in time and frequency domains differently than OFDM symbols, which have a rectangular shape in time domain and a Sinc-shaped response in frequency domain. With a matched receive filter, orthogonality is preserved at the receiver side and no inter-ference is experienced by adjacent symbols in the TFL. The idea of filter hopping is that with a mismatch of the receive filter, the orthogonality conditions are violated and the energy of the trans-mit symbols is widely spread to adjacent symbols and subcarriers. This effect is even significant with a slight mismatch of the transmit and receive filters. In a practical system, Alice and Bob can exchange the filter design using key exchange techniques or a preshared sequence of filters. Then, the filters are varied over the TFL, either continuously or blockwise. A possible eavesdrop-per is then required to try a large number of filter sets and designs in order to get an error-free reception.

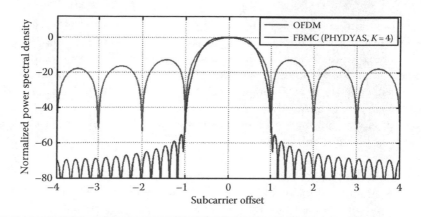

Figure 10.6 Spectral comparison of OFDM and filter bank multicarrier (with PHYDYAS filter).

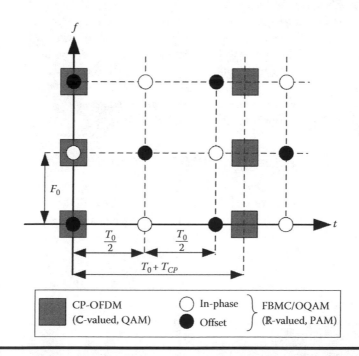

Figure 10.7 Time-frequency-lattices of CP-OFDM and filter bank multicarrier/OQAM. (From Schellmann, M. et al., FBMC-based air interface for 5G mobile: Challenges and proposed solutions, in *Ninth International Conference on Cognitive Radio Oriented Wireless Networks and Communications (CROWNCOM)*, June 2014, pp. 102–107.)

A proposed transceiver chain design for an FBMC filter hopping system is shown in Figure 10.8. It features the time-varying transmit and matched receive filter at Alice's and Bob's side and the eavesdropper, who is using a quasistatic receive filter, as he is not able to follow the sequence of filter changes. The complexity increase due to the variation of the filters in the filter bank is negligible; however, the filter design techniques might inhere a higher amount of processing.

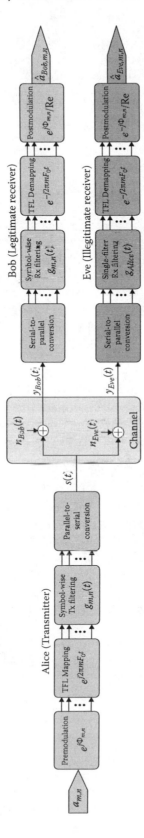

Figure 10.8 Filter bank multicarrier filter hopping transmission chain with eavesdropper. (From Luecken, V. et al., Filter hopping: Physical layer secrecy based on fbmc, in *IEEE Wireless Communications and Networking Conference*, New Orleans, LA, March 2015.)

Then, Eve will experience interference on its receive signal, leading to a degradation of her SINR in a large number of possible channel conditions. By that, a secrecy capacity increase is yielded.

The technique of filter hopping with FBMC is still in a preliminary phase before practical use in a real system. This is mainly due to the limitations in filter design, which are still part of further research. However, with the promising initial results [46], the filter hopping technique might prove suitable for an enhanced secrecy in 5G physical layer applications.

10.5.3 Physical Layer Key Generation

Conventional symmetric key negotiation techniques, like the Diffie–Hellman key exchange [32] or techniques based on the RSA algorithms [63], were not specifically developed for wireless communications. They generally use a secret knowledge at both partner's (Alice's and Bob's) side and then negotiate a shared secret key (in case of Diffie–Hellman) over an authenticated public channel, or employ a public–private key pair for securely exchanging a shared secret (in case of RSA). The resulting key on both sides can then be used, for example, for symmetric encryption techniques. Instead of relying on the computational complexity of a mathematical problem in the case of conventional encryption techniques [64,65], physical layer key generation techniques rely on the wireless channel as a common and reciprocal entropy source for Alice and Bob. The use of these novel techniques can both increase the security of a key exchange and also lower its complexity in comparison to conventional algorithms.

First, considering the conventional techniques, the term *public* for the required public authenticated communication channel in the context of Diffie–Hellman means that an Eve can listen to the communication, for example, by wiretapping the wireless connection or any other signal transmission technique in between. *Authenticated* means that the originators of negotiation messages for the key exchange are verified and the message is unchanged. Otherwise, MIM attacks could possibly compromise the security of the whole key negotiation. The authentication is a necessary prerequisite for the Diffie–Hellman key exchange; otherwise, such an MIM attack is possible. However, extensions and modifications of the Diffie–Hellman key exchange exist, which tackle this problem [66]. The conventional key exchange techniques are well established, but also have several disadvantages, like complexity, the required key distribution techniques with their dependencies, and their focus on the computational complexity of mathematical problems, based on specific functions as a barrier for the eavesdropper [48]. This is one of the reasons why physical layer key generation techniques emerge as an alternative or extension to the classical solutions.

The wireless communication channel as a reciprocal random source for both Alice and Bob can be realized, for example, by using the channel impulse responses or received signal strength [49]. Important physical parameters for this are the mobility, the scattering of the environment, and signal wavelength. Mobility is important for ensuring a significant variation of the channel in order to generate a sufficient amount of random data for generating and renewing the keys [64]. The signal wavelength λ, however, is relevant for the correlation of the channel state in comparison to a close-by position. If an eavesdropper has a distance higher than $\frac{\lambda}{2}$ to the key negotiation partners and the reflections of the environment are sufficiently scattered, he or she will not be able to extract the CSI of the legitimate users [49].

An important aspect for secure communication on a public channel is feedback [52]. Contrary to the definition of Wyner's wiretap channel [10], where the eavesdropper is degraded (having a higher noise level) in comparison to the legitimate users for achieving secure communication, the author of [47] shows that with the so-called public discussion, an advantage over Eve can be gained. This, finally, allows a secret key negotiation even with an initial advantage of the eavesdropper.

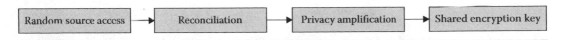

Figure 10.9   Overview of physical layer key generation process.

In the following, the practical structure of the physical layer key generation process is explained, which is also shown in Figure 10.9. The first part of the process is the *random source access*, where the random source is the wireless channel, by accessing it via bidirectional transmissions. This can, for example, be done in a time division duplexing manner, if the channel variation is slow enough. The randomness extracted in this step can consist of characteristics like the channel impulse response or received signal strength [49]. This process is performed until both Alice and Bob have a sufficient amount of secure shared information. This also involves the so-called advantage distillation [48], which seeks to improve the secrecy advantage of Alice's and Bob's observations. For the channel as an entropy source, the authors of [67] presented two possible information-theoretic models, the *source-type* and the *channel-type model*. For the first case, the channel is modeled as a discrete memoryless multiple source, which is basically independent of the communication participants and probed by all terminals. For the *channel-type model* on the contrary, a discrete memoryless channel is used, which features an input from one communication partner and an output to both the other partner and the eavesdropper. Here, the output depends on the input message, meaning that the input to the channel can be controlled by the transmitter, which is practically closer to the model of a real channel. In addition, both models feature the required noiseless public channel in parallel, which is necessary for the subsequent discussion process.

The second stage is the *reconciliation* stage. Here, a mutual agreement between Alice and Bob is achieved by eliminating error factors due to noise and imperfections, for example, due to the type of measurements [48,49]. This can be done using error-correcting codes. The reconciliation may also inhere a partial revelation of information to the eavesdropper. For this reason, the so-called privacy amplification stage is carried out afterward. It has the goal of eliminating the remaining information of Eve over the secret key agreement, which may be left due to the previous random source access and the reconciliation stages [49]. From the *leftover hash lemma* [50,51], we can follow that a secure key can be yielded by using specific reductions functions (hash functions) in a compression process. They yield a close to uniform distribution if the input distribution is sufficiently compressed (with *sufficient*, the reduction by the whole amount of side information of the eavesdropper is meant). Thus, we know that we can get a new key with a reduced length after the compression, where the remaining partial information from the eavesdropper is removed, and thus realize the *privacy amplification* process. This finally leads to the secure *shared encryption key*.

As a conclusion, physical layer key generation is a promising technique for future wireless communications. It can be used as an alternative to conventional techniques, by shifting the secrecy paradigm from the computational complexity of trapdoor functions in current encryption techniques further to an information-theoretic level. Also, physical layer key generation can be used as an additional security measure to classical application layer encryption. By directly integrating it into the physical layer of the communication system, secrecy can already be reached on the lowest layer of the communication structure. Also, for miniature-type devices with power constraints, these techniques can be a promising alternative due to a possible complexity reduction. Still, more research is required to tackle the problem of a low key generation rate in slowly varying channels [64]. Further information on the practical techniques for secure mutual key generation and distillation can be taken from the literature sources [47,48,52,67].

10.5.4 Considerations on Implementation in Future Mobile Networks

The techniques mentioned earlier for physical layer security promise good results; however, some of them come with some constraints, which makes it hard for practical implementation. Interestingly, implementation of physical layer security systems are very limited to this date as most of the physical security techniques require very high computational power for calculations.

Considering the battery life of mobile devices, it is more likely for these techniques to be implemented on the base station side first, hence the adaptive beamforming techniques are a good candidate for future implementation. Artificial noise and interference shaping–based approaches increase secrecy; however, there are more challenges when it comes to practical implementation, for example, it should be ensured that no legitimate user is bothered by the transmitted noise. These techniques should be tested and validated before considered as a practical approach.

However, with software defined radio [68], it is possible to interoperate the radio components with high calculation power and change the radio parameters on the run. This ability is the first step that closes the gap between the research and practice in physical layer security systems. We have conducted a preliminary work in [69], where we show that the artificial noise technique is working in a real-world scenario, and we intend to develop the test bed in order to pioneer the physical layer security implementation in the literature.

10.6 Conclusion

As wireless technologies are a significant part of mobile networks, wireless security is very important for mobile networks of today and the future. In this chapter, the wireless last mile of mobile networks is contextualized, discussing the advantages and disadvantages it brings. The security concerns regarding the usage of wireless technologies are covered. Main concepts of wireless systems and physical layer security are introduced and the relation between the wireless attacks and vulnerabilities is given. The current solutions for security are explained along with the future solutions. Here, we can conclude that the current technologies are not sufficient to meet the security issues; however, the future solutions are very promising. The concerns regarding the implementation of those solutions are valid today and there may be some time needed before these techniques can be implemented in next-generation mobile networks.

References

1. StatCounter. StatCounter global stats, June 2015, online, Available at http://gs.statcounter.com/, accessed August 4, 2016.
2. C. Gonzales. Mobile services business and technology trends. In *IEEE International Conference on Web Services*, Beijing, China, pp. 3–3. September 2008.
3. B. Durkin and I. Lokshina. The impact of integrated wireless and mobile communication technologies on the corporate world. In *Wireless Telecommunications Symposium (WTS), 2015*, New York, NY, pp. 1–5. April 2015.
4. M. Meeker. Kleiner Perkins Caufield Byers 2015 Internet trends, May 2015, online, Available at http://www.kpcb.com/blog/2015-internet-trends, accessed August 4, 2016.
5. P. Mandl, P. Schrotter, and E. Leitgeb. Wireless synchronous broadband last mile access solutions for multimedia applications in license free frequency spectrums. In *Sixth International Symposium on Communication Systems, Networks and Digital Signal Processing*, Graz, Austria, pp. 110–113, July 2008.

6. P. Mandl, E. Leitgeb, M. Loschnigg, T. Plank, and P. Pezzei. FSO and WLAN as backward channel for Internet connections of peripheral regions only covered by DVB-T. In *15th International Conference on Transparent Optical Networks* (ICTON), Cartagena, Colombia, pp. 1–5, June 2013.

7. J. Proakis and D. Manolakis. *Digital Signal Processing*. Pearson Prentice Hall, NJ, 2007.

8. V. Erceg, K. V. S. Hari, M. Smith, and D. S. Baum. Channel models for fixed wireless applications. Contribution to IEEE 802.16.3, July 2001, online, Available at http://www.nari.ee.ethz.ch/commth//pubs/p/EHSB01, accessed August 4, 2016.

9. M. Narandzic, C. Schneider, R. Thomä, T. Jämsä, P. Kyösti, and X. Zhao. Comparison of SCM, SCME, and winner channel models. In *IEEE 65th Vehicular Technology Conference, VTC2007-Spring*, Dublin, Ireland, pp. 413–417. IEEE, 2007.

10. A. D. Wyner. The wire-tap channel. In *Bell System Technical Journal*, 54: 1355–1387, 1975.

11. I. Csiszar and J. Korner. Broadcast channels with confidential messages. *IEEE Transactions on Information Theory*, 24(3): 339–348, May 1978.

12. M. Sarkar and T. Ratnarajah. Secrecy capacity and secure outage performance for Rayleigh fading SIMO channel. In *IEEE International Conference on Acoustic, Speech, Signal Processing* (ICASSP), Prague, Czech Republic, pp. 1900–1903. 2011.

13. S. Yang, P. Piantanida, M. Kobayashi, and S. Shamai. On the secrecy degrees of freedom of multi-antenna wiretap channels with delayed CSIT. In *IEEE International Symposium on Information Theory Proceedings* (ISIT), St. Petersburg, Russia, pp. 2866–2870, 2011.

14. S. Gerbracht, C. Scheunert, and E. Jorswieck. Secrecy outage in MISO systems with partial channel information. *IEEE Transactions on Information Forensics and Security*, 7(2): 704–716, 2012.

15. J. Huang and A. Swindlehurst. Robust secure transmission in MISO channels based on worst-case optimization. *IEEE Transactions on Signal Processing*, 60(4): 1696–1707, 2012.

16. X. Yang and A. Swindlehurst. On the use of artificial interference for secrecy with imperfect CSI. In *IEEE 12th International Workshop on Signal Processing Advances in Wireless Communications* (SPAWC), San Francisco, CA, pp. 476–480, 2011.

17. S.-C. Lin, T.-H. Chang, Y.-L. Liang, Y. Hong, and C.-Y. Chi. On the impact of quantized channel feedback in guaranteeing secrecy with artificial noise: The noise leakage problem. *IEEE Transactions on Wireless Communications*, 10(3): 901–915, 2011.

18. J. Yang, I.-M. Kim, and D. I. Kim. Optimal cooperative jamming for multiuser broadcast channel with multiple eavesdroppers. *IEEE Transactions on Wireless Communications*, 12(6): 2840–2852, 2013.

19. A. Mukherjee and A. Swindlehurst. Ensuring secrecy in MIMO wiretap channels with imperfect CSIT: A beamforming approach. In *2010 IEEE International Conference on Communications* (ICC), Cape Town, South Africa, pp. 1–5. May 2010.

20. A. Goldsmith. *Wireless Communications*. Cambridge University Press, Cambridge, U.K., 2005.

21. W.-C. Liao, T.-H. Chang, W.-K. Ma, and C.-Y. Chi. QoS-based transmit beamforming in the presence of eavesdroppers: An optimized artificial-noise-aided approach. *IEEE Transactions on Signal Processing,*, 59(3): 1202–1216, March 2011.

22. X. Liu, F. Gao, G. Wang, and X. Wang. Joint beamforming and user selection in multicast downlink channel under secrecy-outage constraint. *IEEE Communications Letters*, 18(1): 82–85, January 2014.

23. G. Dartman, O. Cepheli, G. Karabulut Kurt, and G. Ascheid. Beamforming aided interference management with improved secrecy for correlated channels. In *Proceedings of the IEEE 79th Vehicular Technology Conference—VTC'14-Spring*, Seoul, Korea, May 2014.

24. D. Ng, E. Lo, and R. Schober. Robust beamforming for secure communication in systems with wireless information and power transfer. *IEEE Transactions on Wireless Communications*, 13(8): 4599–4615, August 2014.

25. N. Romero-Zurita, D. McLernon, and M. Ghogho. Physical layer security by robust masked beamforming and protected zone optimisation. *IET Communications*, 8(8): 1248–1257, May 2014.

26. S. Leung-Yan-Cheong and M. Hellman. The Gaussian wire-tap channel. *IEEE Transactions on Information Theory*, 24(4): 451–456, 1978.

27. M. Bloch, J. Barros, M. Rodrigues, and S. McLaughlin. Wireless information-theoretic security. *IEEE Transactions on Information Theory*, 54(6): 2515–2534, June 2008.

28. J. Barros and M. Rodrigues. Secrecy capacity of wireless channels. In *IEEE International Symposium on Information Theory*, Seattle, WA, pp. 356–360, July 2006,

29. P. Wang, G. Yu, and Z. Zhang. On the secrecy capacity of fading wireless channel with multiple eavesdroppers. In *IEEE International Symposium on Information Theory*, Nice, France, pp. 1301–1305, 2007.

30. P. Singer and A. Friedman, *Cybersecurity: What Everyone Needs to Know*. Oxford University Press, Oxford, UK, 2014.

31. Z. Lu, W. Wang, and C. Wang. From jammer to gambler: Modeling and detection of jamming attacks against time-critical traffic. In *INFOCOM, 2011 Proceedings IEEE*, Shanghai, China, pp. 1871–1879, April 2011.

32. W. Diffie and M. Hellman. New directions in cryptography. *IEEE Transactions on Information Theory*, 22(6): 644–654, Nov 1976.

33. A. George and M. Kiesler. Secret communication system. US Patent 2,292,387, August 11 1942.

34. B. Gaston. Applications of spread spectrum radio technology for the security market. In *IEEE 28th Annual 1994 International Carnahan Conference on Security Technology*, Albuquerque, NM, pp. 86–91. October 1994.

35. R. Ziemer and R. Peterson, *Digital Communications and Spread Spectrum Systems*. Macmillan, New York, 1985.

36. S. M. Schwartz. Frequency hopping spread spectrum (fhss) vs. direct sequence spread spectrum (dsss) in broadband wireless access (bwa) and wireless lan (wlan). http://sorin-schwartz.com/white_papers/fhvsds.pdf, accessed August 4, 2016.

37. T. Li, J. Ren, Q. Ling, and A. Jain. Physical layer built-in security analysis and enhancement of CDMA systems. In *Military Communications Conference, 2005 (MILCOM 2005) IEEE*, Atlantic City, NJ, Vol. 2, pp. 956–962. October 2005

38. F. Hermanns. Cryptographic CDMA code hopping (CH-CDMA) for signal security and anti-jamming. In *Proceedings EMPS 2004*. Nordwijk, Netherlands, 2004.

39. T. Li, Q. Ling, and J. Ren. Physical layer built-in security analysis and enhancement algorithms for CDMA systems. *EURASIP Journal of Wireless Communication Networks*, 2007(3): 7:1–7:16, July 2007.

40. J. Zhang and M. Gursoy. Relay beamforming strategies for physical-layer security. In *2010 44th Annual Conference on Information Sciences and Systems* (CISS), Princeton, NJ, pp. 1–6. March 2010.

41. W.-C. Liao, T.-H. Chang, W.-K. Ma, and C.-Y. Chi. QoS-based transmit beamforming in the presence of eavesdroppers: An optimized artificial-noise-aided approach. *IEEE Transactions on Signal Processing*, 59(3): 1202–1216, March 2011.

42. S.-H. Lai, P.-H. Lin, S.-C. Lin, and H.-J. Su. On optimal artificial-noise assisted secure beamforming for the fading eavesdropper channel. In *IEEE PIMRC*, Toronto, ON, Canada, pp. 1167–1171. September 2011.

43. G. Dartmann, X. Gong, W. Afzal, and G. Ascheid. On the duality of the max-min beamforming problem with per-antenna and per-antenna-array power constraints. *IEEE Transactions on Vehicular Technology*, 62(2): 606–619, February 2013.

44. X. Gong, M. Jordan, G. Dartmann, and G. Ascheid. Max-min beamforming for multicell downlink systems using long-term channel statistics. In *IEEE PIMRC*, Tokyo, Japan, pp. 803–807. 2009.

45. G. Dartmann, V. Lucken, O. Cepheli, G. K. Kurt, and G. Ascheid. Filter optimization aided interference management with improved secrecy. In *IEEE Vehicular Technology Conference* (VTC Fall), Vancouver, BC, Canada, pp. 1–6. September 2014.

46. V. Luecken, T. Singh, O. Cepheli, G. Karabulut Kurt, G. Ascheid, and G. Dartmann. Filter hopping: Physical layer secrecy based on FBMC. In *IEEE Wireless Communications and Networking Conference*, New Orleans, LA, March 2015.

47. U. Maurer. Secret key agreement by public discussion from common information. *IEEE Transactions on Information Theory*, 39(3): 733–742, May 1993.

48. M. Bloch and J. Barros, *Physical-Layer Security: From Information Theory to Security Engineering*. Cambridge University Press, Cambridge, U.K., 2011.

49. K. Zeng. Physical layer key generation in wireless networks: Challenges and opportunities. *IEEE Communications Magazine*, 53(6): 33–39, June 2015.

50. M. Petkovic and W. Jonker. *Security, Privacy, and Trust in Modern Data Management*, ser. Data-Centric Systems and Applications. Springer, Switzerland, 2007.

51. R. Impagliazzo, L. A. Levin, and M. Luby. Pseudo-random generation from one-way functions. In *Proceedings of the Twenty-first Annual ACM Symposium on Theory of Computing*, ser. STOC'89, pp. 12–24. ACM, New York, 1989.

52. X. Zhou, L. Song, and Y. Zhang. *Physical Layer Security in Wireless Communications*. Taylor & Francis, Boca Raton, FL, 2013.

53. J. Zhu, X. Jiang, Y. Zhou, Y. Zhang, O. Takahashi, and N. Shiratori. Outage performance for secure communication over correlated fading channels with partial CSI. In *IEEE Asia-Pacific Services Computing Conference (APSCC)*, Guilin, China, pp. 257–262.

54. A. Mukherjee and A. Swindlehurst. Robust beamforming for security in MIMO wiretap channels with imperfect CSI. *IEEE Transactions on Signal Processing*, 59(1): 351–361, 2011.

55. H.-M. Wang, M. Luo, X.-G. Xia, and Q. Yin. Joint cooperative beamforming and jamming to secure AF relay systems with individual power constraint and no eavesdropper's CSI. *IEEE Signal Processing Letters*, 20(1): 39–42, 2013.

56. X. He and A. Yener. Providing secrecy irrespective of eavesdropper's channel state. In *2010 IEEE Global Telecommunications Conference (GLOBECOM 2010)*, Miami, FL, pp. 1–5. 2010.

57. W. Li, M. Ghogho, B. Chen, and C. Xiong. Secure communication via sending artificial noise by the receiver: Outage secrecy capacity/region analysis. *IEEE Communication Letters*, 16(10): 1628–1631, October 2012.

58. S. Bayat, R. Louie, Z. Han, B. Vucetic, and Y. Li. Physical-layer security in distributed wireless networks using matching theory. *IEEE Transactions on Information Forensics and Security*, 8(5): 717–732, 2013.

59. M. Ghogho and A. Swami. Characterizing physical-layer secrecy with unknown eavesdropper locations and channels. In *IEEE International Conference on Acoustics, Speech and Signal Processing (ICASSP)*, Prague, Czech Republic, pp. 3432–3435, 2011.

60. S. Anand and R. Chandramouli. On the location of an eavesdropper in multiterminal networks. *IEEE Transactions. on Information Forensics and Security*, 5(1): 148–157, 2010.

61. S. Goel, V. Aggarwal, A. Yener, and A. Calderbank. Modeling location uncertainty for eavesdroppers: A secrecy graph approach. In *IEEE International Symposium on Information Theory Proceedings (ISIT)*, Austin, TX, pp. 2627–2631. 2010.

62. M. Schellmann, Z. Zhao, H. Lin, P. Siohan, N. Rajatheva, V. Luecken, and A. Ishaque. FBMC-based air interface for 5G mobile: Challenges and proposed solutions. In *Ninth International Conference on Cognitive Radio Oriented Wireless Networks and Communications (CROWNCOM)*, Oulu, Finland, pp. 102–107. June 2014.

63. R. L. Rivest, A. Shamir, and L. Adleman. A method for obtaining digital signatures and public-key cryptosystems. *Communication ACM*, 21(2): 120–126, February 1978.

64. S. Gollakota and D. Katabi. Physical layer wireless security made fast and channel independent. In *IEEE INFOCOM Proceedings*, Shanghai, China, pp. 1125–1133. April 2011.

65. Z. Hao, S. Zhong, and L. Li. Towards wireless security without computational assumptions; an oblivious transfer protocol based on an unauthenticated wireless channel. In *IEEE INFOCOM Proceedings*, Shanghai, China, pp. 2156–2164. April 2011.

66. W. Diffie, P. C. Van Oorschot, and M. J. Wiener. Authentication and authenticated key exchanges. *Design Codes Cryptography*, 2(2): 107–125, June 1992.

67. R. Ahlswede and I. Csiszar. Common randomness in information theory and cryptography. I. Secret sharing. *IEEE Transactions on Information Theory*, 39(4): 1121–1132, July 1993.

68. E. Grayver, *Implementing Software Defined Radio*. Springer Science & Business Media, New York, NY, 2012.

69. O. Cepheli and G. Kurt. Analysis on the effects of artificial noise on physical layer security. In *2013 21st Signal Processing and Communications Applications Conference (SIU)*, Haspolat, Turkey, pp. 1–4. April 2013.

Chapter 11

Protecting Mobile Payments Security

A Case Study

Bo Xing

Contents

Abstract

As mobile technologies are becoming more sophisticated and mobile devices are penetrating into our daily life, a brand new type of payment system, that is, mobile payments (m-payments), facilitating users to complete their payment activities while on the move via various wireless devices (in particular mobile phones), has emerged. Currently, mobile network operators, banks, device manufacturers, big Internet titans, and all sorts of other provision institutions (e.g., financial service firms and payment service companies) are playing a part in developing a mature m-payments ecosystem. Therefore, the first part of this chapter is devoted to the discussion of the characteristics of the m-payment system, and an overview of the current security trends related to the m-payments market is also provided. Meanwhile, we have passed the periods of irrational exuberance and equally irrational disillusionment (each championed by involving parties who have yet to fully explore the landscape of m-payments); it's time to stop seeing m-payments as wrecking balls ready to shake many players' conventional business to their foundations. Instead, in this chapter, we think of m-payments from a device manufacturer's perspective via the lens of secure element, a key enabling hardware for m-payments. Although handset manufacturers have been late entrants in the m-payments market, they could still become relevant and play game-changing roles. A principal sticking point in making this role leap is the selection of innovative technology. By selecting an appropriate secure element supplier and integrating this novel hardware to their products, handset manufacturers are expected to be able to find a niche in an expanding m-payments ecosystem, rather than being just a device maker for the financial services: churning out miscellaneous smart widgets while others grab the really valuable parts. The second part of this work illustrates a secure element supplier evaluation scenario. A fuzzy TOPSIS (i.e., technology for order performance by similarity to ideal solution) is also employed to solve the encountered multicriteria decision-making problem. The experimental results demonstrated the feasibility of the utilized methodology in the chosen context.

Keywords: Mobile payments (m-payments), Mobile wallet, Near-field communication (NFC), Eembedded secure element (eSE), Host card emulation (HCE), Fuzzy logic, Technology for order performance by similarity to ideal solution (TOPSIS)

11.1 Introduction

Nowadays, smartphones provide a high level of accessibility (i.e., users have access to everything, all of the time) and change the way we search for information and shop for products, particularly in sectors such as movie, music, games, books, and many kinds of services. For example, it was reported by BIGresearch that, during the 2010 holiday season, consumers were willing to spend $688.87 per head, and a quarter of adult mobile phone possessors have a plan to research or make purchases via their mobile devices [1]. In another survey, majority (or more precisely, 80%) of Canadian industry players are sure that using mobile device for shopping will become mainstream in Canada within the next few years [2,3]. Along with this growth in mobile commerce (i.e., m-commerce), there has been a significant increase in the mobile payments (m-payments), that is, doing shopping for goods or services and finishing the associated payments that originate from a smartphone or a similar device, also called contactless payments. Insight Research Corporation estimated that a $124 billion of mobile financial transactions market value could be generated by about 2.2 billion consumers by 2014 [4]. In fact, 2015 will be the year when m payment platforms become more commonplace [5].

However, there is an inverse relationship between security and accessibility, and finding a lack of consumer confidence in the security (such as authentication and verification) in m-payments can determine its success or failure [6,7]. For example, according to Federal Reserve Board reports, 62% of U.S. consumers with mobile phones and bank accounts do not currently use mobile banking is due to the concern about security [8]. In a similar vein, many researchers (e.g., [9]) pointed out that there is no question that m-payment is going to come; one of the barriers to scale is the security issue since many believe that the consumer buy-in will only occur when systems are secure. For example, how is the user device authenticated? How are data secured on the device? Or who ensures that the card data are not spoofed from one device and replayed on another device? To answer those questions, the most related solution focuses on the card emulation mode (or secure element), that is, an environment in the mobile device where applications and credentials could be securely stored.

With this in mind, the remainder of this chapter is organized as follows: first, some useful terminologies are introduced in Section 11.2; the relevant background and literature review on the m-payments are then provided in Section 11.3, which is followed by a scenario description in Section 11.4; the background of the employed methodologies are detailed in Section 11.5; next, the problems' formulation and numerical study are elaborated in Sections 11.6 and 11.7, respectively; then the future research avenues are highlighted in Section 11.8; finally, Section 11.9 draws the conclusion of this study.

11.2 Useful Terminologies

11.2.1 Mobile Phone–Related Terms

11.2.1.1 2G, 3G, 4G, and 5G

From analog phone calls (i.e., first generation or 1G for short) to third generation (3G) to the current fourth generation (4G) and the forthcoming fifth generation (5G), each generation of mobile technology has been propelled by the need to meet a requirement pinpointed between a certain technology and its predecessor. It is worth mentioning that the main distinct feature of those generations is based on multiradio access technology, which is a concept employed for aggregating data flow of different access. For 5G, it is changed, which means it is not relied on a single dominant technology but will gain interest with other advanced technologies (e.g., employing adjacent and wide spectrum bandwidth, flexible resource allocating and sharing plans, adaptable air interfaces, new waveforms, agile access techniques, advanced multiantenna beamforming and beam tracking and multiple-input multiple-output approaches, and innovative radio resource management algorithms) to construct the 5G infrastructure; the benefits of developing 5G are enormous. Yuan and Zhao [10] pointed out that 5G will significantly increase the peak service rate and the efficiency of other resources, such as energy, frequency, and spectrum expansion. Meanwhile, it will be more flexible and compatible and satisfiable to the users. For example, it will be supported by high-performance systems with capacity three orders of magnitude greater than 4G. In addition, 5G will be a key facilitator for the Internet of things by providing a platform to connect an enormous amount of sensors and actuators with strict energy and transmission limitations. As a result, it will meet the increasing demands of emerging big data, cloud, and vast meshes of connection scenarios, say people-to-people (P2P), people-to-machine, and machine-to-machine.

11.2.1.2 Global System for Mobile Communications

Global system for mobile communications (GSM) is the most widely employed second-generation (2G) wireless telecommunication standard. Originally, it was designed solely for voice, but via short message service, it was later improved to support instant messaging. Since the success of 2G communication systems and at the same time a witness of the early growth of the Internet, a 2.5G system, that is, general packet radio service (GPRS), has been introduced to merge the two ideas for the purpose of enabling packet-based data transmission on GSM networks, or in other words allowing users to download data onto mobile devices. Later, enhanced data rates for GSM evolution was proposed to further forward the GPRS.

11.2.1.3 GSM Association

The Global System for Mobile Communication Association (GSMA) was founded in 1982. The objectives are, on one hand, to create a global working environment for mobile phones and wireless services, and on the other hand, to let operators and their suppliers take the emerging new business opportunities to thrive while transferring the added value to individual consumers and national economies. Nowadays, GSMA has nearly 800 operators with over 250 enterprises existing in a broader mobile ecosystem, which includes handset and device manufacturers, software firms, equipment and Internet service providers, and organizations in contiguous industry sectors. There are many accomplishments achieved by GSMA such as setting GSM standards, enabling roaming interconnection, and establishing industry representation with regulators and government involved.

11.2.1.4 Universal Mobile Telecommunications System

In order to achieve a considerably higher data transmission rate, universal mobile telecommunications system (UMTS) is further introduced as a 3G cellular network standard. It was developed from GSM by introducing innovative technology used on the air interface, while the core network is largely kept unmodified. The system was later improved for data applications by introducing the high-speed packet access, a collective expression that includes two 3.5G technologies, namely, high-speed downlink packet access and high-speed uplink packet access.

11.2.1.5 Long-Term Evolution

With 3G mobile networks, the evolution of mobile communication has reached a new stage. Long-term evolution, aka 4G, has been marketed as a novel high-speed/low-latency service, which provides a long-range, scalable, battery-efficient, and privacy-sensitive platform for end users, applications, and varieties of services. Meanwhile, operators face a serendipitous encounter of becoming a key partner of adjacency awareness development for machine-type communications.

11.2.1.6 Mobile Network Operators

As the name indicates, mobile network operators (MNOs) are the providers that own or control wireless communication services, including radio spectrum allocating, wireless network building, back haul infrastructure maintaining, billing, customer caring, computer systems provisioning, marketing, and repairing [11]. Nowadays, 34 of the world's largest MNOs (serving over 1.3 billion consumers) have been actively collaborating on an m-payment initiative to create and define a global standard for the purpose of realizing mobile near-field communication (NFC) payment services. (Details of NFC are provided in the following texts.)

11.2.1.7 Smart Card

A smart card or, formally, integrated circuit card, is a small chip with embedded integrated circuits that can process information. Normally, there are two broad classes of smart cards, namely, memory card and microprocessor card. The former contains only nonvolatile memory storage components, while the latter could be likened to a simple microcomputer, which means it provides both memory capacity and computational capability, say, user data's storage and secret data's encryption. In addition, smart cards can be categorized into two groups according to their working principles, that is, contact and contactless. Nowadays, one notable feature of a smart card is that it often comes with an embedded tamper-proof chip, which makes the smart card a suitable candidate to be applied in the areas where strong security protection and authentication is required.

11.2.1.8 Subscriber Identity Module

A subscriber identity module (SIM) is a tamper-proof hardware chip (i.e., smart card) inserted inside a mobile handset. Crudely, the core features of a SIM card can be put into two directions. On one hand, it is used to store network parameters securely and identify to the networks. On the other hand, it stores user data and authenticates the subscriber to the MNOs' networks. As a result, when an end user switches from one phone to another, only a SIM card needs to be carried over since it contains all the user's key information.

11.2.1.9 Universal Integrated Circuit Card

The term universal integrated circuit card (UICC) is not commonly used before the era of 3G. It is a new model of smart card that not only contains account information, but also has memory, which can be utilized to enable mobile phones. The main objective of a UICC standard is to define a way of installing and managing multiple applications on an individual mobile handset. This means that it is possible to perform sensitive applications in a tamper-proof manner via the same hardware that is used for the SIM.

11.2.1.10 Universal SIM

The universal SIM (USIM) is a new verification and session key deriving functions for UMTS (i.e., 3G). In general, the USIM application runs on an UICC and offers the same functions as a SIM card.

11.2.1.11 GlobalPlatform

GlobalPlatform, established in 1999, is a standard organization defining smart card interfaces. One of the key advantages of GlobalPlatform is its technology supports different implementation models such as multiapplication, multiactor, and multibusiness. In other words, its infrastructure enhances the interoperability and sustainability, which paves the way for smart card deployments, and in turn, the card issuers, service providers, and technology suppliers all enjoy various benefits.

11.2.1.12 European Technical Standards Institute

The European Technical Standards Institute (ETSI), as a standardization organization, has been working on standards development for various electronics sectors since 1988. They have established many standards for UICC and SIM cards that play a variety of key roles in today's mobile environment such as defining file formats on the card and encrypting information when communicating with the MNO.

11.2.1.13 Near-Field Communication

Technically, NFC is a radio-frequency protocol and proximity-based technology (<10 cm apart) enabling communication between electronic devices, Bluetooth, or infrared technologies for data transfer [12]. According to the difference between operating modes, the NFC protocol can be classified into three categories, that is, reader/writer mode, peer-to-peer mode, and card emulation mode [13]. In the reader/writer mode, an NFC device is capable of reading NFC Forum–mandated tag types, for example, in the scenario of reading an NFC Smart Poster tag or accessing information on the move. In the peer-to-peer mode, data can be exchanged between two NFC devices. Finally, in the card emulation mode, the NFC device appears as an external reader much similar to a traditional contactless smart card. This enables contactless payments (i.e., m-payments) by NFC devices with no need to change the incumbent contactless infrastructure [14].

11.2.1.14 Trusted Service Manager

The trusted service manager (TSM) is an entity in the m-payment ecosystem, and it works as an intermediary institution among m-payment actors, which include card issuers, MNOs, financial institutions, retailers, NFC mobile phone users, and handset manufacturers. The main function of TSM is to manage the customer life cycle of NFC applications securely. In addition, TSM is also responsible for the management of customer service, data center hosting, and quality assurance [14].

11.2.2 Financial Service–Related Terms

11.2.2.1 EuroPay, MasterCard, and Visa

As the combination of initial letters of EuroPay, MasterCard, and Visa, EMV represents the three firms who originally collaborated to develop a standard for smart card interoperation and the corresponding capability of processing devices for financial transactions. In general, an EMV transaction consists of three stages, that is, card authentication, cardholder verification, and transaction authorization. Since the deadline for the transition to the EMV standard is October 1, 2015 in the United States, there is a strong driving force for the widespread availability of m-payments.

11.2.2.2 Electronic Commerce and Mobile Commerce

Generally speaking, electronic commerce (e-commerce) refers to describe conducting business transactions, in particular purchase of goods or services over the Internet. With the ever-broadening acceptance rate of mobile devices, the era of m-commerce is rapidly approaching. Indeed, m-commerce is a natural successor to e-commerce. For example, Starbucks reports that more than 7 million customers now use Starbucks' mobile app for making 4.5 million payments per week, accounting for at least 10% of Starbucks' total U.S. revenue [15]. Moreover, in a detailed analysis of m-commerce, CyberSource research group, a wholly owned subsidiary of Visa Inc., concludes that m-commerce will become a bigger share of overall e-commerce—from 9% in 2012 to a projected 18% in 2014 [16]. The benefits of m-commerce include boosting revenues, reducing costs, enhancing customer experience, and furthering brand proposition.

11.3 Background and Literature Review of the Digital Financial Services

11.3.1 Transformation of Technology-Led Financial Services and Payments Industry

The worldwide payments landscape is undergoing a period of rapid, technology-driven change. Online, mobile, and social technologies have played an important role to revolutionize consumer access information, sparking demand for new services, and finally reflecting concern about the payments attitude. To capitalize on this opportunity, financial services and payments industries must look for ways to position themselves in this rapidly changing market. So what separates the leaders from laggards? One of the innovations that have emerged is the digitization of financial services, which can be roughly divided into two groups, namely, online banking and mobile banking (or payment) [17]. In coming years, however, mobile use will become the epicenter of digital financial services, since smartphones are being the most common starting place for online activities. According to a report issued by GSMA [18], smartphone penetration is already well above 50% in North America and the European Union. In addition, global payments revenues will grow at a CAGR of 5% between 2011 and 2016 [19]. Attracted by those great markets, many financial services and payments industries are adopting a *mobile first* approach.

In general, m-payments can be classified into four main types [20], that is, mobile proximity payments, mobile web payments, mobile remote payments, and mobile point-of-sale (POS). Additionally, m-payment has three distinct models [21], namely, P2P, people-to-business (P2B), and business-to-business (B2B). Compared to PC banking, the main advantage of m-payment is ubiquity. For example, a European banking executive reports that their clients have 11 times as many contacts through mobile than on standard computers [22]. In addition, it provides new opportunities for financial services and payments industries to engage frequently with the customer, providing greater insight into consumers' behavior, increasing the opportunities to build trust, and enhancing loyalty programs and advertising performance [23].

Overall, financial services and payment industries across the world have determined that change must happen in payments. They must not only respond to the competition but are also expected to foot the bill for creating new digital strategies and infrastructures. Among others, advanced technology (e.g., NFC) will help consumers convert to mobile NFC, which provides consumers with a simple, secure method to shop and pay.

11.3.2 Prosperity of Mobile NFC (Contactless) Payments

Today, the discussion about m-payment isn't complete if the NFC technology is omitted. Though there are several different options when building a proxy card emulation device. These options range from building new card emulator hardware from square one to using distinct types of existing hardware. The various choices have different advantages and disadvantages that result in different design and implementation costs and different levels of emulation capabilities. Interested readers should refer to [24] for a detailed comparison. In this study, we will only pay attention to one alternative of using NFC-enabled mobile phones as card emulation devices. Not only has NFC been embedded in most models of smartphones, but also using mobile proximity payments is a natural extension of existing consumer behavior. While BlackBerry mobile phones were the first phones that contained an application program interface (API) for software card emulation, other NFC devices could be adapted to support software card emulation as well. For example, according to ABI research, "global NFC-enable devices will soon exceed 500 million." In addition, 81% of consumers surveyed reported that they find convenience and accessibility to be the most compelling attributes of contactless and mobile technology [25]. Under this circumstance, a blueprint called mobile NFC payments is being gradually accepted worldwide, though the process is not as smooth as one has expected, since it has been used in different ways in a variety of contexts.

Briefly, mobile NFC payments would enable consumers to wave their smartphones in the proximity of a contactless reader (e.g., POS terminals) to conduct transactions efficiently. Examples include mobile ticketing and mobile wallet. Theoretically, there are risks of *man-in-the-middle*, *data corruption*, or *eavesdropping* attacks. To achieve secure m-payments, a set of initiatives has been proposed. At present, a mobile device equipped with two components, namely, an NFC controller and a secure element, is on the verge of broad adoption [26]. Take mobile wallet; the working structure can be summarized as follows:

- *NFC chip component embedding*: At customer checkouts and other POS, contactless payment readers can receive account information sent from NFC chip-enabled mobile devices. From the user point of view, these devices can also be used to read information kept in different contactless-enabled tags, which are often found on objects like advertising collateral and consumer products.

- *Secure element component introducing*: Due to potential man-in-the-middle type of attacks suffered by NFC technology, secure element has been introduced to store and access applications and data safely.
- *Electronic wallet app installing*: In order for a user to manage his or her accounts and initiate desired contactless payments, mobile phones must have a user interface (UI) to facilitate such interaction. Virtually, by possessing different UI apps, a mobile phone can be treated as a digital wallet, which contains many *cards* (from popular ones such as credit, debit, and gift to specific purpose ones, e.g., stored-value accounts, public transit tickets, merchant-specific loyalty cards). End users can choose any suitable card or apps kept in this electronic wallet when finalizing a payment. There are a wide range of digital wallet app collections available in the market, and some manufacturers even have the electronic wallet apps installed before the mobile devices are being delivered.
- *Personalized account information loading*: We all have the experience, say, that a credit card only works after the customized sensitive account information is loaded onto the magnetic strip and the general account holder's information is embossed on the front of the card. In a similar way, before the customer's personal account information is safely stored, a commerce-enabled mobile phone is nothing more than a normal phone.

Nowadays, many people have become accustomed to the contactless worlds and the joy of buying a drink by simply *opening* a mobile wallet (e.g., Google Wallet, Apple Pay, or Samsung Pay) to activate an easy, low-cost transaction. Although this model is still in an early stage of development, there are now dozens of mobile initiatives with different degrees of demonstrated success that provide useful insight for accumulating enough momentum to move this business model forward.

11.3.2.1 Key Component: Secure Element

With advancement of the modern electronic age, to meet an immediate and strong need-to-move everyday transaction requirement, the physical wallet (including a range of valuable items, e.g., coins, paper notes, credit cards, loyalty cards) needs to be transformed into a mobile wallet (e.g., Google Wallet, a digital *container* of different cards and other special offer vouchers). However, as an effective m-commerce solution relies on transparent systems for loading new or updated customer account information into the phone in a dynamic and flexible way, one concern therefore emerges: how will account information be stored inside all those mobile devices? In addition, mobile devices may suffer a great vulnerability to key-logging software or other malwares, which in turn impose a great threat to data inputted via the keypad. So another concern also surfaces: how should m-payments' key stakeholders respond to that challenge? These essential questions have called for the introduction of a new key component to the m-payment area, namely, secure element.

According to GlobalPlatform, a secure element is a tamper-resistant smart card chip with the capability of safely hosting applications and their confidential data, facilitating storage and transaction of payment, and other sensitive credentials [26]. From a technical point of view, it is an independent physical element and designed to permit only trusted payment applications published by a group of well-identified reliable authorities. In addition, they are supported by mature bodies like ETSI, EMV, and GlobalPlatform, which offers a good degree of interoperability [27]. Currently, there are many types of secure element technology available, and each with different pros and cons. In this study, we will focus only on embedded secure element (eSE) situation where a secure element is soldered permanently onto the handset baseband.

11.3.2.2 Simplified NFC–eSE Mobile Device Working Principles

In an NFC-enabled mobile equipment, payment functionality is split into two parts, namely, UI and the payment brand's mobile contactless payment app (see Figure 11.1 for illustration). Although each mobile equipment or operating system may have its own specifically developed UI functionality, the same functional principles must be followed. The UI enables managing app(s), selecting appropriate app, and verifying cardholder, while the payment app itself is responsible for the actual transaction processing. The purpose of eSE is to support different payment apps and keep the customer's payment account information.

11.3.2.3 Secure Element Ecosystem

In general, the m-payments ecosystem involves not only the banking industry but also other members such as TSM and traditional distribution partners like MNOs (e.g., AT&T and T-Mobile), retailers (e.g., Amazon and Alibaba), device manufacturers (e.g., Apple and BlackBerry), and Internet giants (e.g., Google and Baidu). As they all move into the high-margin financial service market, these *actors* are taking different responsibilities depending on the business context. For example, in SIM mode, the major player is the MNOs, while in an embedded SE scenario, the device makers play the main role. Additionally, newcomers and joint ventures have the potential to disrupt the temporarily achieved m-payments ecosystems, which in turn will enrich the payment experience with new value-added services. Consequently, the actual implementation can be challenging and costly as the management of private credentials on a mobile device is complicated.

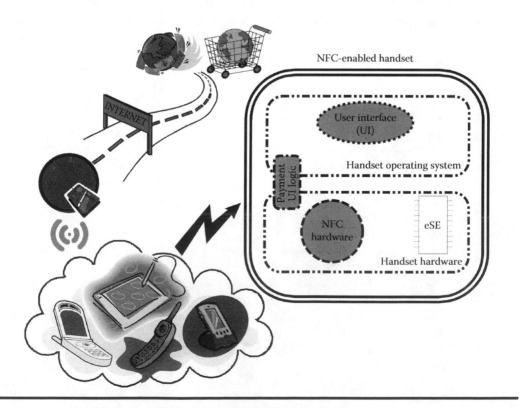

Figure 11.1   Simplified near-field communication mobile device schema.

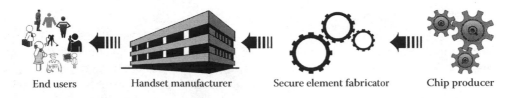

Figure 11.2 Simplified secure element ecosystem.

As shown in Figure 11.2, the chip producer delivers chips to the secure element fabricator. The chip is specifically designed to include a hardware platform that offers the highest levels of security. Then a customized secure operating system will then be installed on the chip by a secure element manufacturer. During this process, each personalized chip (equipped with distinct identification numbers) can be identified by secure element manufacturer with a root key in a secure environment. The secure element is market ready only after this personalization phase. In the case of eSE, handset manufacturers purchase ready-to-use secure elements from their manufacturers and install them in the targeted devices.

11.4 Scenario Description

In this study, we plan to focus on the following scenario: a device manufacturer decides to bake a secure element (i.e., eSE) into its newly introduced mobile devices. A real-world analog case is the Google Nexus S smartphone, which comes with a combination of NFC controller (or PN544 to be precise) and eSE (SmartMX). Normally, a cryptographic processor is included in a secure microprocessor, which is used not only to facilitate authenticating transaction and the associated security issues, but also to provide secure memory space for keeping apps and data. Typically, a secure element has the following general features [14]: (1) it is independent of the handset's operating system and other hardware, (2) enables encrypted protocols to facilitate an enforced access control, (3) allows only the authorized programs' access request to initiate a transaction, and (4) provides the stored data a multilevel of protection through its own snooping- or tamper-free hardware level security.

Apart from all this, the mobile handset manufacturer is eager to implement eSE strategy for many reasons [28–31]:

■ A common architecture for content providers can be provided more freely.
■ Data are encrypted along the entire data travel path.
■ A continuously upgradable security level of devices.
■ More sophisticated technologies are required to provision and manage apps and other stored confidential data.
■ Driven by the requirements of ever-thinner device designs.

We are now at the point where the m-payments are booming and various optimal conditions have been created in many markets. As illustrated in Figure 11.2, a secure element is often supplied by a smart card vendor. In an m-payments scenario, a consumer's credentials are kept on a secure element. Therefore, for a handset manufacturer contemplating eSE solution, selecting a reputable smart card supplier is logical thinking. However, this turns out to be not an easy task considering the recent security scandals suffered by the world's largest smart cards vendor [32]. In this extreme case, spies had reportedly stolen vast numbers of encryption keys hard-coded into the vendor's products. Armed with the keys, decrypting data from the target phones would be trivial. Regardless of its scale, this incident is a good reminder of the vulnerability of mobile phone security, which has mostly been an afterthought in many cases. Bearing this in mind, we set out in this study, from a device manufacturer perspective, to determine how to select an appropriate smart card vendor from a list of possible candidates.

11.5 Background of the Employed Methodology

11.5.1 Fuzzy Logic

In the real world, the solutions to most problems share a common thread that is finding a trade-off between precision and imprecision (or uncertainty). Indeed, uncertainty is a fact of life. To research and deal with those vague phenomena, Zadeh conceived of fuzzy logic in 1965 [33], which is the idea of providing an extension to classical Boolean logic and set theory through introducing fuzzy sets. Compared to precise methods, fuzzy logic refers to a technique of mapping an input space to an output space by means of a list of linguistic rules comprising of the if–then statements [34]. In general, it includes four components, namely, fuzzy sets, membership functions, fuzzy logic operators, and fuzzy rules. From an application viewpoint, every object or value to be allocated with a given fuzzy set to a certain degree within the range of [0, 1] is rather than being a member or not a member of a set (a common classification found in classical set theory). In addition, it allows the evaluation parameters to be expressed as linguistic variables, which works more or less similar to a normal human thinking process.

11.5.1.1 Fuzzy Sets

A set can be called as a fuzzy set if it contains elements that have varying degrees of membership in the set to deal with imprecise or vague data. In contrast with classical (or crisp) sets (i.e., an element's membership to a set is evaluated in binary terms based on a bivalent condition principle: belonging or not belonging to the set), a fuzzy set is specified by a membership function (ranging from 0 to 1) that is deliberately designed to handle the vagueness and imprecision in the context of the application.

For example, let $U = \{x_1, x_2, \ldots, x_n\}$ be a universal set of objects x; we can then define a crisp set A in U as follows:

$$U_A(x) = \begin{cases} 1, & x \in A \\ 0, & x \notin A \end{cases} \tag{11.1}$$

where $U_A(x)$ shows the indication of a clear membership of element x in set A and the conditions of *contained in* or *not contained in* are represented by the symbols $\in$ and $\notin$, respectively. In other words, for crisp sets, an element x in the universe U has only two situations: either belonging to some crisp set A or not, whereas the fuzzy set B in U is extended to accommodate different *degrees of membership* on the real continuous interval [0,1] instead of the notion of binary membership. In this chapter, the notation used to express a fuzzy set is a set symbol with a tilde, $\sim$, on top of it, say $\tilde{B}$, where the functional mapping is thus written as

$$\mu_{\tilde{B}}(x) \in [0, 1] \tag{11.2}$$

where the symbol $\mu_{\tilde{B}}(x)$ is a value on the unit interval that measures the degree to which element x belongs to fuzzy set $\tilde{B}$; equivalently, $\mu_{\tilde{B}}(x)$ is the degree to which $x \in \tilde{B}$.

Overall, as mentioned earlier, the main distinction between crisp and fuzzy sets is their membership functions; a crisp set has a unique membership function, $U_A(x)$, whereas a fuzzy set can have an infinite number of membership functions, $\mu_{\tilde{B}}(x)$, reflecting it from the unit interval [0, 1]. Indeed, the membership function can be represented in many ways, which depends on the application context, such as triangular-, trapezoidal-, exponential-like functions or Gaussian. That means, for fuzzy sets, although one has to sacrifice uniqueness, flexibility can be obtained since the membership function can be adjusted to acquire a maximized utility for a particular application. The author decides not to go to great lengths to discuss this issue since it is out of the scope of this chapter. Interested readers should refer to [35] for more details.

11.5.2 Technology for Order Performance by Similarity to Ideal Solution

The technology for order performance by similarity to ideal solution (TOPSIS) method was initially presented in [36]. It is one of the numerical methods of multiple-criteria decision analysis (MCDA) or multiple-criteria decision-making (MCDM). As we know, the primary goal in MCDA or MCDM is to designate a preferred alternative. To achieve this goal, the TOPSIS helps select the best alternative with a finite number of criteria that simultaneously should have the shortest distance from the positive ideal solution (PIS) and the farthest distance from the negative ideal solution (NIS) [37]. For PIS, we mean that the solution should maximize the benefit criteria and minimize the cost criteria, whereas for NIS, it refers the solution that maximizes the cost criteria and minimizes the benefit criteria. In other words, TOPSIS is a compensatory method that allows a compromise between different criteria. To make use of this technique, attribute values must be numeric and monotonically increasing or decreasing. Briefly, the steps of TOPSIS can be summarized as follows [38]:

- *Step 1*: Gathering the decision-makers' ratings to acquire each decision-maker's aggregated rating and collecting the weight of criteria to compute the accumulated weight
- *Step 2*: Creating the normalized decision matrix
- *Step 3*: Building the weighted normalized decision matrix
- *Step 4*: Determining PIS and NIS, that is, the positive ideal and negative ideal solutions, respectively
- *Step 5*: Calculating the separation measures (or distance) with regard to each weighted alternative
- *Step 6*: Computing the relative closeness coefficient regarding each alternative
- *Step 7*: Defining the rankings of the alternatives based on the closeness coefficient and selecting an alternative with the maximum, that is, closest to the PIS and farthest from the NIS

11.5.3 Fuzzy TOPSIS

Under many conditions, crisp numbered data are inadequate to model complex real-life situations. To cope with that, a natural way is to express the human judgments or comparison ratios as fuzzy sets or fuzzy numbers to designate a preferred alternative, that is, fuzzy versions of TOPSIS. In 2000, the

TOPSIS method was further extended to the fuzzy environment [39], where the ratings and weights are treated as triangular fuzzy numbers due to their simplicity. Since then, it has been applied to a large number of applications, such as supply chain management and logistics [40,41], material selection [42], risk evaluation [43], and airlines efficiency [44,45]. Interested readers should refer to [37] for a thorough analysis of its applications. For the rest of this section, only the basic concepts of fuzzy set, height, and generalized fuzzy number will be briefly reviewed.

Definition 11.1 (Fuzzy Set) A set of pairs $\tilde{A} = (x, \mu_{\tilde{A}}(x) : x \in X)$ can be defined as a fuzzy set $\tilde{A}$ in a universe of discourse X. Here, $\mu_{\tilde{A}}(x) : X \to [0, 1]$ denotes a mapping called the membership function of the fuzzy set $\tilde{A}$, and $\mu_{\tilde{A}}(x)$ represents the membership value or degree of membership of $x \in X$ in the fuzzy set $\tilde{A}$.

Definition 11.2 (Height) The height $h(\tilde{A})$ of a fuzzy set $\tilde{A} = (x, \mu_{\tilde{A}}(x) : x \in X)$ is the largest membership degree possessed by any element in the set, that is, $h(\tilde{A}) = \sup_{x \in X} \mu_{\tilde{A}}(x)$.

Definition 11.3 (Generalized Fuzzy Number) Generalized fuzzy number $\tilde{A}$ is defined as $\tilde{A} = (a, b, c, d; w)$, where, $0 \leq w \leq 1$, and, a, b, c, and $d (a < b < c < d)$ are real numbers. Since the generalized fuzzy number $\tilde{A}$ is a fuzzy subset, its membership function $\mu_{\tilde{A}}(x)$ satisfies the following conditions [46]:

- $\mu_{\tilde{A}}$ is a continuous mapping from the universe of discourse X to the closed interval in [0,1].
- $\mu_{\tilde{A}}(x) = 0$ for $-\infty < x \leq a$.
- $\mu_{\tilde{A}}(x)$ is a strictly increasing function for $a \leq x \leq b$.
- $\mu_{\tilde{A}}(x) = w$ for $b \leq x \leq c$.
- $\mu_{\tilde{A}}(x)$ is strictly decreasing function for $c \leq x \leq d$.
- $\mu_{\tilde{A}}(x) = 0$ for $d \leq x \leq \infty$.

Please note:

- $\tilde{A}$ is a convex fuzzy set with two types of values: when, $w \neq 1$, it is a nonnormalized fuzzy number, whereas for $w = 1$, it is a normalized fuzzy number. In addition, w can also be omitted so that normalized fuzzy number with height equaling one expressed as $\tilde{A} = (a, b, c, d)$.
- If $a = b = c = d$ and $w = 1$, then $\tilde{A}$ can be regarded as a real number a, and $\tilde{A} = (x, \mu_{\tilde{A}}(x))$ with a membership function of

$$\mu_{\tilde{A}}(x) = \begin{cases} 1 & \text{if } x = a \\ 0 & \text{if } x \neq a \end{cases}$$

- If $a = b$, $c = d$, and $w = 1$ then $\tilde{A}$ is treated as crisp interval $[a, d]$, and $\tilde{A} = (x, \mu_{\tilde{A}}(x)$ with a membership function of

$$\mu_{\tilde{A}}(x) = \begin{cases} 1 & \text{if } a \leq x \leq d \\ 0 & \text{otherwise} \end{cases}$$

- If only $b = c$, $\tilde{A}$ then becomes a generalized TFN written as $\tilde{A} = (a, b, d; w)$ or $\tilde{A} = (a, c, d; w)$.
- If $b = c$ and $w = 1$, then $\tilde{A}$ turns out to be a TFN denoted by $\tilde{A} = (a, b, d)$ or $\tilde{A} = (a, c, d)$. For the situations mentioned earlier, the membership function of $\tilde{A} = (x, \mu_{\tilde{A}}(x))$ is expressed as

$$\mu_{\tilde{A}}(x) = \begin{cases} w\frac{x-a}{b-a} & \text{if } a \leq x \leq b \\ w\frac{d-x}{d-b} & \text{if } b \leq x \leq d \\ 0 & \text{otherwise} \end{cases}$$

- If $b \neq c$ then $\tilde{A}$ is called a generalized trapezoidal fuzzy number as $\tilde{A} = (a, b, c, d; w)$.
- If $b \neq c$ and, $w = 1$ the $\tilde{A}$ is called a trapezoidal fuzzy number as $\tilde{A} = (a, b, c, d)$. For the last two cases, the membership function of $\tilde{A} = (x, \mu_{\tilde{A}}(x))$ can be computed via

$$\mu_{\tilde{A}}(x) = \begin{cases} w\frac{x-a}{b-a} & \text{if } a \leq x \leq b \\ w & \text{if } b \leq x < c \\ w\frac{d-x}{d-c} & \text{if } c < x \leq d \\ 0 & \text{otherwise} \end{cases}$$

In general, trapezoidal fuzzy number $\tilde{A} = (a, b, c, d)$ and generalized trapezoidal fuzzy number $\tilde{A} = (a, b, c, d; w)$ indicate different decision-maker's opinions for distinct values of $w(0 < w \leq 1)$. In other words, the value of w represents the degree of confidence of the opinion of the decision-maker. To simplify the problem, in this chapter we use trapezoidal fuzzy number, that is, $\tilde{A} = (a, b, c, d)$, where $w = 1$.

Definition 11.4 (Basic Algebraic Operations on Fuzzy Sets) Suppose that we have two generalized trapezoidal fuzzy numbers, which are $\tilde{A} = (a_1, a_2, a_3, a_4; w_a)$ and $\tilde{B} = (b_1, b_2, b_3, b_4; w_b)$; the addition $\oplus$ for the generalized trapezoidal fuzzy numbers can be computed via the following equation:

$$\tilde{A} \oplus \tilde{B} = (a_1, a_2, a_3, a_4; w_a) \oplus (b_1, b_2, b_3, b_4; w_b)$$
$$= (a_1 + b_1, a_2 + b_2, a_3 + b_3, a_4 + b_4; \min(w_a, w_b)) \tag{11.3}$$

And the corresponding subtraction $\ominus$ can be calculated through the following equation:

$$\tilde{A} \ominus \tilde{B} = (a_1, a_2, a_3, a_4; w_a) \ominus (b_1, b_2, b_3, b_4; w_b)$$
$$= (a_1 - b_4, a_2 - b_3, a_3 - b_2, a_4 - b_1; \min(w_a, w_b)) \tag{11.4}$$

11.6 Focal Problem Formulation

In this study, we treat the focal problem (see Section 11.4) as a GMCDM (i.e., group MCDM) problem with m alternatives and n attributes, which can be mathematically expressed by the following sets:

- ◼ A set of K decision-makers called $\tilde{D} = \{\tilde{D}_1, \tilde{D}_2, \ldots, \tilde{D}_k\}$;
- ◼ A set of m possible card emulation suppliers called $A = \{A_1, A_2, \ldots, A_m\}$
- ◼ A set of n criteria called $C = \{C_1, C_2, \ldots, C_n\}$, with which eSE supplier performance is measured
- ◼ A set of performance ratings of A_i (for $i = 1, 2, \ldots, m$) with regard to criteria C_j (for $j = 1, 2, \ldots, n$) called $\tilde{X} = \{\tilde{x}_{ij}\}$ (for; $i = 1, 2, \ldots, m$; and $j = 1, 2, \ldots, n$), that is, the numeric data of our problem

In matrix format, the targeted problem can be expressed as

$$
\tilde{D} = \begin{array}{c} \\ A_1 \\ A_2 \\ A_3 \\ \vdots \\ A_m \end{array}
\begin{array}{cccccc}
C_1 & C_2 & C_3 & \cdots & C_n \\
\left[\begin{array}{ccccc}
\tilde{x}_{11} & \tilde{x}_{12} & \tilde{x}_{13} & \cdots & \tilde{x}_{1n} \\
\tilde{x}_{21} & \tilde{x}_{22} & \tilde{x}_{23} & \cdots & \tilde{x}_{2n} \\
\tilde{x}_{31} & \tilde{x}_{32} & \tilde{x}_{33} & \cdots & \tilde{x}_{3n} \\
\vdots & \vdots & \vdots & \ddots & \vdots \\
\tilde{x}_{m1} & \tilde{x}_{m2} & \tilde{x}_{m3} & \cdots & \tilde{x}_{mn}
\end{array}\right]
\end{array}
\tag{11.5}
$$

In a similar vein, the aggregated fuzzy weight ($\tilde{W}$) of each criterion can be calculated via the following equation:

$$
\tilde{w} = (\tilde{w}_1, \tilde{w}_2, \ldots, \tilde{w}_n)
\tag{11.6}
$$

For the rest of this section, the outlines of the extent analysis method on fuzzy TOPSIS are first given, and then the method is applied to the eSE supplier selection problem.

- ◼ *Step 1*: Gathering the decision-makers' ratings to acquire each decision-maker's aggregated rating and collecting the weight of criteria to compute the accumulated weight.
  - – To choose the linguistic rating values, we assume the performance ratings of each decision-maker $\tilde{D}_k$ for each alternative A_i with respect to criteria C_j are denoted by $\tilde{R}_k = \tilde{x}_{ij}^k$ (for $i = 1, 2, \ldots, m; j = 1, 2, \ldots, n;$ and $k = 1, 2, \ldots, K$) with a membership function of $\mu_{\tilde{x}_{ij}^k}(x)$. In this chapter, the importance weights of various criteria are treated as linguistic variables (e.g., low, medium, high, very low, medium low), and the same principle applies to the ratings of qualitative criteria as well (e.g., poor, fair, good, very poor, medium poor). We further use linear trapezoidal membership functions to address the vagueness of these linguistic assessments due to the fact that the subjective judgment of a decision-maker can only be approximately represented by linguistic assessments. In particular, these linguistic variables are expressed in positive trapezoidal fuzzy numbers, as shown in Figures 11.3 and 11.4.

    The importance weight of each criterion can be assigned in two ways: either directly or indirectly (using pairwise comparison). For instance, the linguistic variables medium

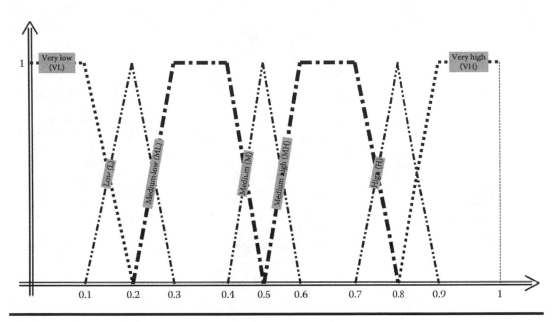

Figure 11.3   Linguistic variables for importance weight of each criterion.

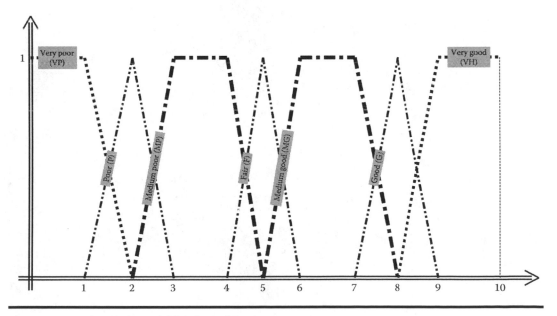

Figure 11.4   Linguistic variables for ratings.

good and good are denoted by (5, 6, 7, 8) and (7, 8, 8, 9), respectively. Membership functions can thus be written as

$$\mu_{\text{Medium Good}}(x) = \begin{cases} 0 & x < 5 \\ \frac{x-5}{6-5} & 5 \le x \le 6 \\ 1 & 6 \le x \le; \\ \frac{8-x}{8-7} & 7 \le x \le 8 \\ 0 & x > 8 \end{cases} \quad \mu_{\text{Good}}(x) = \begin{cases} 0 & x < 7 \\ \frac{x-7}{8-7} & 7 \le x \le 8 \\ 1 & x = 8 \\ \frac{9-x}{9-8} & 8 \le x \le 9 \\ 0 & x > 9 \end{cases} \quad (11.7)$$

— To calculate aggregated fuzzy ratings of all decision-makers, which can be described as trapezoidal fuzzy numbers, $\tilde{R}_k = (a_k, b_k, c_k, d_k)$, for $k = 1, 2, \ldots, K$. Then the aggregated fuzzy rating is given as

$$\tilde{R} = (a, b, c, d) \quad (11.8)$$

subject to the constraints expressed in

$$a = \min_k\{a_k\}, \quad b = \frac{1}{K} \sum_{k=1}^{K} b_k$$

$$c = \frac{1}{K} \sum_{k=1}^{K} c_k, \quad d = \max_k\{d_k\} \quad (11.9)$$

In a similar way, the fuzzy ratings and the importance weights of the kth decision-maker (represented as trapezoidal fuzzy numbers) are described as follows:

$$\tilde{x}_{ij}^k = (d_{ij}^k, b_{ij}^k, c_{ij}^k, d_{ij}^k) \quad (11.10)$$

$$\tilde{w}_{ij}^k = (w_{j1}^k, w_{j2}^k, w_{j3}^k, w_{j4}^k) \quad (11.11)$$

And thus, the aggregated fuzzy ratings and fuzzy weight of alternatives regarding each criterion can be computed as

$$\tilde{x}_{ij} = (a_{ij}, b_{ij}, c_{ij}, d_{ij}) \quad (11.12)$$

$$\tilde{w}_j = (w_{j1}, w_{j2}, w_{j3}, w_{j4}) \quad (11.13)$$

subject to the constraints given as

$$a_{ij} = \min_k\{a_{ij}^k\}, \quad b_{ij} = \frac{1}{K} \sum_{k=1}^{K} b_{ij}^k$$

$$c_{ij} = \frac{1}{K} \sum_{k=1}^{K} c_{ij}^k, \quad d_{ij} = \max_k\{d_{ij}^k\} \quad (11.14)$$

$$w_{j1} = \min_k\{w_{jk1}\}, \quad w_{j2} = \frac{1}{K} \sum_{k=1}^{K} w_{jk2}$$

$$w_{j3} = \frac{1}{K} \sum_{k=1}^{K} w_{jk3}, \quad w_{j4} = \max_k\{w_{jk4}\} \quad (11.15)$$

■ *Step 2*: Creating the normalized decision matrix. In general, this step transforms various attribute dimensions into linear scale, which allows comparisons across criteria. Additionally, as the set of criteria can be divided into two groups, that is, the benefit criteria (the larger the rating, the greater the preference) and the cost criteria (the smaller the rating, the greater the preference), the fuzzy decision matrices need to be normalized. The normalized fuzzy decision matrix is written as

$$\tilde{R} = [\tilde{r}_{ij}]_{m \times n}, \quad \text{for} \quad i = 1, 2, \ldots, m; j = 1, 2, \ldots, n \tag{11.16}$$

where $\tilde{r}_{ij}$ is computed as

$$\tilde{r}_{ij} = \begin{cases} \left(\dfrac{a_{ij}}{u_j^+}, \dfrac{b_{ij}}{u_j^+}, \dfrac{c_{ij}}{u_j^+}, \dfrac{d_{ij}}{u_j^+} \right), & \text{if } u_j^+ = \max_i(u_{ij}), j \in B \text{ (benefit criteria)} \\ \left(\dfrac{l_j^-}{a_{ij}}, \dfrac{l_j^-}{b_{ij}}, \dfrac{l_j^-}{c_{ij}}, \dfrac{l_j^-}{d_{ij}} \right), & \text{if } l_j^- = \min_i(l_{ij}), j \in C \text{ (cost criteria)} \end{cases} \tag{11.17}$$

■ *Step 3*: Building the weighted normalized decision matrix. Considering the importance of each criterion, the weighted normalized decision matrix ($\tilde{V}$) is given by

$$\hat{V} = [\hat{v}_{ij}]_{m \times n} \tag{11.18}$$

where $\tilde{v}_{ij}$ can be obtained via

$$\hat{v}_{ij} = \tilde{r}_{ij}(\cdot)\tilde{w}_j \tag{11.19}$$

■ *Step 4*: Determining PIS and NIS, respectively, we get the following equations:

$$\text{FPIS} = \left(\tilde{v}_1^{\oplus}, \tilde{v}_2^{\oplus}, \ldots, \tilde{v}_m^{\oplus} \right) \tag{11.20}$$

$$\text{FNIS} = \left(\tilde{v}_1^{\oplus}, \tilde{v}_2^{\oplus}, \ldots, \tilde{v}_m^{\oplus} \right) \tag{11.21}$$

Criteria for choosing these can follow the following equation:

$$\hat{v}_j^{\oplus} = \max_i(v_{ij4}) \quad \text{and} \quad \hat{v}_j^{\oplus} = \max_i(v_{ij1}) \tag{11.22}$$

Here, for example, one can employ fuzzy positive ideal to find the maximum values from fuzzy decision matrix with regard to eSE suppliers. Meanwhile, the fourth value from a fuzzy number, for example, d from $\tilde{A} = (a, b, c, d)$, can be utilized to substitute all other values, that is, a, b, and c. In a similar vein, if fuzzy negative ideal is selected, the values of b, c, and d will be replaced with a. By following this kind of logic, we can propose a clear and simple way of choosing fuzzy positive/negative ideals would be (1,1,1,1) or (0,0,0,0), respectively. However, another proposal also makes a lot of sense in practice would be simply identifying the max and min fuzzy numbers from fuzzy decision matrix regarding eSE suppliers as expressed in the following equation:

$$\hat{v}_j^{\oplus} = \max_i(v_{ij}) \quad \text{and} \quad \hat{v}_j^{\oplus} = \min_i(v_{ij}) \tag{11.23}$$

■ *Step 5*: Calculating the separation measures (or distance) with regard to each weighted alternative. The distance matrix can be computed from fuzzy positive ideal solution (FPIS) and fuzzy negative ideal solution (FNIS) to each alternative supplier:

$$d_i^{FPIS} = \sum_{j=1}^{n} d_v(\tilde{v}_{ij}, \tilde{v}_j^+), \quad i = 1, 2, \ldots, m$$

$$d_i^{FNIS} = \sum_{j=1}^{n} d_v(\tilde{v}_{ij}, \tilde{v}_j^-), \quad i = 1, 2, \ldots, m \qquad (11.24)$$

where $d_v(\,,\,)$ is the distance measurement between two fuzzy numbers according to the vertex method. In this chapter, we used the vertex method (adapted from [39]) as expressed in the following equation:

$$d_v(\tilde{m}, \tilde{n}) = \sqrt{\frac{1}{4}[(m_1 - n_1)^2 + (m_2 - n_2)^2 + (m_3 - n_3)^2 + (m_4 - n_4)^2]} \qquad (11.25)$$

■ *Step 6*: Computing the relative closeness coefficient regarding each alternative. The closeness coefficient (CC_i) is then calculated:

$$CC_i = \frac{d_i^{FNIS}}{d_i^{FPIS} + d_i^{FNIS}}, \quad i = 1, 2, \ldots, m, \qquad (11.26)$$

■ *Step 7*: Defining the rankings of the alternatives based on the closeness coefficient and selecting an alternative, that is, closest to the PIS and farthest from the NIS.

Assessment status (outlined in Table 11.1) has also been selected to evaluate the status of eSE suppliers. For this case study, five linguistic variables regarding the subintervals are defined to classify the assessment status of supplier into five categories.

According to Table 11.1, the value of CC_i has the following meanings:
- If $CC_i \in [0, 0.2)$, then *Supplier$_i$* belongs to Class I, and the assessment status of *Supplier$_i$* is *no recommendation*.
- If $CC_i \in [0.2, 0.4)$, then *Supplier$_i$* belongs to Class II, and the assessment status of *Supplier$_i$* is *high risk and recommended*.

Table 11.1 Approval Status

Class	Value of Closeness Coefficient (CC_i)	Criteria
I	$CC_i \in [0, 0.2)$	No recommendation
II	$CC_i \in [0.2, 0.4)$	High risk and recommended
III	$CC_i \in [0.4, 0.6)$	Low risk and recommended
IV	$CC_i \in [0.6, 0.8)$	Approved
V	$CC_i \in [0.8, 0.10]$	Approved and preferred

  - If $CC_i \in [0.4, 0.6)$, then *Supplier$_i$* belongs to Class III, and the assessment status of *Supplier$_i$* is *low risk and recommended.*
  - If $CC_i \in [0.6, 0.8)$, then *Supplier$_i$* belongs to Class IV, and the assessment status of *Supplier$_i$* is *approved.*
  - If $CC_i \in [0.8, 1.0]$, then *Supplier$_i$* belongs to Class I, and the assessment status of *Supplier$_i$* is *approved and preferred.*

11.7 Numerical Study

A smartphone manufacturer wants to select a suitable hardware supplier to purchase the key components (i.e., eSE) of new headsets. After preliminary screening, two candidates (*Supplier$_1$* and *Supplier$_2$*) remain for further evaluation. A committee of three decision-makers, *Decision-Maker$_1$*, *Decision-Maker$_2$*, and *Decision-Maker$_3$*, has been formed to select the most suitable eSE supplier. Two benefit criteria are initially considered, namely, profitability of supplier (*Criterion$_1$*) and technological capability (*Criterion$_2$*).

The proposed method is applied to attack the problem, and the computational process is summarized as follows:

▪ *Step 1*: Three decision-makers assess the importance of these criteria according to the linguistic weighting variables shown in Figure 11.3. The importance weights of each criterion are tabulated in Table 11.2.
▪ *Step 2*: Three decision-makers evaluate the ratings of suppliers with respect to each criterion based on the linguistic rating variables depicted in Figure 11.4. The resulting ratings are outlined in Table 11.3.
▪ *Step 3*: Transform the linguistic evaluations (as shown in Tables 11.2 and 11.3) into trapezoidal fuzzy numbers and calculate the average fuzzy weight of each criterion (see Table 11.4).
▪ *Step 4*: Normalize the fuzzy decision matrix (see Table 11.5).
▪ *Step 5*: Build the weighted normalized fuzzy decision matrix as shown in Table 11.6.
▪ *Step 6*: Compute FPIS and FNIS. Now according to two different criteria, we get two different FPIS and FNIS: FPIS = [(0.9, 0.9, 0.9, 0.9), (1, 1, 1, 1)] and FNIS = [(0.35, 0.35, 0.35, 0.35), (0.35, 0.35, 0.35, 0.35)].
▪ *Step 7*: Calculate the distance of each supplier from FPIS and FNIS with respect to each criterion, respectively. See Tables 11.7 and 11.8 for details.

Table 11.2 Importance Weight of Criteria from Three Decision-Makers

		Criteria	
		Criterion$_1$ and Fuzzy Weights	*Criterion$_2$ and* Fuzzy Weights
Decision-Makers	*Decision-Maker$_1$*	High (H): (0.7, 0.8, 0.8, 0.9)	Very High (VH): (0.8, 0.9, 1.0, 1.0)
	Decision-Maker$_2$	High (H): (0.7, 0.8, 0.8, 0.9)	Very High (VH): (0.8, 0.9, 1.0, 1.0)
	Decision-Maker$_3$	High (H): (0.7, 0.8, 0.8, 0.9)	High (H): (0.7, 0.8, 0.8, 0.9)

Table 11.3 Ratings of the Two Suppliers by Three Decision-Makers under Five Criteria

		Criteria	
Decision-Maker₁		*Criterion₁ and* Fuzzy Ratings	*Criterion₂ and* Fuzzy Ratings
eSE	*Suppliers₁*	Medium good (MG): (5, 6, 7, 8)	Good (G): (7, 8, 8, 9)
Suppliers	*Supplier₂*	Good (G): (7, 8, 8, 9)	Very good (VG): (8, 9, 10, 10)

		Criteria	
Decision-Maker₂		*Criterion₁*	*Criterion₂*
eSE	*Supplier₁*	Medium good (MG): (5, 6, 7, 8)	Good (G): (7, 8, 8, 9)
suppliers	*Supplier₂*	Good (G): (7, 8, 8, 9)	Very good (VG): (8, 9, 10, 10)

		Criteria	
Decision-Maker₃		*Criterion₁*	*Criterion₂*
eSE	*Supplier₁*	Medium good (MG): (5, 6, 7, 8)	Good (G): (7, 8, 8, 9)
suppliers	*Supplier₂*	Good (G): (7, 8, 8, 9)	Very good (VG): (8, 9, 10, 10)

Table 11.4 Fuzzy Decision Matrix and Averaged Fuzzy Weights of Two Candidates

		Criteria	
		Criterion₁	*Criterion₂*
eSE suppliers	*Supplier₁*	(5, 6, 7, 8)	(7, 8, 8, 9)
	Supplier₂	(7, 8, 8, 9)	(8, 9, 10, 10)
Weight		(0.7, 0.8, 0.8, 0.9)	(0.77, 0.87, 0.93, 0.97)

Table 11.5 Normalized Fuzzy Decision Matrix

		Criteria	
		Criterion₁	*Criterion₂*
eSE suppliers	*Supplier₁*	(0.5, 0.6, 0.7, 0.8)	(0.7, 0.8, 0.8, 0.9)
	Supplier₂	(0.7, 0.8, 0.8, 0.9)	(0.8, 0.9, 1.0, 1.0)

- *Step 8*: Compute d_i^{FPIS} and d_i^{FNIS} of candidate eSE suppliers and the associated closeness coefficient as shown in Table 11.9.
- *Step 9*: Based on the obtained results, we know that Supplier₂ belongs to Class IV with the assessment status of *approved*, while Supplier₁ belongs to Class III with the assessment status of *recommended with low risk*. Nevertheless, according the value of CC_i, Supplier₂ is preferred to Supplier₁.

Table 11.6 Weighted Normalized Fuzzy Decision Matrix

		Criteria	
		Criterion$_1$	Criterion$_2$
eSE suppliers	Supplier$_1$	(0.35, 0.48, 0.56, 0.72)	(0.54, 0.7, 0.74, 0.87)
	Supplier$_2$	(0.49, 0.64, 0.64, 0.81)	(0.62, 0.78, 0.93, 0.97)

Table 11.7 Distances between Suppliers and FPIS with Respect to Each Criterion

	Criteria	
	Criterion$_1$	Criterion$_2$
d(Supplier$_1$, FPIS)	0.4	0.33
d(Supplier$_1$, FPIS)	0.28	0.25

Table 11.8 Distances between Suppliers and FNIS with Respect to Each Criterion

	Criteria	
	Criterion$_1$	Criterion$_2$
d(Supplier$_1$, FNIS)	0.22	0.39
d(Supplier$_1$, FNIS)	0.32	0.5

Table 11.9 Calculations of d_i^{FPIS}, d_i^{FNIS}, and CC_i

		d_i^{FPIS}	d_i^{FNIS}		Closeness Coefficient (CC_i)
eSE suppliers	Supplier$_1$	0.73	0.61	1.34	0.46
	Supplier$_2$	0.53	0.82	1.35	0.61

This simple experiment clearly shows that the proposed methodology is very effective in dealing with the focal eSE supplier evaluation and selection problem.

11.8 Future Work

11.8.1 Broader M-Payments Ecosystem Viewpoint

As the next revolution in payments sector, there are huge potential benefits associated with a successful adoption of m-payments for companies who can get the technology innovation right. There are many significant uncertainties involved in the investment and adoption decision-making for m-payments techniques such as hidden technological risks, rapid changing customer demand

and expectations, fierce competition in the marketplace, and ill-defined/over-supplied technology standards [48]. Before a common consensus can be reached among multiple stakeholders, various technological solutions will coexist and, thus, create great uncertainty for many adopters. In addition, the m-payments ecosystem exhibits a high dimension of complexity in its structure, spanning multiple sectors such as banking, retailing, telecoms, consortia, device manufacturers, and various service providers.

The current work was conducted solely from a device manufacturers' perspective, which is based on the fact that the mobile handset is only one element of an end-to-end m-payments system [49]. Nevertheless, the success of m-payments also relies on a fruitful collaboration between different stakeholders for provision. Though such hard-to-realize collaboration does impose additional difficulty on m-payments investment and adoption decision-making, it is critically important for future research to be carried out from a more holistic viewpoint so that the sustainability of certain technologies can be better estimated, the future state of the m-payments market can be comprehensively envisioned, and eventually, more appropriate investment decisions can be made.

11.8.2 Deeper M-Payments Technology Innovation

According to Liu et al. [48], the technology innovations in the m-payments ecosystem can be divided into three groups, namely, component, service, and business infrastructure. The scope of the current study falls within the component level. In addition to the focal eSE technology presented in the current work, there are also some other alternative approaches available for keeping critical information. The rest of this section will highlight several promising candidates for other interested researchers' information.

11.8.2.1 Hybrid Secure Element

Another trend regarding the innovation of secure element technology also includes a combined version of SIM and secure element (SIM-SE). According to a recent report [50], there are over a hundred million SIM-SE–based NFC handsets that were purchased globally during the period of April 2012–January 2013. Moreover, the report predicts that this number will keeping growing to reach about 1.5 billion by 2016. More formally, in the era of 3G, 4G, and the forthcoming 5G, the SIM-SE solution should be referred to as UICC solution since SIM will soon be replaced by UICC due to its sophistication, enhanced processing power, increased memory space, and extended security management functionality [51]. The advantages of employing the UICC solution lie in that the advancements of innovative technology allow the partitioning of the UICC into several security domains, which is emerging as a means to provide a secure hosting environment for various third-party applications. No doubt that the UICC solution enjoys many unique advantages such as a high level of security, universal deployment, portability, remote management, mature standards, and long operational lifecycle [51]. Take the last merit, the design of UICC changes very slowly, in particular compared with the high turnover of mobile phones. The forward compatibility of UICC solution is thus high, which in turn can reduce various operational and replacement costs.

Despite this fact, from the ecosystem perspective, the sole ownership of UICC (by MNO) causes many concerns. One of the big questions is for those early adopters who are planning to fully embrace UICC: do they really assure that all applications can run safely in such new environment? In other words, the current MNO-dominated UICC model favors one party heavily but potentially harms the much needed cooperation and competition. To solve this, future research is needed to come up with more secured models for generating both security confidence and revenue incentive to all stakeholders.

11.8.2.2 Variant SIM

Another logical step in the evolution of the role of SIM and its form factor (long coveted by other parties in the m-payments ecosystem) is to move toward a removable reprogrammable SIM. On October 16, 2014, Apple launched its Apple SIM, which was barely mentioned in the literature [31]. Actually, Apple filed a patent way back in 2010 for an eSE (the main scenario discussed in this study) and has since worked closely with SIM giant Gemalto in this regard. It is well known that Apple has been trying to eliminate the SIM form factor completely since the first iPhone was launched, though the journey is not that quite smooth after being threatened by a number of European MNOs. As people may gradually find out, the Apple SIM concept is presumably the compromised way to realize this ultimate goal, and other alternative solutions, for example, soft SIM (a virtualized SIM residing in a mobile phone's memory), is also in the pipeline to bridge the gap.

In a word, any handset manufacturers that want to expand and enhance their visibility in the m-payments ecosystem should consider the intermediate removable reprogrammable SIM. By doing so, the potential needed to tie the customer on the MNOs would be largely eliminated. However, since most handset manufacturers entering the mobile phone arena do not enjoy Apple's high-end value privilege, a rigorous analysis is still requiring them to find out a balance between duplicating this model and building business deals with the MNOs. This conundrum is witnessed by the slow uptake of the Apple SIM among various MNOs.

11.8.2.3 Emerging Host Card Emulation Technique

In addition to conventional secure element (either eSE or SIM-SE)-facilitated smart card emulation, host card emulation (HCE) is gaining momentum in recent years. As the name suggests, with HCE, the payment applications are held in the mobile phone operating system, which serve as a *host*. HCE is currently backed by several operating systems such as Android 4.4 KitKat and Blackberry OS 10. Take Android 4.4 KitKat, it allows a payment app being stored in the mobile phone operating system (in other words, held in software) to *talk* directly with the NFC controller/antenna. The emergence of HCE opens up the new possibility of performing NFC-based m-payment without using a secure element. It thus may potentially reduce the complexity associated with secure element-based framework. Since HCE paves the way for app providers to load payment apps directly into the target customers' handset via an app store, one may happily expect its prosperity in the near future. This is, however, only part of the picture. Although HCE does simplify the dimension of NFC m-payments ecosystem, it requires a new security treatment, which is either premature at present or does not exist at all. More works are therefore needed to address this challenge.

11.8.2.4 Advanced Authentication Technologies

Additionally, significant progress has also been made on innovative authentication technologies. By 2014, biometric sensors had been implemented in over 300 million mobile devices, which further improved the security level of using such device in executing mobile payment transaction.

Among various technologies in this area, fingerprint recognition is no doubt the most popular one and thus widely integrated by handset manufacturers. Another means that is also attracting a lot of attention in this field is facial recognition technique, which was recently demonstrated by Alibaba, an e-commerce giant from China [52].

To summarize, with the rapid technology advancement, the ecosystem of m-payments is dynamically evolving with different new players moving in and various old participants wiping out. The consequent MCDM problem faced by not only mobile device producers but many other parties will become ever more complex. Though some information predictions such as for mature markets, taking a SIM-SE should be quicker and a lower risk than HCE [49], can be used as a rule of thumb, a more in-depth analysis will be required with the advance of different technological innovations. Therefore, future studies that could extend the current work toward the previously highlighted directions would benefit both academic and industry practitioners in many aspects.

11.9 Conclusion

Traditional financial services are not short of security breaches. In a recent credit card data heist scandal [53], 18 types of customer data were stolen from 3 big South Korean credit card firms (not without their notice), and luckily, no PIN or card verification codes were included. The scale of this event revealed the potential vulnerability of the financial sector and a worry about credit or debit card data safety is thus sweeping many countries. However, the various perks that come with many credit cards (e.g., free air miles and cosmetics) still make keeping a wallet-full worthwhile. With an innovative mobile wallet concept being gradually introduced into the market, it can reasonably expect that people's passion for traditional plastic will soon switch to m-payments.

In the rapidly proliferating mobile payment market, although handset manufacturers have been late entrants, they could still become relevant and play game-changing roles thanks to the development of novel technologies. Since the inherent dilemma of mass market deployment of mobile financial services lies in that who issues and controls the secure element (for storing a set of key information), the device manufacturers can benefit far more than ever before by baking a secure element into the handset (i.e., eSE). The main challenge of realizing this goal is therefore not only to gain a critical mass of customers and ensure sufficient user loyalty, but also to keep eSE truly secure. Only through this, the device manufacturers can become important players in the m-payment ecosystem, and an effective partnership and collaboration with other stakeholders can thus be formed.

Bearing this in mind, in this chapter a new methodology is introduced to address the eSE supplier selection problem. Generally speaking, supplier selection problems are often accompanied by uncertainty and imprecise/missing data issues, and the theory of fuzzy-set is sufficient to address them. In the process of decision-making, the utilization of linguistic variables is highly helpful when classic numerical values cannot be used to express the performance values clearly. Putting it in another way, it is often found that linguistic variables can substitute for numerical values when it comes to evaluating possible eSE supplier candidates (based on some selected criteria and the allocated importance weights). Since decision-makers are often subject to some distractions (e.g., previous experience, personal feeling, and subjective guesswork) in the process evaluating eSE suppliers, this chapter employs a fuzzy version of TOPSIS. It appears from the numerical example that the fuzzy TOPSIS approach can resolve the focal problem when both quantitative and qualitative criteria are involved. We can thus draw a preliminary conclusion that the proposed methodology can effectively assist handset manufacturers in selecting the suitable eSE suppliers.

References

1. D. A. McIntyre. Holiday spending expected to rise to $688.87 per person, 2010, retrieved from http://247wallst.com/retail/2010/10/19/holiday-spending-expected-to-rise-to-688-87-per-person/, accessed on August 28, 2015.
2. R. Lefebvre, E. Simonova, and K. Girdharry. Mobile payments in Canada—The demand side of the equation. CGA-Canada, 2013. Retrieved from http://www.cga-canada.org/en-ca/researchreports/ca_rep_2013-03_informed-view.pdf, accessed August 4, 2016.
3. KPMG. Mobile payments—Is Canada ready? KPMG, 2011. http://www.kpmg.com/Ca/en/IssuesAndInsights/ArticlesPublications/Documents/5803_Mobile%20Payments_Brochure%20Canada%20EN_web.pdf, accessed August 4, 2016.
4. Insight Research Corporation. Insight research says financial applications on cell phones to attract 2.2 billion users. Insight Research, *eNewsChannels*, 2009. http://enewschannels.com/2009/04/22/enc6821_192943.php, accessed August 4, 2016.
5. Trend Watching. 10 trends for 2015. *Trend Watching*, 2014. Retrieved from http://trendwatching.com/trends/10-trends-for-2015/, accessed August 4, 2016.
6. KPMG. Mobile payments outlook. KPMG, 2011. http://www.kpmg.com/Ca/en/IssuesAndInsights/ArticlesPublications/Documents/2011-mobile-payments-outlookv2.pdf, accessed August 4, 2016.
7. KPMG. Monetizing mobile. KPMG International Cooperative (KPMG International), 2011. Retrieved from http://www.mobilepaymentsworld.com/kpmg-banks-race-to-monetize-mobile-payments, accessed August 4, 2016.
8. Board of Governors of the Federal Reserve System. Consumers and mobile financial services 2015. Board of Governors of the Federal Reserve System, 2015. Retrieved from http://www.federalreserve.gov/econresdata/consumers-and-mobile-financial-services-report-201503.pdf, accessed August 4, 2016.
9. M. Jovanovic and M. M. Organero. Analysis of the latest trends in mobile commerce using the NFC technology. *Journal of Selected Areas in Telecommunications*, May Edition:1–12, 2011.
10. Y. Yuan and X. Zhao. 5G: Vision, scenarios and enabling technologies. *ZTE Communications*, 13:3–10, 2015.
11. EMV Migration Forum. Near-term solutions to address the growing threat of card-not-present fraud. EMV Migration Forum, 2015.
12. European Payments Council. White paper on mobile payments, Brussels, Belgium, 2010.
13. NFC Forum. The keys to truly interoperable communications. NFC Forum, Wakefield, MA, 2007. Retrieved from http://www.portech.co.uk/go_files/whitepapers/214793-nfc-forum_marketing_white_paper.pdf, accessed August 4, 2016.
14. C. Guidobaldi. Mobile proximity payment: Ecosystem and overview of NFC technology. *Altran Italia Technology Review*, 7:25–40, 2011.
15. CashStar. Fast-track success in mPayments, loyalty and omni-channel engagement. CashStar, 2013. Retrieved from http://www.cashstar.com/uploads/fast-track-success-in-mpayments-loyalty-omni-channel-engagement-2013.pdf, accessed August 4, 2016.
16. CyberSource. Online fraud report. CyberSource, 2013.
17. Y. Audette. Overview of mobile payments for IT can. KPMG LLP, 2014.
18. GSMA. Mobile and online commerce: Opportunities provided by the SIM. GSMA, 2013.
19. S. Bansal, P. Bruno, F. Istace, and M. Niederkorn. Global payments trends: Challenges amid rebounding revenues. *McKinsey on Payments*, pp. 34–39, September 2013.
20. MasterCard. Mastercard best practices for mobile point of sale acceptance. MasterCard, 2013.
21. B. Andiva. Mobile financial services and regulation in Kenya. Competition Authority of Kenya, Nairobi, Kenya, 2013.

22. M. Jaubert, M. Ullrich, R. Dela, S. Marcu, and J.-B. Malbate. Going digital: The banking transformation road map. A.T. Kearney & Efma, 2014. Retrieved from https://www.atkearney.com/digital-business/ideas-insights/featured-article/-/asset_publisher/Su8nWSQlHtbB/content/going-digital-the-banking-transformation-road-map/10192, accessed August 4, 2016.

23. R. Byrne and J. Hanson. Innovation and disruption in U.S. merchant payments. *McKinsey on Payments*, pp. 33–41, May 2014.

24. M. Roland. *Security Issues in Mobile NFC Devices*. Springer International Publishing, New York, 2015.

25. KPMG International. Mobile payments seen as "mainstream" for consumers within four years. KPMG, 2011. Retrieved from https://www.icma.com/ArticleArchives/MobilePayments_SE2-11.pdf, accessed August 4, 2016.

26. C. Cox and R. Musfeldt. NFC-enabled payments and the role of the trusted service manager. First Data Corporation, 2013. Retrieved from https://www.firstdata.com/downloads/thought-leadership/4529_TSM_WP.pdf, accessed August 4, 2016.

27. SiMalliance. Secure element deployment and host card emulation. SIMalliance Ltd., 2014. Retrieved from http://simalliance.org/wp-content/uploads/2015/03/Secure-Element-Deployment-Host-Card-Emulation-v1.0.pdf, accessed August 4, 2016.

28. C. Abraham. Mobile payments: A study of the emerging payments ecosystem and its inhabitants while building a business case, 2011. Mobile Commerce & Payments Practice/Co-Founder—DROP Labs. Retrieved from http://docplayer.net/15637998-Mobile-payments-a-study-of-the-emerging-payments-ecosystem-and-its-inhabitants-while-building-a-business-case.html, accessed August 4, 2016.

29. S. Kadhiwal and M. A. U. S. Zulfiquar. Analysis of mobile payment security measures and different standards. *Computer Fraud & Security*, 6:12–16, June 2007.

30. C. Cox. Trusted service manager: The key to accelerating mobile commerce. First Data Corporation, 2009. Retrieved from https://www.firstdata.com/downloads/thought.../fd_mobiletsm_whitepaper.pdf, accessed August 4, 2016.

31. UL Transaction Security Division. The future of SIM, 2014. www.ul-ts.com, accessed August 4, 2016.

32. Anonymous. Smartphone security: The spy in your pocket. *The Economist*, 414(8927):21, 2015.

33. L. A. Zadeh. Fuzzy set. *Information and Control*, 8:338–353, 1965.

34. J. Bih. Paradigm shift—An introduction to fuzzy logic. *IEEE Potential*, 25:6–21, 2006.

35. B. Xing, W.-J. Gao, F. V. Nelwamondo, K. Battle, and T. Marwala. Cellular manufacturing system scheduling under fuzzy constraints: A group technology perspective. In *Annual IEEE International Conference on Fuzzy Systems* (*FUZZ-IEEE*), July 18–23, CCIB, Barcelona, Spain, pp. 887–894, 2010.

36. C. L. Hwang and K. Yoon. *Multiple Attribute Decision Making: Methods and Applications*. Springer-Verlag, New York, 1981.

37. M. Behzadian, S. K. Otaghsara, M. Yazdani, and J. Ignatius. A state-of the-art survey of TOPSIS applications. *Expert Systems with Applications*, 39:13051–13069, 2012.

38. F. R. L. Junior, L. Osiro, and L. C. R. Carpinetti. A comparison between fuzzy AHP and fuzzy TOPSIS methods to supplier selection. *Applied Soft Computing*, 21:194–209, 2014.

39. C. T. Chen. Extensions of the TOPSIS for group decision-making under fuzzy environment. *Fuzzy Sets and Systems*, 114:1–9, 2000.

40. G. Kannan, S. Pokharel, and P. S. Kumar. A hybrid approach using ISM and fuzzy TOPSIS for the selection of reverse logistics provider. *Resources, Conservation and Recycling*, 54:28–36, 2009.

41. D. Kannan, A. B. L. D. S. Jabbour, and C. J. C. Jabbour. Selecting green suppliers based on GSCM practices: Using fuzzy TOPSIS applied to a Brazilian electronics company. *European Journal of Operational Research*, 23:432–447, 2014.

42. M. F. Aly, H. A. Attia, and A. M. Mohammed. Integrated fuzzy (GMM)—TOPSIS model for best design concept and material selection process. *International Journal of Innovative Research in Science, Engineering and Technology*, 2:6464–6486, 2013.

43. X. Zhou and M. Lu. Risk evaluation of dynamic alliance based on fuzzy analytic network process and fuzzy TOPSIS. *Journal of Service Science and Management*, 5:230–240, 2012.
44. C. P. Barros and P. Wanke. An analysis of African airlines efficiency with two-stage TOPSIS and neural networks. *Journal of Air Transport Management*, 44–45:90–102, 2015.
45. P. Wanke, C. P. Barros, and Z. Chen. An analysis of Asian airlines efficiency with two-stage TOPSIS and MCMC generalized linear mixed models. *International Journal of Production Economics*, 169:110–126, 2015.
46. G. S. Mahapatra and T. K. Roy. Optimal redundancy allocation in series-parallel system using generalized fuzzy number. *Tamsui Oxford Journal of Information and Mathematical Sciences*, 27:1–20, 2011.
47. C. T. Chen, C. T. Lin, and S. F. Huang. A fuzzy approach for supplier evaluation and selection in supply chain management. *International Journal of Production Economics*, 102:289–301, 2006.
48. J. Liu, R. J. Kauffman, and D. Ma. Competition, cooperation, and regulation: Understanding the evolution of the mobile payments technology ecosystem. *Electronic Commerce Research and Applications*, 14:372–391, 2015.
49. S. Pannifer, D. Clark, and D. Birch. HCE and SIM secure element: It's not black and white. Consult Hyperion, Guildford, U.K., 2014.
50. GSMA. Mobile commerce in retail: A look at the opportunities provided by mobile commerce in the retail industry. GSMA: Mobile Commerce, 2013. Retrieved from http://www.gsma.com/digital commerce/wp-content/uploads/2013/08/GSMA-Mobile-Commerce-in-Retail-White-Paper-V2.pdf, accessed August 4, 2016.
51. GSMA. Pay-buy-mobile: Business opportunity analysis. GSMA, London, U.K., 2007.
52. J. Duvaud-Schelnast and M. Born. Mobile payment: Is this the turning point? Arthur D. Little, 2015. Retrieved from www.adl.com/MobilePayment, accessed on November 19, 2015. Retrieved from www.adlittle.com/downloads/tx_adlreports/ADL_M-payment.pdf, accessed August 4, 2016.
53. Anonymous. South Korean credit-card data: Card sharps. *The Economist*, 410(8871):62, January 25, 2014.

Chapter 12

Cyberattack Surface of Next-Generation Mobile Networks

Filipo Sharevski

Contents

Abstract

This chapter presents the cyberattacking surface of the next-generation mobile networks. Drawing on native security approaches in the mobile environment, this chapter renders the present state of mobile network resilience and identifies the areas prone to cyberattacks. Extending toward the cloudification and in-software definition concepts, this chapter provides an overview of the next-generation mobile network architecture. Next, this chapter presents the associated cyberattack surface relative to the threat models, the security challenges, and the attacking vectors. Specifically, this chapter elaborates the characteristics of all the security-related threats for the evolved packet core segment, the software-defined backbone, and the cloud-based radio access network using the spoofing, tampering, repudiation, information disclosure, denial-of-service, and elevation of privilege (STRIDE) categorization and dynamic flow representation. This chapter extends the cyberattacking surface elaboration to include discussion on the existing defensive mechanisms or the segment-specific ones that can be adopted for the purpose of an integral software-defined mobile network (SDMN) protection. In conclusion, this chapter emphasizes the key design components in developing a robust defense solution and provides recommendations for its implementation in protecting the next-generation mobile networks.

12.1 Introduction

At the beginning of the 1980s, the public landline telephone service was extended to accommodate the *mobility* feature to its native service. From today's perspective, this technological breakthrough was crucial in transforming the way we communicate and interact today. The mobile evolution has pushed the frontiers to gigabit data rates in the radio segment, guaranteed quality of service, and 99% global penetration. There will be 9.2 billion network entities, demanding 30.5 exabytes/month of data up to the year 2020 [1]. In response to the exponential growth, the network architecture accommodated the service-oriented paradigm by employing the network functional virtualization (NFV), user/control separation, and the in-software definition of the mobile logic.

The need to protect the mobile infrastructure has increased as a result of its continuously expanding cyberattack surface. At the beginning, the concern was to eliminate the blunt opportunities for user impersonation, eavesdropping, overbilling, and radio jamming. The Global System for Mobile (GSM) introduced user authentication, signaling anonymization, and traffic encryption in its security suite. However, the lack of mutual authentication and the weak encryption made the networks susceptible to man-in-the-middle attacks. A novel, comprehensive security architecture was developed in the following third-generation (3G) upgrade of the network architecture. Organized in several classes, the Universal Mobile Telecommunication System (UMTS) security architecture actively protects the radio interface, the internal signalization traffic, and the network nodes from adverse cyber events [2]. Notwithstanding the long-term evolution (LTE) reinforcement, the deployments still remain vulnerable to targeted eavesdropping, smart jamming, signaling-based attacks, and service logic exploits [3].

Affecting the mobile network resilience are the architectural changes with every generation, the rapid increase in attacking techniques and tools, and the service and device diversification. For an attack to succeed, an adversary need not rely on expensive equipment and technology-specific knowledge. There are numerous malware programs and tools that can launch signalization denial-of-service, mask the wireless signal in a given area, or impersonate a legitimate user. They are able to turn large mobile populations into a cellular botnet and amplify the negative consequences on network availability. This parallels the mobile evolution tendency to accumulate threats native to other domains, such as the common operating system and IP-related attacks.

The next generation of mobile architecture is equally prone to adverse cyber events, given the characteristic of numerous vulnerabilities for the *cloudification* [4]. Focused on the future threat landscape, this chapter consolidates the elements relevant for the associated cyberattack surface. First, the known cyber events are overviewed to introduce the main concepts of mobile subversion. The service-oriented technological enablers are detailed next. Subsequently, the cyberattack surface of the next-generation architecture is elaborated relative to the threat models, security challenges, and attacking vectors. The concluding segments elaborate on the associated defense mechanisms and emphasize the key security implications for the next-generation mobile networks.

12.2 Mobile Network Cyberattack Landscape

12.2.1 Service and Security Architectures

The traditional network organization distinguishes between *radio access* and *core* network segments. The latest third-generation partnership project (3GPP) reference model includes GSM, UMTS, and the evolved radio access networks next to the evolved packet core (EPC) as depicted in Figure 12.1. On the user side, the major components are the subscriber identification module (SIM) and the terminal equipment. The main fronthaul elements are the base stations (or NodeB and evolved NodeB) and the radio controllers (base station and radio network controller [RNC]). The core segment involves the mobile multimedia entity (MME), serving/packet data gateway (S/PGW), serving GPRS support node (SGSN), home subscriber system (HSS), policy and charging rule function (PCRF), and the Internet multimedia subsystem (IMS). The SIM card stores a globally unique international mobile subscriber identity (IMSI) number, authentication key (Ki), and additional card identification required for a secure interconnection. The network registration and authentication are coordinated with the HSS and MME, providing traffic encryption and interconnection toward other mobile entities and packet data networks (PDNs).

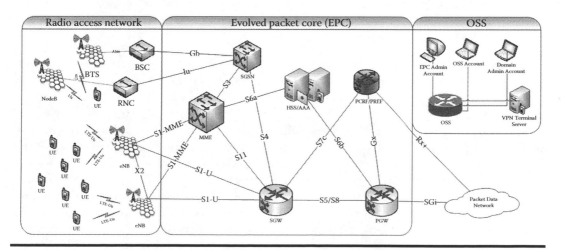

Figure 12.1 Third-generation partnership project evolved packet core and evolved universal terrestrial radio access network with a support for a backward radio access network access.

With the comprehensive security architecture in UMTS, the mobile service protection is extended beyond the GSM user authentication and data encryption. The novel security features involve mutual authentication, signaling protection, and Advanced Encryption Standard (AES) implementation with longer, 128 bit key sizes [5,6]. In response to strict security requirements, UMTS protection is enforced with security mechanisms organized in five classes. The objective of *network access* mechanisms is to guard against attacks on the radio interface, while the objective of *network domain* mechanisms is to protect the signalization and user traffic in the core segment. There are also *user domain* security class for SIM and terminal equipment protection and *application domain* class for secure traffic exchange on the application level. A *visibility and configurability of security* is a class that enables mobile users to have control over the protection measures enforced by the network. LTE protection provides further key extension, reinforces the encryption for the user and control traffic, and provides additional data re-encryption (IPSec) for an external access [7].

12.2.2 GSM Cyber Exploits

The main GSM security drawbacks are the absence of mutual authentication, obscure and weak encryption, only the first-hop protection, and the lack of strict identification scheme. Due to the inability of the mobile user to authenticate the base station he or she communicates with, a malicious cyber adversary is able to launch a man-in-the-middle attack. By impersonating the serving base station to the user, the adversary is able to listen, track, and tamper with the data traffic from a given geolocation [8]. This attack can be further amplified by performing a cryptanalysis of the weak encryption algorithm and *eavesdrop* on the entire communication. The same technique can be applied on the transmission trunks, having them often realized over microwave links or leased lines. Next, there are several possibilities for launching a denial-of-service attack against the radio access network, a user, or a core network function. An adversary can jam the radio frequencies, drop all the traffic for a certain user, or modify the signaling messages to saturate a given core control function. Finally, a SIM card can be cloned and used to impersonate a user for further attacks of overbilling and roaming abuse [2,9].

12.2.3 UMTS Cyber Exploits

The UMTS introduced a comprehensive security architecture that aims to minimize the present attacking surface and improve overall network robustness [10]. The UMTS Authentication and Key Agreement (AKA) was introduced to protect against network impersonation and ensure safe key derivation. Service logic was upgraded to protect from basic signaling denial-of-service attacks, user impersonation, and eavesdropping-related attacks. However, the architecture by itself has not proven to be sufficient for a practical protection of the real network deployments due to several reasons. First, in enabling backward compatibility, the rogue base station attack remained plausible, given the option for a radio access fallback to a GERAN-type access. Second, the convergence toward IP-based architecture opened a door for cross-domain attacks beyond the radio and core segments. Third, a flawed implementation of the security features has also enabled network domain security subversion. Fourth, the cheap equipment together with the open software implementations of critical network elements enabled easy tampering for a wide range of attackers. As a result, novel attacks types and techniques surfaced against operational UMTS networks.

In regard to cross-infrastructure attacks, adversaries are able to subvert the call forwarding and client billing services and intercept user traffic or cause denial-of-service attacks. Similarly, they are able to bypass the billing system and the location-based instant messaging service to track certain user location or flood the messaging server. Further, a session-mix-up attack can be launched against the AKA and its associate carrying protocol that enables user authentication subversion within the core segment. Essentially, a skilled attacker can leverage the session-mix-up attack to proceed with the rogue base station exploitations [11]. In this context, a novel set of UMTS denial-of-service attacks have emerged, exploiting the signaling plane and the IMS service. A malicious user can overflow the RNC, combining the periodic radio access bearer requests and small traffic bursts. Similarly, the HSS function can be flooded with simultaneous registration requests with different IMSIs. It is also possible to inject non-SIM-originated signalization, variate the session initiation protocol, and create cellular botnets as to diversify these attacks [12–16]. In aggregate, the UMTS cyberattack surface, in fact, extends over the one already associated with the second generation of mobile networks

12.2.4 LTE Cyber Exploits

In the fourth generation, sophisticated attack vectors emerged with the increase of mobile device capabilities, free tools for exploitation (i.e., Metasploit, Angler, or Blackhole), and interest in cyber probing. Within the first wireless hop, user equipment was found to be vulnerable to location tracking attacks based on the cell radio temporal identifier and packet sequencing [17]. In terms of evolved UTRAN-related DoS and distributed DoS attacks, a myriad of techniques have been reported as successful in tampering with the radio resource control mechanisms. It is possible to falsify the buffer status reports and in that way to progressively consume bandwidth until the resources of given cell are depleted [18,19]. Interestingly, the illegitimate base station attack persisted even for the EUTRAN-supported access. Though scaled to a femtocell level, the disruptive man-in-the-middle power of the attack can still be leveraged for tracking, impersonation, traffic interception, and service tampering [20,21].

Based on the signaling amplification approach, many of these attack vectors are tailored to propagate into the EPC. A cellular botnet is capable of saturating the HSS with an authentication request, while a smart attacker is capable of bypassing the charging policies and causing accounting volume inaccuracies [22,23]. Due to the extensive reliance on the mobile communication access, the propagation effect spans beyond the core segment, affecting the IP-related services. As such, the LTE cyberattack surface includes the IP-spoofing, address and session hijacking, service abuse and theft, eavesdropping, and DoS attacks targeting the IMS [24]. Again, the technological improvements in LTE do little to prevent networks' susceptibility to user-initiated attacks, internet-related exploits, and insider threats [25]. Table 12.1 depicts the general attacking surface according to its threat dimensions in each of the mobile communication generations up until today.

12.3 Next-Generation Mobile Networks: Concepts and Solutions

The key technology enablers of the next-generation networks are the software-defined networks and radio (SDN and SDR) and the NFV [26]. Featuring powerful attributes, they promise to eliminate the traditional problems of optimal network capacity planning, vendor lock-ins, and the high operational expenditures. Being in a development stage, the design priorities are set on the network harmonization, interoperability, and service logic adaptations in accordance with the

Table 12.1 Cyberattack Surface of the Operational Mobile Networks

Threat Category	GSM	UMTS	LTE/LTE-Advanced
Spoofing	Illegitimate base station, SIM card cloning	Illegitimate base station with a forced service downgrade, session-mix-up attack	Rogue femtocell, IP spoofing
Tampering	Malicious traffic injection, packet sequence reordering/dropping	Call forwarding service abuse, malicious traffic injection	Overbilling, insider threat, IMS service abuse, and theft
Repudiation	SIM card cloning	Client charging bypass, SIM-less network presence	Charging policies bypass
Information disclosure	IMSI, T-IMSI interception, man-in-the-middle eavesdropping, Kasumi cryptanalysis, nonprotected core signaling	IMSI-targeted interception, man-in-the-middle eavesdropping	IMS session eavesdropping, location tracking, topology fingerprinting, session hijacking, network location exposure
Denial-of-service	Jamming, packet dropping, SMS flooding, signalization storms	Smart jamming, RNC overflow, HSS, IMS, and message services flooding, signalization storms	Distributed cellular botnets support, RRC/MAC signaling subversion, HSS/IMS flooding
Escalation of privilege	Unauthorized roaming traffic insertion	Unauthorized roaming traffic insertion	Unauthorized roaming traffic insertion

5G vision [27]. The future core cloudification assumes complete separation of the data and control flows, offering several implementation variants for virtual deployment of the core service nodes over an NFV infrastructure (NFVI). Analogously, the fronthaul cloudification concentrates the radio resource management to improve the wireless data rates and service quality. The overall envisioned architecture of the next-generation mobile network is depicted in Figure 12.2.

12.3.1 EPC as a Service

EPC-as-a-Service (EPCaaS) emerged as a concept that combines the advantages of SDNs and the functional vitalization to offer a better implementation flexibility. Assuming a full 3GPP compliance [28,29], it offers flexible and on-demand scaling, easier operational maintenance, virtual machine (VM) modularity, bounded latency, maximized throughput, and high availability. EPCaaS can be implemented in two main variants, either with a full or partial virtualization of the core network functions. The fully virtualized variant has both the data and the control plane implemented in VMs, while in the partial variant only the control plane is virtualized [26]. In the fully virtualized

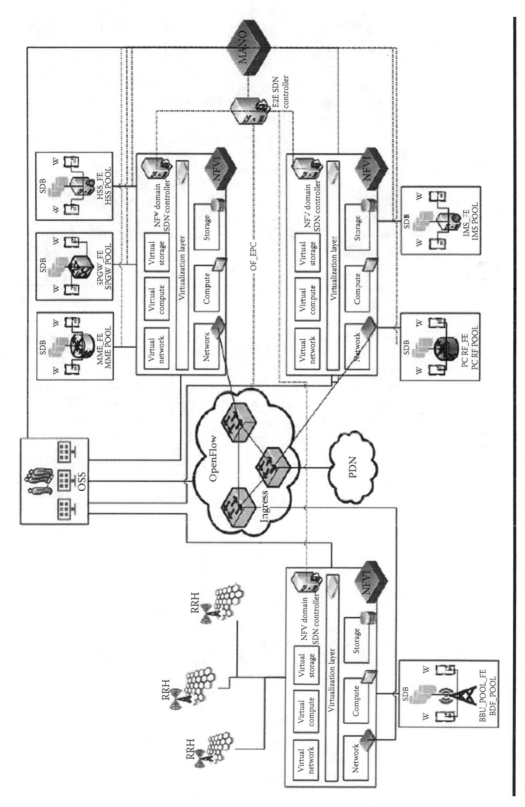

Figure 12.2   The next-generation mobile network architecture.

EPC variant, there are four implementation alternatives with respect to the 3GPP functionalities and the virtualization elements:

- *1:1 mapping*: Each 3GPP EPC functionality is mapped in a separate and state-full VM. Since this is a straightforward EPC translation, the scalability, and the operational maintenance remain an issue for the practical implementation.
- *1:N mapping*: EPC functionality is decomposed into a *front end, service logic*, and a *state database*, each implemented in a virtual component pool, as depicted in Figure 12.2. The front end handles all the protocol interfaces toward the other network entities. The service logic is organized in *worker* components that actually implement the logic of that specific EPC functional entity. Since the workers are stateless, the state database contains the user session state, as well as logging and operational information. This implementation variant offers greater flexibility and responsiveness to fluctuating traffic demands. Possible problem with this variant is the synchronization between the components and large processing delays.
- *N:1 mapping*: All the EPC functional entities are merged into one virtual component. The advantage of such a concentration is that the control plane interactions can be optimized on a network level. However, the scalability and the functional concentration as a single point of failure are serious issues posed in this implementation variant.
- *N:2 mapping*: This is similar to the previous mapping, only the control and data planes are split here. The state database is used as in the 1:N variant, while all the control functions are collapsed into a one *control* component. The user data are handled by a separate *switch* component that is responsible for packet forwarding and policy enforcement. This organization is useful for a fast deployment of a core network logic, however, on the account of an increased control plane complexity.

Technically, the first two mappings extend over the NFV dimension, while the later ones extend over the SDN dimension of the EPCaaS. A central service manager (Management and Orchestration; MANO) coordinates the service orchestrators and cloud controllers for a dynamic EPCaaS configuration according to an agreed infrastructure template graph.

12.3.2 Software-Defined Backbone

The EPCaaS variants assume full virtualization of the MME, HSS, S/PGW, PCRF, and IMS functions and complete realization of the backbone and the traffic toward the external PDNs. With a realistic traffic load, the virtualized network functions become a performance bottleneck due to the protocol interactions and the control–data plane dependencies [30]. Therefore, it might be useful to organize the forwarding actions in a separate virtualization suite to ensure proper service execution. This approach enables flexibility and robustness by enforcing the EPC policies in a flow-based fashion, while overcoming the scalability and quality of service (QoS)–related limitations. On this account, Figure 12.2 includes a separate SDN as a backbone between the EPCaaS and the fronthaul, implementing the OpenFlow concept [31]. In the all-IP 3GPP architecture, the OpenFlow enables parametrization as of the user profile and traffic behavior, offloading the user plane processing burden of the central EPCaaS infrastructure. Thus, the OpenFlow segment has an ingress switch handling the gateway tunneling protocol (GTP) sessions with the fronthaul, PDN interface, and backhaul interface toward the EPCaaS components.

12.3.3 Cloud-Based Radio Access Network

In the wireless segment, the *cloud radio access network* (C-RAN) is the fronthaul architecture providing the advantages of a virtualized and cloudified solution. Basically, the C-RAN is a concept where the baseband units (BBUs) are virtualized and accordingly shared between various radio resource units (RRUs) or base stations. The benefits are the smooth peak traffic amortization, statistical multiplexing gain, increased throughput and decreased delays, easier operation, and decreased cost [32]. The RAN-as-a-service is enabled with the virtualized management and orchestration of the BBU pools/base stations [33]. The SDN concept comes into play as a control play solution for the dynamic resource allocation, traffic load balancing, and fast self-organization and reconfiguration.

There are several coexisting scenarios, scaling the coverage from a femto to a macro level. The elastic cell segmentation rests on a virtualization-wise topological reorganization and software-wise functional definition, aiming to tackle enormous traffic demand through a smart spectrum utilization. An example deployment scenario has the RRUs colocated with the antennas on the cell site and connected to the coordinating BBH with a high-bandwidth optical link. As depicted in Figure 12.2, the associated BBU resources are realized over an NFVI. The functionalities provided by the BBUs include EPCaaS interface, radio resource control, media access control, sampling, coding, modulation, MIMO mapping, and enhanced intercell interference coordination. This leaves only the common public radio interface on the RRU side, which in aggregate improves the efficiency, scalability, and the dynamic traffic adaptation.

12.4 Cyberattack Surface of the Next-Generation Mobile Networks

The cyberattack surface of the next-generation mobile networks encompasses all threats that would undermine normal service delivery. Being modular in nature, the future mobile architecture is prone to both service-related and technology-related attacks. The threat landscape of the GSM, UMTS, and LTE networks in this case is extended toward virtualization and SDN related threats. Accordingly, the cyberattack surface is decomposed on a logical domain level, and the threats are analyzed in the context of a convergent network deployment. Following the abstraction in Table 12.1, the interactions within each logical domain are summarized using data flow diagrams* (DFDs). The respective threat model is built over this information using the spoofing, tampering, repudiation, information disclosure, denial-of-service, and elevation of privilege (STRIDE) analytical approach, yielding the integral cybersecurity profile of the cloudified mobile network depicted in Figure 12.2.

12.4.1 OpenFlow-Related Threats

The OpenFlow DFD is depicted in Figure 12.3. A trusted boundary exists between the EPCaaS functions and the E2E domain controller given the split between the responsibilities in administering the mobile service logic and the virtual infrastructure parametrization. To account for the hierarchical orchestration, another boundary rests between the local and the E2E SDN domain

---

* The term *data flow* in the context of the threat analysis corresponds with the logical interactions between the entities under consideration. In the OpenFlow terminology, *data flow* refers to the actual user sessions realized over the network.

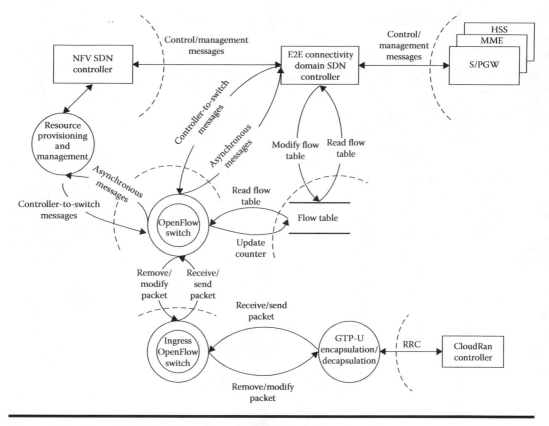

Figure 12.3 Simplified data flow diagram of the OpenFlow domain.

controller. A similar distinction exists toward the OpenFlow switches and the forwarding tables, representing the control and data plane interactions. The last threat boundary is set between the OpenFlow and the C-RAN controller, completing the software-defined perimeter. Next to the controllers, the OpenFlow topology includes switches as multiprocess entities, handling the GTP en-/decapsulation, the resource provisioning and management, and the traffic behavior parametrization. The logical interactions represent the EPC control traffic, NFV management and orchestration traffic, and the OpenFlow administration.

As for the spoofing-related threats, the OpenFlow segment is susceptible to the so-called *topology poisoning* attacks. An adversary can bypass the SDN perimeter boundaries and launch either the *host location hijacking* or the *link fabrication* attack [34]. In case of a VM-level modification/migration in the EPCaaS/C-RAN domains, the host profile of the associated function changes, entailing updates in the OpenFlow host tracking service. Since the update mechanism requires no authentication, the E2E controller can be tricked to accept an update from a spoofed host address, in case the spoofed packet reaches it first. Assuming the adversary has already established an OpenFlow foothold, all it needs is to craft and send OF packets with the same identifier of the target host. Given the stateless implementation of the EPC control plane [33], the control flows can be hijacked and leveraged for further user-targeted exploitations. The host hijacking attack resembles the same attributes of the ARP cache poisoning attack characteristic for the traditional networks, bearing similar detrimental effects for the OpenFlow domain.

Another possibility for a topology poisoning is to exploit the logical link discovery protocol (LLDP). In this case, the E2E controller checks the "integrity/origin" and the "path" LLDP invariants to ensure it communicates with legitimate switches. A deception is possible if these invariants are subverted; that is, the adversary needs to inject fake internal links on compromised hosts or VM into the network topology. To do so, he or she needs to obtain a valid LLDP packet, fabricate it, and tune the injecting rate as the targeted link to trick the controller or can simply relay a valid LLDP stream to fake the OpenFlow topology. The link fabrication can be further leveraged for a denial-of-service or man-in-the-middle attack. By injecting a fake link into the exiting topology, the adversary can impersonate the OF spanning tree service and remove operational switch ports. With that, the legitimate traffic is blackholed, causing both control- and user-related denial-of-service. A similar possibility exists when the shortest path service is impersonated by an adversary. In this case, the fake link is injected into the deceive E2E to calculate the shortest path between the target switches. The traffic traversing the fake route is under direct control of the adversary who is able to tamper with it or disclose valuable information.

The E2E controller spoofing opens further possibilities for DoS, information disclosure, and tampering/nonrepudiation with the controller-to-switch and modify/read message streams. Instead of forging a target link, service denial can be caused by sending bogus user streams to the controller that result in the installation of new flow rule. Since the OF forwarding tables have limited capacity, the number of different flow rules (corresponding to different user profiles) needs to be slightly bigger so as to overflow them. This can be achieved by varying the source and destination session ports, knowing that the OpenFlow logic installs rules matching only the header fields exactly. With an indefinite flow rule expiration, a new legitimate rules cannot be installed, leading to severe packet losses for the rest of the user population [35]. The denial-of-service attack can be mounted in a revised fashion against the OpenFlow switches. Instead of flooding, the adversary with a network presence can flush the flow tables on a regular interval before valid flows are populated [36].

Assuming a passive adversary monitoring the OpenFlow traffic, there is a possibility for disclosing information on the present network topology or the E2E behavior. To learn which flow rules are aggregated for certain user flows, the adversary needs to impersonate a legitimate mobile user or elicits response for a target flow. Knowing that the E2E connectivity controller will install new flow rule for a nonmatching packet stream, the adversary can observe the round trip time of the forged flow and fingerprint the OpenFlow parametrization or charging policy enforcement information. Likewise, an *inference* attack can yield the flow table capacity and usage, valuable for the fine tuning of the aforementioned *resource consumption* attack [37]. Exploiting the control/management messages to trigger a controller-to-switch interaction, the adversary needs to combine the number of forged flows with the round trip time for their installation. Another opportunity for information disclosure exists due to a faulty implementation of the *netconf* protocol on the controller side. Exploiting the *XML eXternal Entity vulnerability* (XXE) could exfiltrate configuration details or plaintext credentials stored locally on the E2E controller [38]. Overall, the disclosed information set in the EPCaaS or C-RAN context may reveal the network capabilities and its internal structure. Examples include fingerprinting distinction of the ingress OF switches handling GTP protocol conversion, load-balancing policy identification, and lawful interception recognition on the network side. If an attacker has a direct access to the OF infrastructure, this information can be utilized for user-level information disclosure, flow duplication, or even flow modification [36].

Table 12.2 summarizes the OpenFlow-related threats as related to their relative subversion means. Considering a skilled adversary profile (malicious insiders, cellular botnets, etc.), the Open-Flow domain is susceptible to spoofing, tampering, and DoS-related attacks [39]. However, without an exact instantiation, the repudiation-related attacks in the OpenFlow environment are relevant

Table 12.2 OpenFlow-Related Threats

Domain	Threat Category	Proof-of-Concept/Related Attacks
OpenFlow	S	Host location hijacking, link fabrication
	T	Topology poisoning attacks, man-in-the-middle attacks
	R	n/a
	I	Inference, network fingerprinting, XML eXternal Entity
	D	Resource consumption, controller flooding, flow table flooding/flushing
	E	n/a

threats to the accountability of the internal interactions. If an insider is able to eliminate any logging information set at the E2E or MANO nodes, there will be no forensic evidence of the incident. The same rationale holds for the privilege escalation threats, given that in this case an EPCaaS or C-RAN functionality needs to be fully under adversary control [40,41].

12.4.2 NFV-Related Threats

Concordant to the ETSI security guidance [26], Figure 12.4 depicts the simplified DFD for the NFV segment. According to the mobile cloudification principle, the EPCaaS and C-RAN functions are abstracted in a telecom (application) domain on the NFVI. The organization reflects on a practical infrastructure-as-a-service (IaaS) cloud deployment model, where the virtualization resources do not necessarily have to be within the operator's perimeter. That is, the mobile network functions are potentially colocated with other nonrelated tenants. In the telecom stratum, there is a central virtual management function (VMF) that interacts with the orchestrator (MANO) and the virtual entities. Basically, the VMF performs the NFV domain SDN controller functions, extending beyond the simple *hypervisor* functionality in the traditional virtualization context. There are two corresponding trust boundaries related to the entity management and orchestration message flow. The VMF manager coordinates the dynamic resource allocation with another two virtualized infrastructure managers (VIMs), the transport and the service one. The transport VIM further organizes the networking resources needed for the EPCaaS/C-RAN control traffic, while the service VIM provisions the necessary processing and storage capabilities. Finally, the underlying NFVI interfaces the OpenFlow domain as of the control traffic execution and routing of the user flows. The remaining trust boundaries delimit the VIM interaction, NFVI resource allocation, and the NFVI–OpenFlow interaction.

Assuming a remote access, VMF's management consoles are susceptible to application programming interface (API) or injection attacks aiming to tamper the orchestration and entity management interactions [42]. If an adversary is able to hijack the controls over the VMF console, a virtual resource–targeted denial-of-service attack can be launched on the entire NFVI platform. EPCaaS or C-Ran control traffic can be blackholed, the dedicated storage capacity can be eliminated for targeted mobile functions, or the processing limited to delay the service delivery. Another DoS possibility is to turn off or suspend a network function, given their implementation in a separate

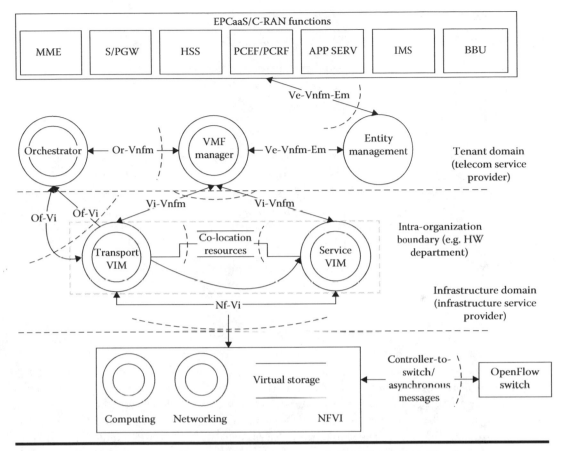

Figure 12.4 **Simplified data flow diagram of the network functional virtualization segment.**

machine. The central NFVI controllers are not immune to local exploitations, too. A malicious guest can launch several *VM-to-VMF* attacks as to arbitrary access the physical resources dedicated for the EPCaaS or C-RAN functions. Targeting the Peripheral Component Interconnect (PCI) configuration space, the adversary can perform *unauthorized memory access* or generate *device memory collisions*, interrupting the VMF interactions to form a guest VM location [43]. Additionally, the adversary can exfiltrate EPCaaS/C-RAN data to the one under his control. This elevated privilege can be utilized to execute arbitrary code with VMF privileges, inject, or manipulate the related interrupt vectors.

Equally dangerous are the *cross-VM* attacks, where a malicious guest VM is able to successfully recover the AES encryption keys from the NFVI cache using Bernstein's correlation technique [44]. By impersonating a virtualized component by ARP poisoning, a malicious guest is able to gain access to a targeted VM traffic [45]. Regardless of the VM traffic encryption, the EPCaaS or BBU behavior can be derived by analyzing the traffic packet sizes and their time distribution. Having the raw controller-to-switch hop counts, the telecom topology structure can be disclosed, learning the transport VIM and orchestration configuration. Like in the traditional networking environments, this opportunity can be leveraged for a network denial-of-service (resource wastage, flooding) or transport VIM saturation.

Table 12.3   Network Functional Virtualization–Related Threats

Domain	Threat Category	Proof-of-Concept/Related Attacks
Network functional virtualization	S	API-related attacks, ARP poisoning, VMF, and transport/service VIM impersonation
	T	VMF/NFVI compromise, physical resource tampering
	R	VMF–VIM interaction forgery
	I	Unauthorized memory access, Bernstein's correlation attack, network traffic analysis, minimum disclosure routing
	D	Device memory collisions, resource wastage attacks, network flooding, transport VIM saturation, economic denial-of-service
	E	Unauthorized memory access, device memory collisions, interrupt-related attacks

Table 12.3 summarizes the threats characteristic for the NFV domain. Beyond the operational benefits, virtualization introduces serious risks in regard to the mobile service delivery. Prone to imperfect isolation, both the core and radio access control is susceptible to various spoofing, tampering, and information leakage threats. The centralization brings a single-point-of-failure threat pertaining the VMF management and orchestration. As in the OpenFlow case, the NFV domain SDN controller's subversion has a cascading effect on the entire NFVI. Able to elevate its privileges, a malicious VM guest can invoke time delays, data loss, or data manipulation and with that to degrade the mobile delivery logic. Realistically, the cross-VM attacks can fingerprint the NFVI cluster or extract plain text passwords, encryption keys and memory data [46–48]. This can lead to physical denial-of-service attack, network usurpation attacks on the IaaS data centers, or a sophisticated, Economic Denial-of-Sustainability (EDoS) attack [45,49,50]. The logical separation entails security policies that might violate the main mobile security principles, extending the cyberattacking surface with a large set of virtualization-related exploitation points [51,52].

12.4.3 EPC-Related Threats

The EPC segment is vulnerable to attacks targeting the service delivery logic. The mobile botnets were proven to cause DDoS by flooding the HSS/S/PGW functions with a large volume of signalization and authorization traffic [12]. In addition, the service delivery logic is proven susceptible to IP spoofing, mobile-to-mobile probing, and delay pattern recovery attacks [53]. Overall, the availability-related threats and the signaling storms have the strongest inhibiting potential for any practical software-defined mobile network (SDMN) deployment [22]. The EPC threats also relate to AKA vulnerabilities [54], weakness in the integrity protection [55], and the handover handling procedures [54]. As noted in [56], the EPC has weak points as well in the IMS and the mobile charging function. Though addressed in the latest 3GPP security architecture [57], the plausibility of the STRIDE threats associated with the conventional IMS implementation [58] remains valid for the convergent SDMN variant. The server/media redirection attacks, the man-in-the-middle P-CSCF attacks [59], or the SIP-related exploits can be applied with existing attack vectors [60].

Table 12.4 Evolved Packet Core–Related Threats

Domain	Threat Category	Proof-of-Concept/Related Attacks
EPC	S	IP spoofing, server/media redirection
	T	Charging/billing bypassing, overbilling, AKA tampering, man-in-the middle P-CSCF attacks, authentication loophole
	R	Charging/billing bypassing, overbilling
	I	Delay pattern recovery attacks
	D	HSS/PGW signaling/authorization storms, roaming-initiated HSS flooding, mobile-to-mobile probing
	E	Charging/billing bypassing, overbilling

As for the mobile charging–related threats, Reference [61] identifies an authentication loophole enabling traffic charging bypass or overcharging. Table 12.4 summarizes the EPC-related threats.

12.4.4 C-RAN-Related Threats

Employing the principles of virtualized service distinction, Figure 12.5 depicts the simplified DFD of the software-defined fronthaul. Overarching the telecom and infrastructural stratums is the C-RAN controller, bearing all the VMF responsibilities as in the general NFV scenario. One trust boundary exists between the RRUs and mobile users delimiting the network edge. Compliant with the open radio interface (ORI) standard, the software-defined fronthaul implements a distributed base station architecture where multiple operators place their BBU pools on the same virtualized infrastructure [62]. Therefore, a separate boundary aligns with the Ir and X2 interfaces. Another trust boundary exists as of the resource and traffic management, complementing the one between the C-RAN and the OpenFlow segments.

On the network edge, the traditional threats of *receiver jamming* and *media access control (MAC) layer* attacks remain applicable also for the C-RAN environment. The adversary can decrease the received signal-to-noise ratio (SNR) below the required threshold by transmitting noise over the receiving thereby jamming the control/data traffic for a given RRU. Beyond the basic *barrage* jamming, the adversary can jam a certain bend of interest or can send asynchronous off-tones to cause interchannel or coexistence interference for a multiradio access technology (RAT)-supported devices [63]. Another jamming technique is to spoof the primary and secondary synchronization channels and with that force mobile users to synchronize with a bogus reference [64]. Several adversaries can cooperate to exploit the MAC scheduling by sending simultaneous uplink requests with a high QoS requirements [19]. Next to the standardized 3GPP access, the radio cognition brings another set of attacks that the future cellular architecture needs to account for. Opportunistically utilizing the licensed white spaces, the primary user emulation attacks (PUEAs) mimic the spectral characteristics of primary users to gain priority access to the wireless channels available. Spectrum sensing data falsification attacks (SSDFs) can be launched to cause an incorrect perception of the radio environment, enabling greedy white spaces occupation. There is a viable opportunity for corrupting the cognition control channel (CCC), leading to a wireless performance degradation.

The internal C-RAN interactions are subject to the same security violations associated with the virtualization domain. The X2 and Ir communication may be affected by unauthorized memory

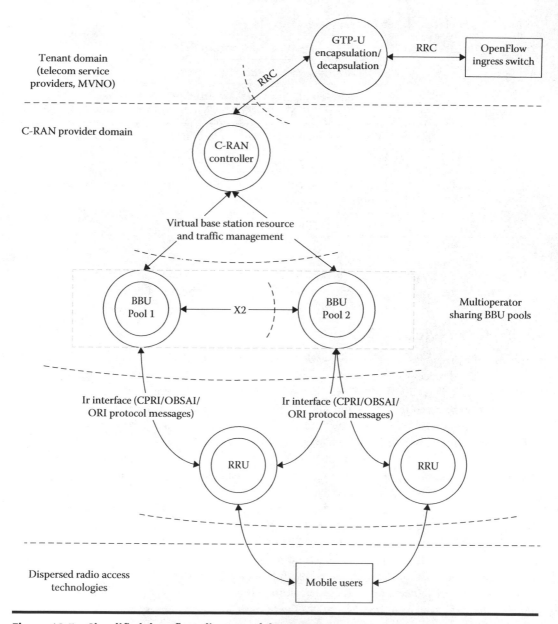

Figure 12.5 Simplified data flow diagram of the C-RAN segment.

access, memory collisions, or resource wastage attacks. Fronthaul performance degradation can be as well caused by a C-RAN controller saturation, RRC traffic forgery, or malicious C-RAN recon-figuration. Also, the traffic distribution on a cell site level can be fingerprinted by monitoring the BBU activity, resource allocation, and processing load [65]. The aggregate SDF cyberattack space depicted in Table 12.5 reinforces the general threat of virtual network management centralization. Affecting the mobile logic, an adversary can reuse the same attack vectors and subvert one of the critical controllers in each of the service delivery domains. In the C-RAN case, the user-initiated

Table 12.5 C-RAN–Related Threats

Domain	Threat Category	Proof-of-Concept/Related Attacks
C-RAN	S	Synchronization spoofing, PUE, SSDF, CCC attacks
	T	MAC-layer attacks, RRC traffic forgery
	R	PUE, SSDF, CCC attacks, malicious C-RAN reconfiguration
	I	Radio channel eavesdropping, PUE attacks, virtual traffic fingerprinting
	D	Receiver jamming, cognition control channel, PUE, SSDF, CCC attacks, C-RAN controller saturation
	E	NFV exploits

attacks aggravate the already difficult virtualization conditions toward implementing an integral fronthaul protection.

12.5 Cyber Defense Design Requirements and Recommendations

Various solutions exist to safeguard the SDMN attack surface. The stateful SDN processing [66] and the concept of centralized SDN/NFV integration [67] have the potential for a more holistic topology view, thus, enabling cross-domain anomaly detection. From a security perspective, the flow-based network access control solution presented in [66] can provide the opportunity for flow definitions that restrict behavior that is previously identified and categorized as one of the EPC-related threats enlisted in Table 12.4. The flow definitions and the proposed IEEE 802.1x can be used for early detection and proactive reconfiguration in face of a coordinated denial-of-service attack from a distributed cellular botnet. This proactive defense/detection can be implemented in a separate management and control entity responsible for enforcing the network resilience policies in the SDMN architecture from Reference [67]. Moreover, the resilience design can also be organized in several section-specific management and control entities, that is, one responsible for monitoring, detection, and reconfiguration of the C-RAN segment and another one responsible for the EPC segment. In that way, the integral network protection can eliminate the logic-related C-RAN threats from Table 12.5.

Essentially, these entities serve the role of the infrastructural support nodes for the intrusion detection solutions proposed in References [68–71]. Coupled with the cell-pot infrastructure, the proposed mechanisms can be extended to aid the overall moving target defense mechanism, as proposed in Reference [72]. In effect, the proposed evolved defense mechanism (EDM) can scan and fingerprint each of the logical SDMN segments, categorize the attack according to its predefined attack pattern from the current network resilience policy, and try to eliminate with a preconfigured variation strategy. If there is a novel or modified attack vector, the sector-specific management and control entities will report the deviation from the most close pattern (or set of patterns) to the central resilience controller. In case no automated strategy is found, a novel strategy is developed and reconfigured manually on these nodes and immediately implemented in response to the novel attack. Again, enhancements to this logic are possible by organizing the network reconfiguration strategies based on not only the attack pattern but also the targeted domain, thwarting malware-initiated SDMN attacks with a strategy similar to the one in [73], or protecting the edge of the network as elaborated in [74–76].

Following this premise, the software-defined backbone resilience can be established in accordance with the security monitoring and policy enforcement logic as proposed in [77]. For example, the control loop actions for the service and control planes can be abstracted to control-related attack recognition strategies, while the application and data plane can be abstracted to mobile user-related attack recognition strategies. The associated reconfiguration strategies can be incorporated in the network resilience policy by combining the self-healing recovery actions from [77] with the deep packet inspection feature from [78].

Based on this discussion, the *key design requirements* recommended for a robust SDMN defense solution are summarized as follows:

- ■ *Infrastructure*: Management and control entities
  - ● *Central resilience controller*: Network-wide and cross-segment interactions monitoring, analysis, and resilience strategy evolution (feedback loops, human-assisted strategy modification)
  - ● *Segment-level resilience controllers*: Function-related monitoring, fingerprinting, and aggregated event analysis for an anomaly detection in the software-defined backbone, the EPCaaS and C-RAN segments
- ■ *Cyberattack surface snapshot*: Current set of attacking vectors and attacking patterns used for a preliminary reconfiguration strategy selection; updated as per any novel attack vector detected or reported as detected elsewhere (this refers also to attacks with an implicit potential to be applied in the SDMN setting)
- ■ *Resilience policies*: Variation strategies for control-level and user-level protection
- ■ *Reconfiguration-based*: Fast self-healing, attack eradication, traffic rerouting
- ■ *Flow-based*: Access control, service filtering and re-provisioning, SLA enforcement

Enabling proactive management and control, the resilience infrastructure needs to be parametrized to respond to the current processing and performance capabilities to each of the nodes in order not to affect the normal mobile service delivery. It is recommended for the cyberattack surface snapshot to be regularly updated not just with SDMN-targeted attack vectors but also with vectors known to have a potential to be applied against the operative infrastructure. By the same token, the recommendation for the reconfiguration-based and flow-based policies is to be refined depending on the information on the malicious user behavior together with the information on important societal events and the observed mobile service utilization by the legitimate users over time.

12.6 Summary

The cyberattacking surface of the next-generation mobile networks equally draws on the vulnerabilities native to the mobile networking segments as well as the ones characteristic for the cloud-based, virtualized, and software-defined environments. Extended in this way, a complex set of threats relate to the confidentially, integrity, and availability of the next-generation mobile services. Taking this into consideration, the existing defensive mechanisms are discussed together with the segment-specific ones toward an integral SDMN defense. Respectively, several key design components and recommendations for practical implementation are included to set the main course in developing a robust and efficient SDMN cyberprotection.

References

1. Ericcson. Ericsson Mobility Report, Stockholm, Sweden, 2015.
2. N. Boudriga. *Security of Mobile Communication*. Taylor & Francis, Boca Raton, FL, 2009.
3. J. T. J. Penttinen. *LTE/SAE Security*, John Wiley & Sons Ltd, Chichester, West Sussex, UK, 2012.
4. D. Kreutz and F. Ramos. Software-defined networking: A comprehensive survey. *Proceedings of the IEEE*, 103(1):1–61, 2015.
5. 3rd Generation Partnership Project. Specification of the MILENAGE algorithm set: An example algorithm set for the 3GPP authentication and key generation functions f1, f1*, f2, f3, f4, f5 and f5*; Document 1: General (Release 12). 3GPP, Sophia Antipolis, France, p. 10, 2014.
6. 3rd Generation Partnership Project. Specification of the 3GPP confidentiality and integrity algorithms; Document 1: f8 and f9 specification (Release 12). 3GPP, Sophia Antipolis, France, p. 10, 2014.
7. 3rd Generation Partnership Project. 3GPP TS 35.202—3G security—Specification of the 3GPP confidentiality and integrity algorithms—Document 2: KASUMI specification (Release 12). 3GPP, Sophia Antipolis, France, p. 24, 2014.
8. U. Meyer and S. Wetze. On the impact of GSM encryption and man-in-the-middle attacks on the security of interoperating GSM/UMTS networks. In *IEEE International Symposium on Personal, Indoor and Mobile Radio Communications*, Sophia Antipolis, France, pp. 2876–2883, 2004.
9. I. I. Androulidakis. *Mobile Phone Security and Forensics: A Practical Approach*. Springer Science and Business Media, New York, 2012.
10. 3rd Generation Partnership Project. Specification of the A5/3 encryption algorithms for GSM and ECSD, and the GEA3 encryption algorithm for GPRS; Document 1: A5/3 and GEA3 specifications (Release 12). 3GPP, Sophia Antipolis, France, 2014.
11. J.-K. Tsay and S. F. Mjolsnes. Computational security analysis of the UMTS and LTE authentication and key agreement protocols. *Lecture Notes in Computer Science*, 7531(1):65076, 2012.
12. A. Merlo, M. Migliardi, N. Gobbo, F. Palmieri, and A. Castiglione. A denial of service attack to UMTS networks using SIM-less devices. *IEEE Transactions on Dependable and Secure Computing*, 11(3): 280–291, 2014.
13. G. Kambourakis, C. Kolias, S. Gritzalis, and J. H. Park. DoS attacks exploiting signaling in UMTS and IMS. *Computer and Communications*, 34(3):226–235, 2011.
14. C. Xenakis and C. Ntantogian. An advanced persistent threat in 3G networks: Attacking the home network from roaming networks. *Computers & Security*, 40(1):84–94, 2014.
15. Z. Quian, Z. Wang, Q. Xu, Z. M. Mao, M. Zhang, and Y.-M. Wang. You can run, but you can't hide: Exposing network location for targeted DoS attacks in cellular networks. In *17th Annual Network & Distributed System Security Symposium (NDSS)*, San Diego, CA, 2012.
16. G. Suarez-tangil, J. E. Tapiador, P. Peris-lopez, and A. Ribagorda. Evolution, detection and analysis of malware for smart devices. *IEEE Communications Magazine*, 16(2):961–987, 2014.
17. G. Mantas, N. Komninos, J. Rodriguez, E. Logota, and H. Marques. Security for 5G communications. In J. Rodriguez, ed., *Fundamentals of 5G Mobile Networks*. John Wiley & Sons, Chichester, U.K., pp. 207–220, 2015.
18. G. Gorbil, O. Abdelrahman, M. Pavloski, and E. Gelenbe. Modeling and analysis of RRC-based signalling storms in 3G networks. *IEEE Transactions on Emerging Topics in Computing*, 4(1):113–127, 2015.
19. J. Jermyn, G. Salles-Loustau, and S. Zonouz. An analysis of DoS attack strategies against the LTE RAN. *Journal of Cyber Security and Mobility*, 3(2):159–180, 2014.
20. N. Golde, K. Redon, and R. Borgaonkar. Weaponizing femtocells: The effect of rogue devices on mobile telecommunications. In *Network and Distributed System Security Symposium*, San Diego, CA, 2012.
21. J. Beekman and C. Thompson. Breaking cell phone authentication: Vulnerabilities in AKA, IMS, and Android. Presented as part of the *Seventh USENIX Workshop on Offensive Technology*, Washington DC, pp. 1–10, 2013.

22. R. P. Jover. Security attacks against the availability of LTE mobility networks: Overview and research directions. In *2013 16th International Symposium on Wireless Personal Multimedia Communications (WPMC)*, Atlantic City, NJ, pp. 1–9, 2013.

23. N. Vallina-rodriguez, S. Sundaresan, C. Kreibich, U. C. Berkeley, and C. Nat. Beyond the radio: Illuminating the higher layers of mobile networks access point name. In *The 13th International Conference on Mobile Systems, Applications, and Services*, Florence, Italy, pp. 1–15, 2015.

24. S. Natouri, C. Lac, and A. Serhrouchni. A model-based resilience analysis for IMS. *Journal of Networks*, 9(3):588–603, 2014.

25. G. Lenzini, S. Mauw, and S. Ouchani. Security analysis of socio-technical physical systems. *Computers & Electrical Engineering*, 43(1):258–274, 2015.

26. M. R. Sama, L. M. Contreras, J. Kaippallimalil, I. Akiyoshi, H. Qian, and H. Ni. Software-defined control of the virtualized mobile packet core. *IEEE Communications Magazine*, 53:107–115, February 2015.

27. L. Gavrilovska, V. Rakovic, and V. Atanasovski. Visions towards 5G: Technical requirements and potential enablers. *Wireless Personal Communications*, 87(3):731–757, 2015.

28. 3rd Generation Partnership Project. 3GPP TS 23.002 V13.1.0—Technical specification group services and systems aspects—Network architecture (Release 13). 3GPP, Sophia Antipolis, France, 2014.

29. European Telecommunications Standards Institute. GS NFV 002—V1.2.1—Network Functions Virtualisation (NFV); architectural framework, Sophia Antipolis, France, Vol. 1. ETSI, pp. 1–21, 2015.

30. A. S. Rajan, S. Gobriel, C. Maciocco, and K. B. Ramia. Understanding the bottlenecks in virtualizing cellular core network functions. In *2015 IEEE International Workshop on Local and Metropolitan Area Networks (LANMAN)*, Beijing, China, pp. 1–6, 2015.

31. M. R. Sama, S. B. H. Said, K. Guillouard, and L. Suciu. Enabling network programmability in LTE/EPC architecture using OpenFlow. In *2014 12th International Symposium on Modeling and Optimization in Mobile, Ad Hoc, and Wireless Networks (WiOpt 2014)*, Hammamet, Tunisia, pp. 389–396, 2014.

32. A. Checko, H. L. Christiansen, Y. Yan, L. Scolari, G. Kardaras, M. S. Berger, and L. Dittmann. Cloud RAN for mobile networks—A technology overview. *IEEE Communications Surveys & Tutorials*, 17(1):405–426, 2015.

33. T. Taleb, M. Corici, C. Parada, A. Jamakovic, S. Ruffino, G. Karagiannis, and T. Magedanz. EASE: EPC as a service to ease mobile core network deployment over cloud. *IEEE Network*, 29(2):78–88, 2015.

34. S. Hong and H. Wang. Poisoning network visibility in software-defined networks: New attacks and countermeasures. In *NDSS*, San Diego, CA, pp. 8–11, February 2015.

35. R. Klöti, V. Kotronis, and P. Smith. OpenFlow: A security analysis. In *Proceedings of the International Conference on Network Protocols ICNP*, Göttingen, Germany, 2013.

36. M. Antikainen, T. Aura, and S. Mikko. Spook in your network: Attacking an SDN with a compromised OpenFlow switch. *Secure IT Systems*, 7617(2014):229–244, 2014.

37. J. Leng, Y. Zhou, J. Zhang, and C. Hu. An inference attack model for flow table capacity and usage: Exploiting the vulnerability of flow table overflow in software-defined network, arXiv preprint arXiv:1504.03095, 2015.

38. D. Jorm. SDN and security. *ONOS Project*, 2015 [Online]. Available: http://onosproject.org/2015/04/03/sdn-and-security-david-jorm/, accessed April 26, 2015.

39. M. Dabbagh, B. Hamdaoui, M. Guizani, and A. Rayes. Software-defined networking security: Pros and cons. *IEEE Communications Magazine*, 53(6):73–79, 2015.

40. I. Alsmadi and D. Xu. Security of software defined networks: A survey. *Computers & Security*, 53(9): 79–108, 2015.

41. European Telecommunications Standards Institute. Network Functions Virtualisation (NFV)—Security and trust guidance, Vol. 1, pp. 1–21, ETSI, Sophia Antipolis, France, 2014.

42. M. Ali, S. U. Khan, and A. V. Vasilakos. Security in cloud computing: Opportunities and challenges. *Information Sciences* (*New York*), 305:357–383, 2015.

43. G. Pék, A. Lanzi, D. Balzarotti, S. Anitpolis, S. Antipolis, and C. Neumann. On the feasibility of software attacks on commodity virtual machine monitors via direct device assignment. In *ASIACCS*, Kyoto, Japan, pp. 305–316, 2014.

44. G. I. Apecechea, M. S. Inci, T. Eisenbarth, and B. Sunar. Fine grain cross-VM attacks on Xen and VMware are possible! Cryptology ePrint Archive Report 2014/248, IACR Cryptology ePrint Archive, 2014.

45. L. R. Bays, R. R. Oliveira, M. P. Barcellos, L. P. Gaspary, and E. R. Mauro Madeira. Virtual network security: Threats, countermeasures, and challenges. *Journal of Internet Services and Applications*, 6(1): 1–19, 2015.

46. D. A. B. Fernandes, L. F. B. Soares, J. V. Gomes, M. M. Freire, and P. R. M. Inácio. Security issues in cloud environments: A survey. *International Journal of Information Security*, 13(2):113–170, 2014.

47. M.-D. Nguyen, N.-T. Chau, S. Jung, and S. Jung. A demonstration of malicious insider attacks inside cloud IaaS vendor. *International Journal of Information and Education Technology*, 4(6):483–486, 2014.

48. F. Rocha and M. Correia. Lucy in the sky without diamonds: Stealing confidential data in the cloud. In *2011 IEEE/IFIP 41st International Conference on in Dependable Systems and Networks Workshops* (*DSN-W*), Hong Kong, China, pp. 129–134, June 2011.

49. D. Zissis and D. Lekkas. Addressing cloud computing security issues. *Future Generation Computer Systems*, 28(3):583–592, 2012.

50. Q. Yan and F. R. Yu. Distributed denial of service attacks in software-defined networking with cloud computing. *IEEE Communications Magazine*, 53:52–59, April 2015.

51. W. E. I. Huang, A. Ganjali, B. H. Kim, S. Oh, and D. Lie. The state of public infrastructure-as-a-service cloud security. *ACM Computing Surveys*, 47(4):68, 2015.

52. H. Hawilo, A. Shami, M. Mirahmadi, and R. Asal. NFV: State of the art, challenges and implementation in next generation mobile networks. *IEEE Network Magazine*, 28(6):18–26, 2014.

53. V. C. Perta, M. V. Barbera, and A. Mei. Exploiting delay patterns for user IPs identification in cellular networks. *Lecture Notes in Computer Science* (including subseries *Lecture Notes in Artificial Intelligence and Lecture Notes in Bioinformatics*), 8555:224–243, 2014.

54. J. Cao, M. Ma, S. Member, I. H. Li, and Y. Zhang. A survey on security aspects for LTE and LTE-A networks. *IEEE Communications Surveys & Tutorials*, 16(4):1–20, 2014.

55. T. Wu and G. Gong. The weakness of integrity protection for LTE. In *Proceedings of the Sixth ACM Conference on Security and Privacy in Wireless and Mobile Networks* (*WiSec'13*), Budapest, Hungary, p. 79, 2013.

56. B. Van Leeuwen, V. Urias, C. Glatter, and A. Interrante-Grant. Testbed for cellular telecommunications cyber vulnerability analysis. In *2013 IEEE Military Communications Conference* (*MILCOM 2013*), San Diego, CA, pp. 1391–1397, 2013.

57. 3rd Generation Partnership Project. 3GPP TS 33.401—3GPP System Architecture Evolution (SAE); security architecture (Release 12). 3GPP, Sophia Antipolis, France, p. 18, 2015.

58. K. Shuang, S. Wang, B. Zhang, and S. Su. IMS security analysis using multi-attribute model. *Journal of Networks*, 6(2):263–271, 2011.

59. N. Vrakas, D. Geneiatakis, and C. Lambrinoudakis. Obscuring users' identity in VoIP/IMS environments. *Computers & Security*, 43:145–158, 2014.

60. E. Belmekki, M. Bellafkih, and A. Bellafkih. Enhances security for IMS client. In *2014 Fifth International Conference on Next Generation Networks and Services*, Casablanca, Morocco, pp. 231–237, 2014.

61. C. Peng, C. Li, H. Wang, G. Tu, and S. Lu. Real threats to your data bills: Security loopholes and defenses in mobile data charging. In *CCS*, Scottsdale, AZ, pp. 727–738, 2014.

62. European Telecommunications Standards Institute. Open Radio equipment Interface (ORI), Vol. 1, Release 4, pp. 1–16, ETSI, Sophia Antipolis, France, 2014.

63. Rohde & Schwarz. Vulnerabilities of LTE and LTE-advanced communication. White paper. Rohde & Schwarz, Munich, Germany, p. 35, 2014.

64. M. Lichtman, J. H. Reed, T. C. Clancy, and M. Norton. Vulnerability of LTE to hostile interference. In *Proceedings of the 2013 IEEE Global Conference on Signal and Information Processing (GlobalSIP 2013)*, Austin, TX, pp. 285–288, 2013.

65. G. Baldini, T. Sturman, A. R. Biswas, R. Leschhorn, G. Godor, and M. Street. Security aspects in software defined radio and cognitive radio networks: A survey and a way ahead. *IEEE Communications Surveys & Tutorials*, 14(2):355–379, 2012.

66. J. Matias, J. Garay, N. Toledo, J. Unzilla, and E. Jacob. Toward an SDN-enabled NFV architecture. *IEEE Communications Magazine*, 53:187–193, April 2015.

67. L. Mamatas, S. Clayman, and A. Galis. A service-aware virtualized software-defined infrastructure. *IEEE Communications Magazine*, 53:166–174, April 2015.

68. E. F. El Gaml, H. Elattar, and H. M. El Badawy. Evaluation of intrusion prevention technique in LTE based network. *International Journal of Scientific and Engineering Research*, 5(12):1395–1400, 2014.

69. S. Liebergeld, M. Lange, and R. Borgaonkar. Cellpot: A concept for next generation cellular network honeypots. In *Proceedings of the 2014 Workshop on Security of Emerging Networking Technologies*, San Diego, CA, February 2014.

70. A. J. Alzahrani and A. A. Ghorbani. A multi-agent system for smartphone intrusion detection framework. In *Proceedings of the 18th Asia Pacific Symposium on Intelligent and Evolutionary Systems*, Singapore, Vol. 1, pp. 101–113, 2015.

71. Z. Yan, P. Zhang, and A. V. Vasilakos. A security and trust framework for virtualized networks and software-defined networking. *Security and Communication Networks*, 2(1):71–81, 2015.

72. H. Zhou, C. Wu, M. Jiang, B. Zhou, W. Gao, T. Pan, and M. Huang. Evolving defense mechanism for future network security. *IEEE Communications Magazine*, 53:45–51, April 2015.

73. J. François and O. Festor. Anomaly traceback using software defined networking. In *International Workshop on Information Forensics and Security*, Atlanta, GA, 2014.

74. X. Duan and X. Wang. Authentication handover and privacy protection in 5G HetNets using software-defined networking. *IEEE Communications Magazine*, 53:28–35, April 2015.

75. N. Yang, L. Wang, G. Geraci, M. Elkashlan, J. Yuan, and M. Di Renzo. Safeguarding 5G wireless communication networks using physical layer security. *IEEE Communications Magazine*, 53(4):20–27, 2015.

76. D. Montero, M. Yannuzzi, A. Shaw, L. Jacquin, and A. Pastor. Virtualized security at the network edge: A user-centric approach. *IEEE Communications Magazine*, 53:1–10, April 2015.

77. I. Grida, B. Yahia, T. Rasheed, I. Grida, B. Yahia, T. Rasheed, and D. Siracusa. Softwarized 5G networks resiliency with self-healing softwarized 5G networks resiliency with self-healing. In *First International Conference on 5G for Ubiquitous Connectivity*, Levi, Finland, pp. 229–233, 2014.

78. P. Yasrebi, M. Sina, H. Bannazadeh, and A. Leon-Garcia. Security function virtualization in software defined infrastructure. In *2015 IFIP/IEEE International Symposium on Integrated Network Management (IM)*, pp. 778–781, 2015.

Index